THE
BIOLOGY
OF PARASITIC
FLOWERING
PLANTS

THE
BIOLOGY OF
PARASITIC FLOWERING
PLANTS

JOB KUIJT

UNIVERSITY OF CALIFORNIA PRESS
BERKELEY AND LOS ANGELES 1969

University of California Press
Berkeley and Los Angeles, California

University of California Press, Ltd.
London, England

To Jeannie

"Ueberhaupt, man findet an
diesen Gewächsen so etwas
Sonderbares, so etwas Räthsel-
haftes, ein solch geheimnisvolles
Wesen, das mit dem Schatten,
in dem sie gewöhnlich wuchern,
so vertraut zu sein scheint, dass
man sich gedrungen fühlt, ihre
Bedeutungen zu enträthseln"

(MEYEN, 1829).

Preface

BOTANISTS have written treatises on aquatic plants, on carnivorous plants, on climbing plants, on succulent plants, but a general book covering parasitic phenomena in higher plants is new in conception, or at least in execution.

This book, like most books, was conceived and completed in too short a time. Nothing would have pleased me more than to repeat or extend many studies referred to in this work. The working botanist faces these temptations continually, and must make compromises whenever committing material to paper. It is useful and, indeed, necessary to survey the scientific scenery from certain vantage points, even though recognizing that its topography obscures some areas and that its horizons are imaginary. This book represents such a survey, and is similarly defective. I would issue a plea to those who discover such defects, large or small, to do me the honor of communicating them to me so that, if a later edition materializes, I can make amends.

The plan of this book is a simple one. Its backbone is provided by what I should like to think of as up-to-date biological monographs of the parasitic groups. Aside from these chapters there is an introductory chapter dealing with matters of historical, medicinal, and folkloristic interest. Chapter seven represents as comprehensive a treatment of the haustorium as space would permit. The final two chapters are general ones in which I have drawn together, from all parasitic groups, items relating to the physiology and evolution of parasitic angiosperms.

I have based this book largely on the splendid resources of the library of the University of California at Berkeley. I have been able to add information or illustrations from the University Herbarium in nearly all parasitic groups.

The use of various facilities of the Department of Botany at Berkeley is gratefully acknowledged. By far the largest share of this book was written during the tenure of a Guggenheim Fellowship for which I am very thankful. My wife and Dr. Robert Ornduff read the entire manuscript, Dr. W. M. Laetsch parts of chapters 6, 7, and 8, and Peggy Bloom chapter 1. Their numerous suggestions are much appreciated.

In dedicating this book to my wife I do not merely follow an accepted tradition. Only those who have been close to my family during the work reflected in these pages can appreciate how much sacrifice on her part has been necessary, and how willingly her help was given. If this book deserves any credit, a very large part is hers.

J. K.

Contents

Parasitic Flowering Plants and Man

PARASITES have enthralled the naturalist for centuries. The aura of fascination surrounding them is due in part to an intuitive fear, however unjustified, of all things parasitic. Thus Erasmus Darwin (1825, p. 171) was moved to compare "the harlot-nymphs, the fair Cuscutas" with mythological monsters:

> Round sire and sons the scaly monsters roll'd,
> Ring above ring, in many a tangled fold,
> Close and more close their writhing limbs surround,
> And fix with foamy teeth the envenom'd wound.

This feeling of apprehension is a natural extension from animal parasitology. The biologist, however, sees parasites as superbly successful species in the great mosaic of living things. Indeed, there is genuine admiration in his attitude toward them as species which have, during their evolutionary history, gradually constructed so precise a relationship to other living things as to loosen many of the physiological shackles oppressing nonparasitic forms. That a price has been paid for this security need not diminish his admiration. If, as has been said, evolution is opportunistic, parasites are the greatest opportunists of all.

The vast scientific literature on parasites is dominated by work on animal parasitism and, more recently, parasitism by lower plants, bacteria, and so on. Innumerable are the books and scientific articles in which "parasitism" refers only to human or animal parasitism, or only to parasitism by cryptogamic plants. When Caullery (1952) writes his treatise on "Parasitism and Symbiosis," and Yarwood (1956) gives an account of "Obligate Parasitism," they ignore parasites among higher plants. There is little sense in taking issue with this well-established meaning. Such a conception of parasitism is, nevertheless, incomplete. I hope that the present work will in some degree restore a sense of balance in regarding parasitism as a more general biological phenomenon.

There is no single predecessor for a book on parasitic flowering plants. The vast and disunited scientific literature dealing with angiospermous parasites is largely unsorted. No serious attempt has been made to compare and harmonize the various modes of parasitism in flowering plants. The early work of Solms-Laubach may come closest to it, but antedated the discovery of parasitism in some cases. One or two more recent discussions are helpful but not sufficiently comprehensive. Perhaps the most glaring defect of Christmann's (1960) little book is that it ignores several families in which parasitism is known or suspected. The discussion of parasitic plants by Schmucker (1959) may be the best modern treatment, although it omits the families Myzodendraceae, Olacaceae *sensu lato*, and Krameriaceae. The literature review by Subramanian and Srinivasan (1960) also neglects Myzodendraceae, and is little more than an uncritical compilation.

The most important concept in this book, and possibly the most intangible, is of course parasitism. Definitions abound, and the most commonly accepted are not necessarily the most precise. For a working concept of parasitic relationships in higher plants let us, for the purposes of this book, insist on the criterion of a natural and specialized physiological bridge (haustorium), composed at least in part of living tissues, through which nutrients and water are transported from one organism to another. That this concept is sometimes fallible need not diminish its present usefulness.

In order to unify the resulting notion of parasitism, we must consider a number of special cases. An intimate fusion between a plant of *Passiflora* and one of *Evonymus* was reported from France (Peé-Laby, 1904). A spreading, haustorium-like body of the *Passiflora* stem extended into the host wood. Remarkable as this report is, it represents an extremely rare occurrence. We may wish to think of this relationship as parasitic.

Similar observations were made in North America (MacDougal, 1911). Both *Cissus laciniata* and a species of *Opuntia* were discovered growing on a *Yucca*, and *Opuntia toumeyi* on *Cercidium microphyllum*. Moran (1966) found a vigorous plant of *Opuntia* growing on *Idria columnaris* (fig. 1-1). No true organic fusion has been demonstrated in these American reports. It is possible that these were no more than root developments in a host cavity, even if the roots had something to do with the formation of this cavity. I shall return to these reports when discussing the origin of parasitism (chap. 9).

The suggested view of parasitism in higher plants excludes a number of other mutual interactions between different species which are sometimes looked upon as parasitic. The stems or roots of neighboring forest trees are occasionally so closely associated that their continuing secondary growth brings about a degree of fusion. The unrelenting pressure in both cambial layers makes such a structural union inevitable where the two organs are confined to a small space. It is not impossible that some materials pass from one into the other partner. Kadambi (1954) noted that in a *Eugenia-Santalum* fusion (nonhaus-

Fig. 1-1. *Opuntia* in a possibly parasitic relationship with a tree of *Idria columnaris*, Baja California (courtesy Reid Moran, after Moran, 1966).

torial) the fruit of the former had the distinctive flavor associated with sandal heartwood. A similar fusion of the trunks of *Ficus infectoria* and *Vateria indica* seemed to have resulted in transfer of materials (Kadambi, 1954). In coniferous forests, root grafting can be very common, and large groups of trees can be organically united by a complex of interspecific grafts (Kozlowski, 1964). Sizable amounts of nutrients are thus transmitted (Woods and Brock, 1964), as can be seen by the continued growth of callus on the stumps of certain trees which have been felled. While I do not deny the important ecological implications of these observations, they fall largely outside the scope of this book. Root grafts seem to be manifestations of secondary growth, and involve no haustorial organs. (On this phenomenon as a possible origin of the parasitic mode of life see chap. 9.) Aside from actual root grafts, mycorrhizal connections between adjoining individuals may also transmit nutrients.

It is held in some quarters that true saprophytism does not occur in angiosperms; that such plants as *Sarcodes, Monotropa,* or *Corallorhiza* in reality are parasitic on the fungus intimately associated with their roots. There is little doubt that various materials are transported from one partner to another in such plants. If so liberal a concept of parasitism be pursued, it becomes even more impossible to exclude the vast number of vascular plants which, while not holosaprophytic, are nevertheless mycorrhizal. Mac-Dougal (1911) has estimated that about half of all seed plants use complex food materials derived from other organisms through a mycorrhizal or parasitic arrangement. Definitions of this sort are arbitrary, but I feel that the most natural limits for a treat-

ment of parasitism separates the plants of this book from those having mycorrhizal associates. I agree with Caullery's (1952, p. 285) conclusion: "Commensalism, parasitism and symbiosis are only categories created by us and as soon as they are thoroughly analyzed it is impossible to delimit them." My definition of parasitism, if fallible, is more a pragmatic than a biological definition.

The truly symbiotic relationship which is often thought to represent the evolutionary maturation of a parasitic one is totally lacking in the mutual relationships of higher plants. This is not to say that the attacked individual might not suffer visible injury or, if it does, might not still reproduce successfully. But even in the most advanced angiospermous parasites the part of the host which is parasitized is not normally reproductive (e.g., in *Pilostyles* and *Arceuthobium* parasitism). If the evidence of nutrient transfer from the parasite to the host happens to be particularly striking (Tubeuf, 1923; Docters v. Leeuwen, 1954; Kuijt, 1964c) this should not deceive us, for such benefits are illusory, and the host's reproductive capacity is in fact curtailed or eliminated. A situation in which both partners draw significant benefits from an intimate nutritional contact is unknown among higher plants.

The difficulties in separating holosaprophytes from true parasites have caused a number of erroneous statements to creep into the botanical literature. Some Burmanniaceae, as *Mamorea*, are greatly reduced saprophytes, and it is no wonder that at least two students of this group refer to them as parasites or probably so (Jonker, 1938; Sota, 1960). More striking is the case of *Salomonia* of Polygalaceae, which was said by both Blume (1823) and Griffith (1845) to be parasitic on roots. Subsequent authors have repeated this assertion. Blume even placed

Salomonia next to Orobanchaceae! As Penzig (1901) has pointed out, there is no evidence for the parasitic nature of this genus, the members of which are mycorrhizal saprophytes. Monotropaceae also carry the stigma of parasitism at times (Hutchinson, 1959).

In a separate category, understandably, are parasitic manifestations associated with the reproductive cycle of flowering plants. These phenomena, interestingly enough, find some of their most extraordinary expressions in the ovaries of certain parasitic plants, especially in the Santalalean lineage.

Following the suggested concept of parasitism, we might ask where, in higher plants, parasites may be encountered. The first fact we encounter in this initial survey is that no parasites are known from any vascular cryptogams. In the gymnosperms there is a solitary report of a possibly parasitic conifer from New Caledonia, *Podocarpus ustus,* a report which requires further documentation (De Laubenfels, 1959). Except for this possible instance it would seem that there are no parasitic gymnosperms. Perhaps more surprising is the absence of parasites from the monocotyledonous ranks. It is a startling and unexplained fact that the known parasitic vascular plants are limited to the dicotyledons.

When we focus on the latter group and try to discover where parasitism has arisen we are faced with taxonomic difficulties. These can scarcely be anticipated in the present chapter, and the reader will need to refer to the various groups to find justification for the systematic notions underlying the following statement. *It is my thesis that parasitism has arisen at least eight different times in unrelated groups of dicotyledons:* (1) Santalales; (2) Scrophulariaceae and Orobanchaceae; (3) Rafflesiaceae and Hydnoraceae; (4) Balanophoraceae; (5) *Cuscuta;* (6) *Cassytha;* (7) Lennoaceae; (8) *Krameria.*

From the illustrations in this book it can readily be seen that the various parasitic groups have little in common in their general appearance. They include trees and shrubs and herbaceous perennials and annuals. Many do not betray their parasitic nature to the onlooker. Other parasites are among the most exotic productions of the plant kingdom, so extraordinary in appearance and mode of life that for years botanists could not believe them to be flowering plants.

Whether parasitism, at least within each of the first three groups, might have evolved more than once may never be known. The problem is particularly interesting in the first two groups, where undoubted parasites are closely related to nonparasitic forms. Are plants of both modes of life derived from parasitic or from autotrophic ancestors? If the latter, we might expect a closer affinity between parasitic members than between them and autotrophic ones. This may, in fact, not be true in Santalales. It would be difficult to maintain that the three known parasitic Olacacean genera (*Olax, Cansjera, Ximenia*) are more closely related to the mistletoes than to *Heisteria*

longipes, which is apparently independent. If the immediate ancestors uniting parasitic and nonparasitic Santalales were autotrophic, it is possible that parasitism within the order is of multiple origin. This is not too far-fetched, perhaps, if we consider the insectivorous habits of Utriculariaceae. The insect-trapping apparatus of *Utricularia* could have had a different evolutionary origin from that of *Pinguicula.* These seem to be two distinct manifestations of an underlying predisposition. Is it not conceivable that the genetic make-up of Santalalean plants includes a degree of predisposition to parasitism? This would allow for the origin of this habit at various times and in various groups.

If the common ancestors of plants having these different modes of life were parasitic forms, on the contrary, the consequence would be different: the loss of haustorial organs, the conversion from parasitism to the autotrophic state. This possibility, as far as I am aware, has not been mentioned in the botanical literature, perhaps because it seems unlikely. It is significant, in this context, that the roots of the autotrophic *Heisteria longipes* (Olacaceae) should be devoid of root hairs, a condition often thought to be related to a parasitic way of life (see chap. 9). Even though a parasitic ancestry of such independent forms is at present nearly insupportable, the possibility must be kept in mind.

Our awareness of parasitism in higher plants has been slow in coming. Rural people in many parts of the world have probably been acquainted with the phenomenon much longer than the surviving records indicate. The effect of parasites such as *Striga, Orobanche,* and *Cuscuta* on crop plants can be devastating and obvious, and must have been particularly so to people eking out a living by primitive methods.

The earliest reference to parasitic plants might well be that of Theophrastus (*De Causis Plantarum,* XVII), who gives a vague report on a dodder-like plant covering bushes in the fields of Babylon, and even reaching up to small trees. This information, repeated later by Pliny, is not precise enough to identify the species of *Cuscuta* (Mirande, 1900). The information might conceivably have reached Theophrastus from seafarers, and have come instead from *Cassytha,* a coastal plant much like dodder but belonging to Lauraceae. Pliny's reference to the plant's aromatic odor makes Lauraceae even more likely. It may be, therefore, that Dioscorides in his Materia Medica is the first writer who clearly speaks of *Cuscuta* (*C. epithymum;* Mirande, 1900).

Theophrastus was aware of mistletoes also, and seems to have known of both *Viscum album* (fig. 1-2) and *Loranthus europaeus.* A full citation of the original texts and an interpretation of Theophrastus' statements are found in Tubeuf (1923).

The writings of a group of Arabian scholars of the tenth century indicate that these observers had a reasonably accurate conception of the parasitic nature of some plants. I give a translation of the relevant passage in Dieterici (1861) probably based on

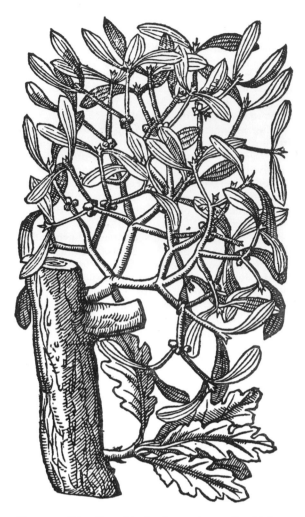

FIG. 1-2 Two parasites illustrated in herbals: left, *Viscum album,* common European mistletoe, on oak (M. de l'Obel, Kruydtboeck, 1581); right, *Pedicularis sylvatica,* a lousewort (M. de l'Obel, Plantarum seu Stirpium Icones, 1581).

Cuscuta: "Another kind, the actions of which correspond to those of the animal soul while its body remains that of a plant, is the parasitic plant, for it attaches itself to trees, seeds and thorns, and feeds itself as the worm from the juices of its host plant, thus with its soul carrying out the actions of animals." If we substitute for "the actions of the animal soul" the notion of heterotrophic life, we obtain a statement of parasitism with which it is difficult to quarrel.

As far as the Western world is concerned, the earliest recognition of parasitism in several groups is difficult to pinpoint. I have found no clear statement of the parasitism of Rafflesiaceae prior to Robert Brown's momentous publication on *Rafflesia arnoldii* (1822), but I would not be surprised to know that such a statement exists for *Cytinus.* No exact starting point can be given for the notion of parasitism of *Cuscuta, Cassytha, Orobanche,* and mistletoes. Probably the first person to realize that *Cynomorium* of Balanophoraceae was not a fungus but a parasitic plant on the roots of other plants was Micheli (1729). A century later a new parasitic group, the genus *Lennoa,* was added by La Llave and Lexarza (1824). Myzodendraceae was added to the list by Hooker (1846). The botanical world was taken by surprise when Mitten (1847) revealed that *Thesium linophyllum,* a plant known to many generations of

botanists, was a parasite on roots of neighboring plants. Mitten's note reminded Decaisne of the difficulties of growing *Pedicularis* and *Melampyrum* ornamentally. He examined the roots of several Rhinanthoideae (*Melampyrum, Odontites,* and *Rhinanthus*) and made the exciting discovery that these plants, indeed, had haustorial attachments to other plant roots (Decaisne, 1847). *Lathraea* was a known parasite long before that time, however (Meyen, 1829; Bowman, 1833; Unger, 1833), but was not regarded as a member of that subfamily until much later. *Krameria* completes the present list of angiospermic parasites (Cannon, 1910). The possibility of parasitism in *Podocarpus ustus* was raised more recently (De Laubenfels, 1959). The parasitism of individual genera of Olacaceae, Santalaceae, and Scrophulariaceae is discussed in the relevant chapters below.

It must not be thought, however, that the existence of discrete parasitic plants was easily accepted by the general botanical public. Trattinick (1828) could not bring himself to believe that the various Balanophoraceae and Rafflesiaceae which were then becoming known were plants in the proper sense of the word, and he called them "phanerogamisirende Schwämme," which demonstrated a type of "vegetabilische Verrücktheit." Trattinick proposes the erection of a special category for these grotesque vegetable productions, Sarcophytae. His rationale is so hilarious that I cite the original:

. . . so bleibt uns nichts übrig, als sie in eine eigene Categorie, als Sonderlinge zusammen zu werfen, etwa wie man in einem Irrenhause die Geisteskranken zusammenbringt, deren Manien höchst verschieden sind, von denen jedoch keiner das ist, was er zu seyn vorgiebt, oder sich einbildet. Auch unsere Sarcophytae . . . ahmen auf eine barocke Weise etwas nach, was sie nicht sind, so die Aphyteja ein Hydnum, das Cynomorium eine Typha, die Rafflesia eine Brassica capitata, der Cytinus eine Cotyledon, die Balanophora, eine Arum, das Corallophyllum eine Clavaria, die Sarcophyte eine Brassica Botrytis u.s.w.[1]

---

[1]". . . we therefore have no choice but to cast them together, as oddities, into their own category, much as in an asylum we bring together the mentally ill, whose mania are extremely varied, but of whom no one is really what he pretends or imagines to be. Our Sarcophytae also imitate, in a baroque fashion, something which they are not; *Aphyteja* (*Hydnora*) imitates a *Hydnum, Cynomorium* a *Typha, Rafflesia* a *Brassica capitata, Cytinus* a *Cotyledon, Balanophora* an *Arum, Corallophyllum* (roots of *Lennoa*) a *Clavaria, Sarcophyte* a *Brassica botrytis*, etc."

The fact that this opinion was published shows that it was taken seriously. When Thunberg (1775) published his description and figures of *Hydnora* he considered it a fungus related to *Lycoperdon* and *Clavaria*. As late as 1830, in his *Flora Javae*, a botanist as eminent as Blume thought that Rafflesiaceae and Balanophoraceae were closely allied to fungi. Junghuhn, "the Humboldt of the East Indies," could not believe that Balanophoraceae would grow from seed (Goeppert, 1847). Meyen (1829) had tried to prove that parasites such as *Lathraea, Orobanche,* and Rafflesiaceae are malformations emanating from the roots of other plants, without the necessity of seeds. Yet he must have recognized a degree of distinctness in the *Rafflesia* plant when he held that its "calyx" (bracts) is transitional between the parasite proper and the "mother plant." He felt that, taxonomically, such organisms are to be aligned with the mother plant wherever possible.

Another interesting misconception with regard to parasitic angiosperms was that based on dodder growing on the grapevine. Dodder sometimes becomes established on the stalks of the fruits and even on the fruits themselves. A profuse, beardlike mass of filamentous stems then grows out from the bunch of grapes. This "uva barbata" of early western writers became associated with bizarre legends and botanical mistakes (Mirande, 1900). No less a botanist than Tabernaemontanus erected a separate genus for it, illustrating it in his *Kräuterbuch,* where it remained even in editions up to 1687. To another famous early botanist, J. Bauhin, "uva barbata" was a monstrosity of the grapevine. John Gerarde (1633, p. 226), may have been the first to realize and describe the exact nature of the "bearded grape." Even a century later Guettard (1744) thought the phenomenon worthy of illustration (fig. 1-3). Although Guettard and his

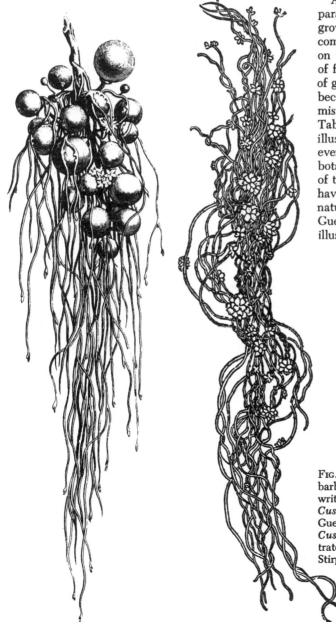

FIG. 1-3. Left, the "Uva barbata" of early botanical writers, being parasitism of *Cuscuta* on grapes (after Guettard, 1744); right, *Cuscuta europaea* as illustrated in Fuchs (De Historia Stirpium, 1542).

more sophisticated contemporaries were in doubt about the situation, the rural superstitions connected with it probably survived long after this time. As late as 1831 the finding of such a "monstrosity" was thought to be related to the appearance of a comet in the sky of the previous year! One of the striking features of the "uva barbata" is the laxness of the dodder branches, as is well illustrated in Guettard's and Tabernaemontanus' figures. Zietz (1954) has shown that the attenuated reaching out of the dodder stem is related to an abundant supply of nutrients. It is therefore conceivable that the nutritive wealth of the grapes and their supporting stems brings about this attenuated and rapid growth.

The superb monograph on Balanophoraceae by Hooker (1856) cleared the way for a more rational appreciation of the parasitism of higher plants. There could be no doubt of the individuality of Balanophoraceae as parasitic plants. If this were true for these most fungoid of parasites, the older notion was even more difficult to defend for the others, especially after the early haustorial study of *Lathraea* (Bowman, 1833) and the demonstration of seed germination in *Cynomorium* (Weddell, 1858-1861). When Solms-Laubach (1867-1868) published his comprehensive account of the haustorial organs of angiospermic parasites, the apparent blending of parasite and host turned out to be merely an illusion.

THE USES OF PARASITIC PLANTS

The local abundance and occasional prominence of some parasitic plants have provided an invitation to man to make use of them. Throughout the world, many parasites are utilized in a variety of ways most of which are unrelated to the parasitism of the plants. Indeed, the parasitism of many terrestrial parasitic plants has been discovered only fairly recently and cannot have been known to many people in earlier times.

The succulent nature of some parasitic plants in arid regions has made them a source of food for both man and animals. The fleshy stems of *Ammobroma,* growing in one of the most arid deserts of the world, were baked and eaten by Papago Indians (Gray, 1854). *Lennoa* is eaten as a vegetable even in present-day Sinaloa (see chap. 6). In Central Asia and Afghanistan, two species of *Orobanche* are said to have been eaten like asparagus (Kirk, 1887), although on the herbarium labels of *Orobanche* collected by Dr. R. W. Chaney in Mongolia we read that these plants, when eaten by camels, make the animals ill. *Boschniakia* was eaten by Indians in British Columbia (Henry, 1915). The pulpy fruits of the two species of *Prosopanche* in Argentina are eaten, either raw or fried, by Indians (Cocucci, 1965), although the numerous hard seeds give the flesh an unpleasant, sandy texture. The fruits of *Hydnora* are consumed, after having been baked, by Bushmen, Hottentots, and Somalis (Vaccaneo, 1934;

Story, 1958), who must compete with jackals, foxes, and porcupines. On Madagascar *Hydnora* is said to produce one of the best fruits locally available (Jumelle and Perrier de la Bâthie, 1912).

In less arid regions other parasites occasionally enter into the local food economy. According to Hooker (1856), at least three genera of Balanophoraceae are eaten: *Cynomorium coccineum,* on the Canary Islands; *Lophophytum;* and *Langsdorffia,* in the mountains around Bogotá. He states that the "soft receptacle of *Langsdorffia* is eaten when ripe, and considered stimulating and refreshing." One wonders if this is Hooker's discreet Victorian way of reporting on the use of these plants as aphrodisiacs (see below). Two further genera of Balanophoraceae, not then known to Hooker, have been added to the list of comestibles. The natives of New Caledonia eat young inflorescences of *Hachettea* (Rendle *et al.,* 1921), and the curious subterranean *Juelia* of the Andes is also edible (Asplund, 1928). In Scrophulariaceae it is said that the root of *Pedicularis* is edible (Szczawinski and Hardy, 1962), although its actual use does not seem to have been established.

Various mistletoes are cut from their hosts and fed to cattle during times of drought in Africa (Marloth, 1913), a practice known even in the time of Theophrastus (*De Causis Plantarum,* II.17). In Texas the poor grades of *Phoradendron serotinum* intended for the Christmas market are often fed to sheep and cattle (Griffis, 1956). Tubeuf (1923) reports that in contemporary Europe the same thing was known, though mistletoe (*Viscum album*) is sometimes locally regarded as a poisonous plant. In fact, the recurrent reports of toxic qualities of mistletoes should give cattle-breeders pause for thought. Twisselmann (1956) states that abortion may follow when pregnant cattle feed on *Phoradendron.* In the more rustic days of the American West, a decoction of mistletoe was rumored to be similarly effective in human pregnancy (Twisselmann, pers. comm.). Kingsbury (1965) refers to the presence of toxic amines in *Viscum* and *Phoradendron,* and states that cattle have been killed by eating *Phoradendron* in California. While such reports have not been substantiated, they are sufficiently persistent to warrant caution.

When we next consider the use of wood we can immediately restrict our concern to a handful of genera of Olacaceae and Santalaceae. In the Indo-Malayan region, *Scorodocarpus, Ochanostachys,* and even *Ximenia* are put to such structural uses as bridge piles, telephone poles, and house construction (Heyne, 1950). *Strombosia javanica* is prominent locally in house construction, as it appears to be resistant to termite attack. The related *Anacolosa,* on the contrary, lacks this resistance. In Africa the genus *Coula* produces a heavy, hard, dense, very durable wood which is immune to termite attack and therefore has many construction uses (Dalziel, 1937). *Ongokea* is used for carpentry work (Louis and Léonard, 1948). In South America the closely related

Minquartia (perhaps the Black Manwood of commerce) has a similarly high reputation, and is used for house posts, railway ties, and the like (Record and Hess, 1943).

Perhaps the most famous of tropical woods is sandalwood, *Santalum album* (fig. 1-4). The economic importance of this precious aromatic wood goes back to the dawn of history (Metcalfe, 1935). So much has the tree been planted through the centuries that its original home is a matter of controversy (Sprague and Summerhayes, 1927). Some of the earliest records seem to point to Timor as the source of the best and most abundant sandalwood, and this has prompted the suggestion that *S. album* is not truly native to the Indian subcontinent. However, natural forests of *S. album* have since been found in the heads of ravines in Maharashthra State, India (Choudhuri, 1963). It is possible, of course, that this tree was widely dispersed through the Indo-Malayan region, though it may have had a scattered distributional pattern and great variations in frequency. A summary of sandalwood geography was produced by Skottsberg (1930), who included the Australian *Mida* and *Eucarya* in his survey.

The high commercial regard for sandalwood has resulted in an intensive search for new stands and for related or other species with similar characteristics (Metcalfe, 1935). Because of the unrelenting demand, a number of island populations of the tree were virtually wiped out early in the nineteenth century (Rock, 1916). The destruction of the Hawaiian *Santalum freycinetianum*, which formerly existed in magnificent groves, began in the late 1700's. The rise of the Hawaiian monarchy under Kamehameha is said to have been based mainly on the income from this precious wood. In a single year Kamehameha and his subjects received $400,000 from this source. Most sandalwood seems to have been sold on Chinese and Polynesian markets. Under the next monarch, however, the supply began to ebb, and after 1825 the wood lost its significance as a national source of income. An account of the recent partial recovery of this tree on Oahu has been given by St. John (1947).

Santalum fernandezianum of the Juan Fernandez Islands seems to have been exterminated in this fashion. On his visit to the islands Skottsberg (1910) found only a single gaunt surviving tree. As far as I am aware, the species has not been collected again.

A number of substitutes for true sandalwood have been introduced, none of which are endowed with the high qualities of *S. album*. Australian sandalwood is derived mostly from *Eucarya spicata*, a close relative of the true sandalwood. Similar wood of high quality comes from Polynesia and New Caledonia. East Africa contributes *Osyris tenuifolia* (Engler and Volkens, 1897), and *Ximenia* is used as a substitute even in India. In addition to these Santalalean substitutes, Metcalfe (1935) gives the following unrelated local substitutes: *Erythroxylum monogynum* (Erythroxylaceae; India), *Eremophila mitchelli*

Fɪɢ. 1-4. *Santalum album,* the famous sandalwood, in National Botanic Gardens, Lucknow, India (courtesy Dr. S. P. Bhatnagar, from Bhatnagar, 1965).

(Myoporaceae; Australia), *Brachylaena* (Compositae; African region). The sandalwood of West Indian and Venezuelan commerce is believed to be *Amyris balsamifera* (Rutaceae). The sandalwood for King Solomon's temple was probably yet another substitute, *Pterocarpus santalinus* (Leguminosae), although the Egyptians knew and used true sandalwood as long ago as 1700 B.C. (Boerhave Beekman, 1949-1955).

None of the substitutes approaches the India sandalwood in quality of wood and oil. The State of Mysore, where its culture even today is a state monopoly, is the center of sandalwood country; Bombay is the main commercial center and port. Sandalwood, a slow crop, takes about thirty years before the precious heartwood has reached sufficient volume to justify felling the tree. The roots are also utilized, and are particularly high in oil content. The pale yellow sandal oil is distilled from this wood in large copper vessels over a period of ten days. Aside from the production of sandal oil, the wood is a favorite material for intricate carvings, and is burned ceremoniously at burials and other religious rituals of Hindus, Moslems, Buddhists, and Parsis. Some of the information above comes from Engler and Volkens (1897), but the recent account of Mohan Ram (1964) shows that the situation has not changed much since the beginning of the century, when sandalwood constituted the main income of the royal house of Mysore, as it had some hundred years earlier in Hawaii. Today every tree of *Santalum album* in Mysore, no matter where it grows, is the property of the state.

Some of the earliest Sanskrit medical works speak of the beneficial uses of sandalwood products. Through the centuries these products have been applied to sufferers from gonorrhea (Heyne, 1950). While modern medicine has, of course, replaced sandalwood in this respect among more sophisticated Indians, the ancient folk medicine persists rurally (Mohan Ram, 1964).

The major use of the precious sandal oil in eastern countries seems to be in anointing the body. India alone uses the equivalent of 3,000 to 4,000 tons of sandalwood annually. Sandal oil is used in the manufacture of soap and perfumes; aside from its own famous aroma, it fixes the more volatile and delicate fragrance of jasmine, rose, and other flowers (Heyne, 1950). The chief export market at present is France. The United States buys much from Australia.

A number of plant products are derived from parasites, products which are not important commercially but nevertheless are of interest. *Cuscuta tinctoria*, as its name indicates, has been used in the preparation of a dye (Gaertner, 1950), and the same is true of some Indian species (Watt, 1889-1896). *Hydnora* in parts of Africa is so abundant and contains so much tannin that it has been gathered for tanning purposes (Vaccaneo, 1934).

The use of birdlime, normally prepared from the berries of mistletoe, is not restricted to Europe. It has been reported from Africa (Dalziel, 1937; Watt and Breyer-Brandwijk, 1962), where the berries are macerated in water and manipulated until the viscous material sticks to the fingers. In Europe, birdliming has been practiced since time immemorial (see chap. 2), but it would not be surprising if it had disappeared in the Europe of today.

In at least two genera of Balanophoraceae an inflammable, waxy material is so abundant that a very useful wax may be manufactured from them. At the time of Hooker (1856) the stems of *Langsdorffia* were sold on Bogotá markets to be used directly as candles on saints' days. It would be interesting to know if this were still true today. Plants of *Balanophora elongata* and *B. globosa* are used in the same way in the Tjibodas region of Java (Heyne, 1950). While the *Balanophora* plants, like *Langsdorffia*, can be used directly as fagots, a small, rapidly burning candle is also manufactured from them (Ultée, 1926). The plants are too rare, however, to represent a commercial commodity.

An ingenious use has been made of *Cuscuta* in the study of plant viruses. This genus of parasites has a very wide host tolerance, and a single individual can be made to parasitize completely different host species. Thus a degree of physiological continuity exists. If one host happens to be infected with a virus, this virus may travel via the haustorial connections not merely into the dodder but even into the second host, thus infecting it. The dodder, as it were, serves as a "virus bridge." This remarkable technique has widened the study of host ranges of viruses. Earlier techniques by means of grafting were severely limited by incompatibility factors. To cite an example: we know, through studies of such virus bridges of *Cuscuta campestris,* that the witches'-broom virus of alfalfa will grow on carrot, tomato, and periwinkle, but not on parsley (Kunkel, 1952). The possibilities are nearly unlimited because of the large number of *Cuscuta* species and their great host latitude. More than a dozen species have now been used successfully as bridges for a great variety of viruses (Hosford, 1967). The ability of a dodder species to transmit a virus, nevertheless, is not a generalized one. There are numerous examples of dodder species which are selective in the viruses transmitted. Also, the viruses which pass through the dodder do not necessarily reproduce within it. The question as to the mechanism of selection has scarcely been raised, let alone answered. It may well be that the great structural variability of the haustorium (see chap. 7) is correlated with its effectiveness as a virus bridge. The genus *Cassytha*, notwithstanding its remarkable similarity to *Cuscuta*, apparently will not serve the same purpose (Sakimura, 1947).

FROM MEDICINE TO MAGIC, AND RELIGION TO FOLKLORE

In his efforts to influence supernatural powers and better his lot, primitive man turned to both animate and inanimate objects in his environment. Here again parasitic flowering plants have not escaped such uses, although, with the possible exception of the European mistletoe, parasitism was not an important element. The great pool of popular, more or less sophisticated beliefs and knowledge in this regard is as varied as it is intractable.

The reports on the medicinal uses of mistletoe have been so persistent through the years that even modern pharmacognosy has shown interest in them. Claims for the beneficial effects of mistletoe preparations have ranged from Sir John Colbatch's "most wonderful specifick remedy for the cure of convulsive distempers" (1719), similar uses in contemporary Africa (Soyer-Poskin and Schmitz, 1962), and many applications in South Africa, by both Africans and Europeans (Watt and Breyer-Brandwijk, 1962), to the relief of digestive distress in California Indians (Chesnut, 1902). Tubeuf (1923) has summarized the medicinal uses of *Viscum album* from the time of the Druids to our century. Even when Tubeuf's monograph appeared, the extent and variety of rural uses of this mistletoe throughout Europe were astounding. Decoctions were used against cramps and hemorrhages; and in certain parts of Austria the berries when eaten were thought to prevent pregnancy. In the early American West a similar tradition, possibily introduced by European immigrants, existed with respect to *Phoradendron*. Even more common in Europe were a great variety of therapeutic uses in cattle and other domesticated animals. Several recent articles on the subject (Thun, 1943; Pora *et al.*, 1957; Samuelsson 1958, 1959; Selawry *et al.*, 1961; Graziano *et al.*, 1967) present evidence for effects on blood pressure, for hemostatic and even carcinostatic activity. Mistletoe preparations were used against cancer by Mayas, and by Mediterranean people even at the time of Christ.

A decoction of *Quinchamalium* in South America is employed in the treatment of open sores; *Osyris* and *Thesium* in Africa provide remedies for many ailments (Pilger, 1935; Watt and Breyer-Brandwijk, 1962). Especially useful is *Ximenia* in Africa, which is used in cosmetics, and against hookworm and syphilis; it also increases the potency of bulls, and keeps witches away! Modern studies have revived the credibility of some indigenous uses: it has been demonstrated that a component of the root of *Thesium* called quercitrin causes dilation of blood vessels and thus a fall in blood pressure. The same substance stimulates the heart, and may regulate its rhythm. Whether the death of sheep that have eaten *Thesium namaquense* is related to these effects is not known. In other Santalales we encounter minor medicinal uses of *Heisteria*, *Ximenia*, *Strombosia*, and others (Pilger, 1935; Dalziel, 1937; Louis and Léonard, 1948; Heyne, 1950; Watt and Breyer-Brandwijk, 1962).

Beyond the Santalales the use of parasites in therapeutics is erratic and localized. *Thonningia* tubercles are said to be effective against headaches when taken in a macerated form (Lecomte, 1896). The bark of *Krameria* at one time was made into a dental powder, and had other medicinal uses in addition to the less reputable one of dilution of wine (Taubert, 1894). *Cynomorium coccineum* was held in high esteem during the Crusades for its hemostatic properties, at which time it was known as *Fungus melitensis*. *Prosopanche* still enjoys some repute among Patagonian Indians as a hemostatic agent, disinfectant, expectorant, and remedy for cardiac diseases (Cocucci, 1965), in the form of pollen, pulverized anthers, or decoctions of floral parts or roots. *Cuscuta*, in various parts of Europe and Asia, has been used as a powerful purgative and diuretic. In China, dodders provide remedies against fevers, angina, and rabies. Some of the minor medicinal uses of this genus in India apparently can be traced to Galen (Watt, 1889-1896; Gaertner, 1950; Watt and Breyer-Brandwijk, 1962). *Lathraea* has been mentioned in relation to epilepsy (Hegi, 1907-1931). A species of *Alectra* in India has many therapeutic uses and seems particularly promising in the treatment of leprosy (Bedi, 1967).

Between such uses and the place of parasitic plants in folklore, magic, and religion there is no clear dividing line. Somewhere near this division are the eyebright or "Augentrost" (*Euphrasia*) and "Zahntrost" (*Odontites*) of the early herbals, which recommended these genera as remedies for eye and tooth difficulties. The early American use of the so-called cancer-root (*Orobanche*) against that disease might point to a late survival of the doctrine of signatures, transplanted to the New World, for the roots of broomrapes have an irregular, coralloid appearance.

Some parasitic plants have been used as aphrodisiacs. The use of *Ximenia* for the benefit of bulls in Africa has been mentioned. Intended for human use is the so-called "muira-puama" from South America which is thought to be a product of *Ptychopetalum* (Anselmino, 1933) or *Liriosma* (Youngken, 1921). One preparation "is employed as a nerve stimulant and aphrodisiac in doses of 3 to 6 pills daily before meals," mainly by the French, reports Youngken. Also from South America, we hear of the use of Balanophoraceae as aphrodisiacs. Of *Lophophytum* it is said that secret consumption of the plants by young men makes it easier for them to attract members of the opposite sex (Harms, 1935). A similar action of the edible *Langsdorffia* in the vicinity of Bogotá is vaguely hinted at by Hooker (1956). Mr. Paul C. Hutchison, of the University of California's Botanical Garden at Berkeley, informs me that similar traditions persist in the Peruvian Andes. In 1957, when visiting Chiclayo, Peru, he saw plants of an unidentified genus of Balanophoraceae sold as an aphrodisiac in the local market.

Some of the vernacular names of parasites tell us that local people have recognized these plants as unusual. *Dactylanthus taylori* of New Zealand was known among Maoris as "Flower of Hades." The Caribs of the Caribbean island Dominica have respectful names such as "Maître Bois" and "Roi Bois" for their mistletoes (Hodge and Taylor, 1957). One of several names of mistletoes in West Africa is "God's Dropping," although no special religious beliefs seem to be connected with these plants in Africa. The Dutch and German names for dodder, "Duivelsnaaigaren" and "Teufelzwirn" (Devil's Yarn) testify to superstitions which long ago lost their power.

The European mistletoe *Viscum album* (fig. 1-5) is the legendary parasite par excellence — one of the very few plants that represents a living legend, and by far the most brilliant among them. Its fame goes back into prehistory, and has been transplanted to the mistletoes of other lands, especially *Phoradendron* of North America.[2] I am indebted to Tubeuf's monu-

---

[2]For information on the mistletoe industry in North America, see Howard and Wood (1955). Dr. Wood reports that most mistletoe found in Northern markets originates in Texas, New Mexico, and Oklahoma. A small amount of the mistletoe offered for sale in San Francisco recently turned out to be *Viscum album*, which had escaped detection for more than fifty years (Howell, 1966).

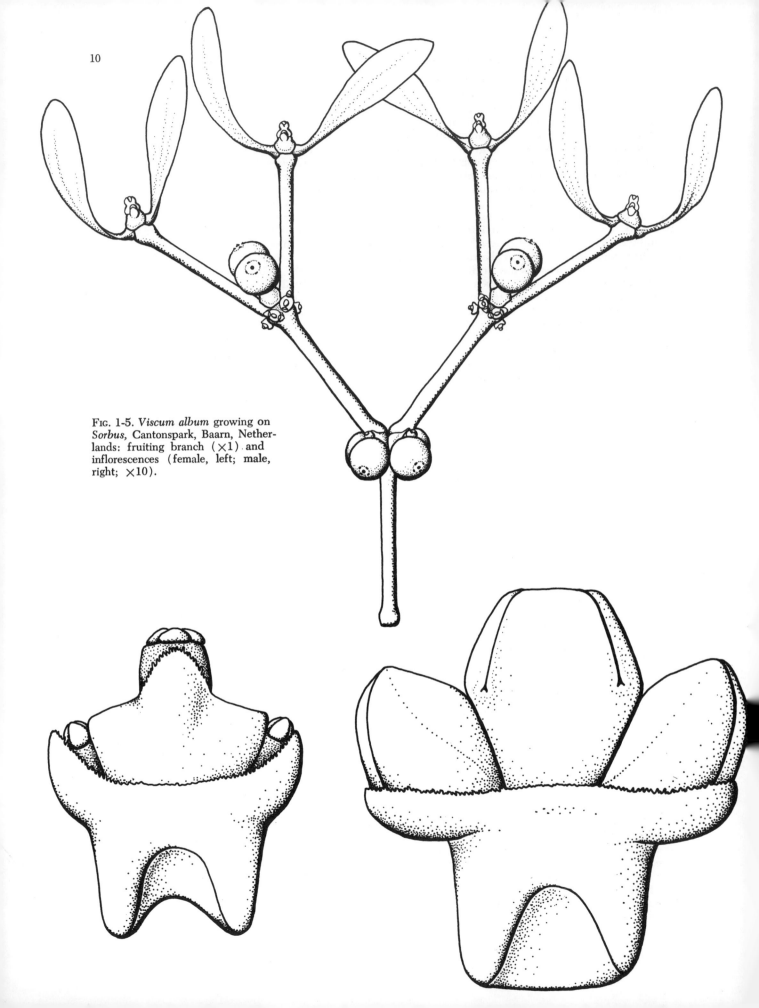

10

Fig. 1-5. *Viscum album* growing on *Sorbus,* Cantonspark, Baarn, Netherlands: fruiting branch (×1) and inflorescences (female, left; male, right; ×10).

mental monograph on *V. album* (1923) for the following account.

The legendary fame of the mistletoe is derived from two European sources which appear separate to us but may spring from the same body of cultural expression. The Golden Bough of Virgil's Aeneas, which provided the title of Frazer's (1900) classical study of magic and religion, is a mistletoe only by exegesis: it is compared to the mistletoe, but not identified as such.

The second source leaves no doubt about the mistletoe's identity, although the poets who wrote of the plant had no clear idea of its appearance. Mistletoe occurs neither on Iceland nor in northern Scandinavia, although it was more widespread in the past (Hafsten, 1957). One of the ancient Eddas of Iceland, dating from about A.D. 1220, represents a retelling of legends even at that time centuries old. There is abundant evidence that the relevant saga is not of medieval origin, but has its roots in the time of the Vikings.

The main elements of this famous saga may be paraphrased as follows. When Balder, the favorite of the gods and son of Odin, has threatening dreams, his mother, Frigg, places all objects and beings under oath not to harm him. But the evil Loki has observed that Frigg neglected to take the oath from one thing which seemed unimportant, the mistletoe, growing west of Valhalla. While the other gods amuse themselves in proving that even spears and stones cannot harm Balder, Loki takes the mistletoe to the god Hödur, Balder's brother, who is unable to take part in the sport because of his blindness. Loki hands Hödur the mistletoe and shows him how to throw it at Balder like a spear. When Hödur does so, the spear pierces his brother, who falls dead.

Frigg sends her son Hermod on Odin's steed Sleipnir to buy Balder back from the Realm of the Dead. Meanwhile, a funeral procession of all the gods, including Odin with his ravens and Valkyries and numerous giants, makes its way to the shores of the sea. Balder's wife Nanna dies of sorrow, and her body is placed next to his on the funeral ship. Many valuables, such as Odin's miraculous ring and Balder's steed, accompany the dead as the mortuary ship with its burning funeral pyre sails out to sea.

After a journey of nine days through deep, dark valleys, Hermod has reached the river Gjöll of the nether world, and is further directed by the guard at the bridge. Eventually he finds the mansion of the dead, where Balder is enthroned. Hermod receives a promise that Balder will be allowed to return if he is so loved that all creation mourns him. Messengers are sent far and wide, and all weep, even stones and metals. But on their return the messengers encounter a she-giant who refuses to shed even a tear for the son of Odin. Thus Balder is forced to remain in the Realm of the Dead.

An additional ending to this legend is given in another Edda, in somewhat apocryphal fashion, which describes the hopes of those remaining behind for Balder's return. But this will occur only at the end of time when the sun has lost its fire and the earth has sunk into the sea. Then a new generation of gods will receive the returning Balder into a new world of plenty where evil has vanished, and Balder will live peacefully with Hödur in the Hall of Odin.

The appeal of the legend is immortal. There is an unmistakable parallel with the events surrounding the life of Christ, but the two traditions are probably related only by descent: it is not a matter of one having been transformed into the other. Similar legends are known from cultures far removed from the European scene. Such tales of love and death and eventual return are part and parcel of the universal craving for immortality. It is conceivable, nevertheless, that the prominence of the Balder saga in the early Teutonic mind helped to pave the way for the eventual acceptance of the Christian religion.

The prominence of mistletoe shows the Balder saga to be an introduction from more moderate climates, for *Viscum album* could not have existed in northern Scandinavia or Iceland even at that early date. The tradition may have been part of the cultural stream which moved from Asia Minor into European regions in pre-Christian times. This does not necessarily mean that the central motif of the mistletoe might not have been incorporated into the legend in transit through Teutonic central Europe, although it is equally possible that even this element emanates from Asia Minor.

It is interesting to speculate on the reasons for the legendary importance of the mistletoe. Might there not be a connection between the parasitism of the plant and the lethal role assigned to it in ancient tradition? Or does it signify the primitive recognition of the mistletoe's power to draw tribute from the host tree?[3] There are many records of the mistletoe having been used as a symbol of strength or protection, as by carrying its branches into battle. The evergreen condition of *Viscum album* has undoubtedly added the appeal of immortality to this image. In the fifteenth century, rosaries of mistletoe wood were in favor, especially in the Rhineland. For these rosaries, most of which were made in the Czechoslovakian and Austrian area, both *V. album* and *Loranthus europaeus* were used.

No wonder the European use of mistletoe as a remedy for a great variety of ills of man and his animals survives into our century! It is a logical extension of the reverence of generations for this legend which lived among them. Even in modern Gaelic the name applied to mistletoe means "All-Heal."

In the mistletoe cult of the Druids the plant occupied a formal religious status. The information on this cult is based on the works of Pliny, who lived in the first century. Again I am guided by Tubeuf's

---

[3]The popular mind often does not distinguish between strangler figs and mistletoes; in many Latin American countries the name "matapalo" applies to both.

monograph (1923) as well as by Frazer's (1900). The Druids represented an order of priests assisting in religious rites in ancient Gaul, Ireland, and Britain. At certain times, probably at Midsummer Eve, a sacrificial ceremony was performed. The mistletoe, preferably growing on oak, was cut from the tree with a golden sickle by a Druid, and caught in a white cloth. Only then was the sacrifice made: animals and even human beings were slain and burned.

Viscum album is not, however, normally parasitic on oak. Many reports of this sort seem to have originated from the hyperparasitism of this mistletoe on *Loranthus europaeus*, which prefers oaks. During the winter, when *V. album* is most conspicuous, the leafless *Loranthus* blends with the leafless oak, and it is easy to confuse them (Tubeuf, 1923). An additional source of confusion lies in the fact that some North American species of *Phoradendron* are virtually limited to oaks. In Druidical times the powers of mistletoe were added to those of the oak, which in Europe was held sacred. That even the contemporary Western world has inherited the ancient esteem for the oak is shown in the use of oak leaves in some of the highest military decorations bestowed by Germany, the United States, and probably other countries.

In contemporary Europe there are many remnants of the Druidical ceremony, although they are rarely recognized for what they are (Frazer, 1900). In German folklore the cutting of mistletoe at the summer solstice was still encountered at the turn of the twentieth century. The Swedish Midsummer Eve celebrations, sometimes complete with mistletoe collecting, are such a relic. *Viscum album* has been legally protected in Sweden since 1910, however, and has enlarged its range considerably (Walldén, 1961). The bonfires until recently actually bore Balder's name, but did not involve the burning of specific objects. It is probably not going too far to say that the Swedish celebrations and similar ones in other parts of Europe, as well as the early Druidical ceremonies, represent the reenactment of the Balder saga in skeletal form, including the gathering of mistletoe and the burning of bonfires with or without the sacrifice symbolic of the death of Balder.

But what does this have to do with our modern tradition of kissing under the mistletoe? Our present usage of the mistletoe should be projected against its historical and legendary backdrop. The Teutonic gods and the rituals of the Druids are gone, and their place has been taken by the newer and more sophisticated mythology of Christianity. Yet, in the manner of reticulate evolution, two long-separated branches of the cultural tree have come together and fused in our modern Christmas. The place of the pagan mistletoe in this most prominent of Christian festivals today is more secure than ever.

The kissing tradition at Christmas seems to be purely English in origin. I cannot say when it originated, but I am under the impression that it is a rather modern extension of the mistletoe myth. It has been exported to North America, but seems to have gained little acceptance elsewhere. This is in contrast to the use of mistletoe for decoration and as a harbinger of good wishes and fortune. Tubeuf (1923) has chronicled the rapid increase in the popularity of mistletoe in the cities of continental Europe in the early decades of this century, and there is reason to believe that its popularity has increased since that time. It would be inaccurate to trace it all to Britain. We can see it also as a revival of ancestral customs which were dormant on the continent, but were still practiced occasionally here and there.

Efforts to trace the "kissing habit" to one specific ritual or custom nearly two thousand years ago are futile. This is not to say that this tradition of the English-speaking world might not be historically connected to the ritual of the Druids. But this is in the same category as a claim to be a descendant of King Solomon. A direct lineage of this sort appears only when looking backward. Neither in evolution nor in cultural history does any "pure line" exist. Ideas and customs are passed on from generation to generation, continuously changing and interbreeding under the influence of cultural cross-currents and modified local usage. The partners involved in the kissing ceremony under the Christmas mistletoe are no less a link in this cultural evolution for being blissfully unaware of it.

2

The Mistletoes

Since the traditional Christmas mistletoe is the *Viscum album* of Europe and, by substitution, *Phoradendron serotinum* of North America, it comes as a surprise to many people that mistletoes are not typically inhabitants of the temperate zones. Nearly all mistletoe genera are exclusively tropical or subtropical, and only a handful of species have successfully established themselves elsewhere. Outside the temperate zones there are more than seven hundred species of mistletoes. A recent estimate which is less conservative doubles this number (Barlow, 1964).

Botanists are used to having this largest of all assemblies of parasitic plants arranged in a single family, Loranthaceae. The evidence which has accumulated over many years has tended to emphasize the systematic distance between the only two subfamilies (Loranthoideae and Viscoideae), so much so that a separation into distinct families (Loranthaceae, *sensu stricto*, and Viscaceae) seems to have become inevitable. Although I accept, at present, the two-family status and feel that the families do have a somewhat different ancestry (see chap. 3), the mistletoes are here treated in a single chapter for the sake of economy.[1]

Loranthaceae, in this stricter sense, include about twice as many genera as Viscaceae. The most obvious difference between the two groups is the much larger size of the Loranthacean flower, except for the American genera *Oryctanthus, Ixocactus, Phthirusa,* and to a lesser extent *Struthanthus* and a couple of Old World mistletoes. Many tropical Loranthaceae have brilliant flowers of exceptional beauty, as do *Psittacanthus* and *Aetanthus*. Viscaceae, on the contrary, have inconspicuous flowers which are small and drab. Several other contrasts between the two families are evident. Loranthaceae are characterized by the rim-like outgrowth (calyculus) on the ovary, a Polygonum-type embryo sac, and a biseriate multicellular suspensor which is the longest one known in angiosperms. Viscaceae lack a calyculus, have an Allium-type of embryo sac but no suspensor, and have strictly unisexual flowers.

There is no single criterion, however, by which to separate the two families. Female flowers of *Viscum* show a calyculus-like rim (Schaeppi and Steindl, 1945); there are a number of dioecious mistletoes in Loranthaceae and a number of inconspicuous ones.

Another distinction frequently cited is the position of the viscid layer, which is said to develop to the inside of the vascular bundles (Viscaceae) or to the outside of them (Loranthaceae), but this distinction is probably an artifact. The various places of overlap do not, however, seem to constitute an embarrassment to the two-family status (see p. 50).

Perhaps it is the similarity of the mistletoe habit in the two families that has made it difficult to regard them as anything but subfamilies. We have actually gone very nearly full circle in these concepts. The idea of two separate families was first expressed by a number of botanists in the 1850's, when Viscaceae were formally set apart form Loranthaceae *sensu stricto* by Miquel (1856). Few botanists, however, paid heed to such opinions until details of a morphological nature were added a century later, through the work of Maheshwari and his students. The unique development of the Loranthacean embryo sac and young embryo has now been traced through many species of the Old World. The two-family concept has been revitalized and currently receives a considerable amount of support (Dixit, 1962; Barlow, 1964). Barlow is explicit in stating that the two families are the result of independent, convergent evolution toward aerial parasitism, from the related family Santalaceae.

Viscacean genera are usually clear-cut groups. The largest genera are *Phoradendron* and *Dendrophthora*. The latter has about fifty-five species of neotropical distribution. The former extends farther north and farther south. One can only guess at the size of *Phoradendron* as a genus. Taxonomists have greatly exaggerated the number of species, but it certainly is considerably larger than *Dendrophthora*. These two genera, by the way, are the only ones which are difficult to separate, for their distinction lies in the number of locules of the exceedingly small anther—hardly a convenient character.

In Loranthaceae the generic limits are often vague, and their circumscription has fluctuated greatly. Few genera are as clearly delimited as their Viscacean counterparts. *Oryctanthus* is one of these, and monotypic genera like *Tupeia* and *Ixocactus* are, of course, discrete entities. But in many other places in Loranthaceae, we discern a veritable spectrum of forms which produce difficulties in classification. The extremes of such a series of forms can easily be seen, so long as no lines need be drawn between adjacent genera. Further studies will undoubtedly clarify many of these problems.

This may be illustrated with an example in a

---

[1]In an article written after but published before this book I have found it necessary to resurrect Van Tieghem's Eremolepidaceae to accommodate *Eremolepis, Antidaphne,* and *Eubrachion* (Kuijt, 1968).

neotropical setting. At first sight there seems little reason for taxonomic confusion in *Phrygilanthus, Struthanthus, Psittacanthus,* and *Gaiadendron. Struthanthus* is a scandent shrub with rather drab flowers. *Phrygilanthus* and *Psittacanthus* usually have brilliant red flowers, but the latter's fruits lack endosperm. *Gaiadendron,* the giant of American mistletoes, forms small trees which may grow in the soil where roots of other plants are attacked, and has sprays of golden-yellow flowers. Unfortunately, *Phrygilanthus* seems to intergrade with *Struthanthus* on one side and with *Gaiadendron* on the other. The endosperm criterion of *Psittacanthus* is hypothetical for most species, as it has scarcely been studied. While some species have no endosperm in the mature fruit, there is no way of telling what happens in mistletoes which seem to form a transition between this genus and *Phrygilanthus.* To add to the confusion, the genus *Aethanthus* is segregated from *Psittacanthus* because it *does* develop endosperm. The fact that, in turn, *Struthanthus* is difficult to separate from *Phthirusa* blurs yet another line. One retains the impression of a galaxy of forms, some forming natural clusters and others serving to bridge adjacent clusters. Although such groups of plants present exasperating problems, they have a peculiar fascination for botanists, who return to them with ever-changing techniques and approaches.

Aside from the obvious preponderance of mistletoes in the tropics, a peculiar distributional feature was noted at a fairly early stage in their history. All genera but one are restricted to either the Western or the Eastern Hemisphere. The genera of New World mistletoes are not encountered in the rest of the world and vice versa. The dwarf mistletoes (*Arceuthobium*) are the only clear exception, and they are limited to the Northern Hemisphere, aside from an occurrence in the equatorial Aberdare Mountains of Kenya. *Phrygilanthus* in the Southern Hemisphere has long been considered the other aberrant genus in this distributional pattern, but Barlow (1962) contends that extra-American members should not be included in the genus. This is not to deny a reasonably close affinity between certain Australian and South American mistletoes. Aside from the mistletoes just mentioned, *Gaiadendron* of Central America and north-Andean America is closely related to *Atkinsonia ligustrina* of Australia.

A knowledge of the host range (conifers only) may help to clarify the present geographical distribution of *Arceuthobium.* A fairly recent continuity across the Bering Strait has allowed much more southerly floral elements than conifers to migrate from one side to the other. Over this reasonably continuous host bridge of conifers *Arceuthobium* encountered no serious barriers. Even today, *A. americanum* occurs in Canada's Northwest Territories (Kuijt, 1960-1961).

A somewhat different view of the geography of Loranthaceae is implied in the work of Balle (1960). She has pointed out that most primitive African Loranthaceae are those bearing the greatest affinities with Indo-Malayan ones. It is true also that Australian Loranthaceae are related more closely to both African and American members than the latter two are mutually. There emerges, therefore, the idea not so much of a hemispheric division but rather of a bipodal distributional pattern finding its extremes in Africa and America. This pattern emphasizes both the importance of the Atlantic dividing line and the historical subantarctic connection between South America and the Australian continent. The best evidence for the latter in Loranthaceae is the geography of the most primitive generic trio *Nuytsia* (W. Australia), *Atkinsonia* (Australia), and *Gaiadendron* (Cordilleran neotropics). Because Viscacean genera are rather isolated from each other, no support for or conflict with the bipolar pattern is evident in that family. In Santalaceae a similar situation is found (cf. chap. 3).

In any consideration of distribution of parasites that of the host species must be weighed. Yet here there is either little factual material or else the evidence points to a wide range in the preference of parasites for their hosts which is rarely helpful in matters of geographic distribution. All the reliable data indicate that tropical mistletoes can attack a large variety of taxonomically unrelated hosts (Docters v. Leeuwen, 1954; Kuijt, 1964c). A well-documented (but perhaps extreme) example is *Dendrophthoe falcata* with its 343 known host species (Narasimha and Rabindranath, 1964). The same situation is even more strikingly demonstrated by a single individual of *Phrygilanthus acutifolius* attacking, by means of different roots, members of Leguminosae, Celastraceae, Euphorbiaceae, and Myrtaceae (Hoehne, 1931). That much is to be learned here is shown by occasional local preferences. Even in the temperate zones host-specificity is often quite low: the number of hosts of *Viscum album* is very large (Tubeuf, 1923; Rao, 1957). Perhaps only in *Arceuthobium* do we encounter something like the host preferences which characterize much of animal parasitology. *Arceuthobium oxycedri* in nature is limited to junipers, and *A. minutissimum* has been collected only on the Himalayan *Pinus griffithii* (*P. excelsa*). In the Pacific provinces and states of North America *A. douglasii* is virtually restricted to *Pseudotsuga menziesii,* and the North American *A. pusillum* to *Picea* spp. Some of the more reduced *Phoradendrons* of California have narrowed their host selection to certain conifers. In the mistletoes such preferences may be associated with an advanced evolutionary status, and are exceptional.

HABIT AND MODE OF PARASITISM

The terrestrial habit sporadically appears in more advanced mistletoes. *Arceuthobium campylopodum* has been found on the roots of *Pinus sabiniana* (Dr. R. Scharpf, pers. comm.), and Dr. J. Steyermark has collected a *Phoradendron* near or just above soil

FIG. 2-1. *Gaiadendron punctatum*, a facultatively epiphytic or terrestrial mistletoe: *a*, young plant with small tuber from which aerial shoots have emerged; *b*, regeneration of shoots from roots under natural conditions; *c*, part of a mature tree, with a hummingbird visiting the flowers (all after Kuijt, 1963*a*, 1965*a*).

level. In many wet tropical forests, roots and stems of potential hosts are available at various levels, and additional rare occurrences on host roots may be anticipated. *Ixocactus hutchisonii* (Kuijt, 1967*a*) may grow on epicortical (external) roots of fellow mistletoes.

In no less than three genera of mistletoes, however, the terrestrial habit is common. The giant of all mistletoes is the West Australian *Nuytsia floribunda*, called the Christmas tree because of its brilliant show of yellow-orange flowers about that time of the year. It reaches a height of 10 meters or more. That this tree is indeed a parasite has been shown by its haustorial connections to the roots of small grasses and even cultivated carrots (Herbert, 1918-1919). Small plants and young trees may sprout from the roots of the mother tree many meters away. The shrub *Atkinsonia ligustrina*, of the Blue Mountains of New South Wales is also a root parasite (Menzies and McKee, 1959). The third genus, *Gaiadendron*, is also known to be a parasite (Kuijt, 1963*a*, fig. 2-1, 2-2). While often terrestrial and matching *N. floribunda* in size, there is evidence that in Costa Rica the terrestrial habit is a result of man's interference with the structure of the forest. In that country's virgin forests *G. punctatum* is most common among the luxurious epiphytic growth on the limbs of forest giants. This mistletoe is unique in being a true epiphyte (with respect to the supporting tree, which it does not parasitize) as well as a true parasite (with regard to its fellow epiphytes). *Gaiadendron* is versatile in that it sprouts leafy shoots from various points on its roots, and thus is able to reproduce itself vegetatively. The taxo-

nomic confusion blurring the outlines of Andean *Gaiadendron* and *Phrygilanthus* probably hides from us a number of species which also tend toward a terrestrial habit. As far as I am aware, the true state of affairs of the *Helixanthera terrestris* and *H. ligustrina* of India has not been detailed; we must accept Hooker's (1886) statement that at least the former mistletoe always grows from the ground and shows no parasitic attachments to the aerial parts of neighboring plants *Amyema scandens* and *Amylotheca pyramidata* germinate on the ground, climb the host tree, and then parasitize the branches (Menzies and McKee, 1959).

A striking liana habit frequently occurs in the Brazilian mistletoe *Phrygilanthus acutifolius* (Hoehne, 1931). This parasite may have a complex and anastomosing network of subterranean roots which may reach lengths of 20 meters or more. Surprisingly, haustorial contacts with roots of other plants seem to be rare. But even more surprising is that, at the end of such a subterranean root, a sheathing system of roots may envelop the base of a neighboring tree and there support many haustorial organs. One individual of this species parasitized five different species of trees in this fashion. A long serpentine woody organ, often closely appressed to the host trunk, connects the unusual root system to the leaf- and flower-bearing branches in the crown, where haustorial contacts are infrequent. It might easily be, however, that *P. acutifolius* germinates in the crown of the tree and there establishes its initial haustorial connections. Subsequent roots, in their search for further host organs, may descend the tree and eventually give rise to a terrestrial root system. This being true, the versatile mode of life of *P. acutifolius* resembles that of *Gaiadendron* more than would at first appear. A more detailed study of germination and establishment of the mistletoes forming an apparent bridge between the two genera would yield interesting data.

Many of the branch-inhabiting tropical mistletoes

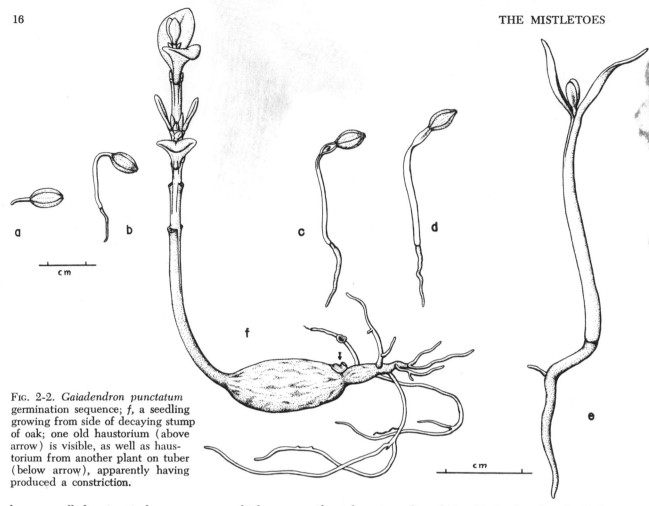

Fig. 2-2. *Gaiadendron punctatum* germination sequence; *f*, a seedling growing from side of decaying stump of oak; one old haustorium (above arrow) is visible, as well as haustorium from another plant on tuber (below arrow), apparently having produced a constriction.

have so-called epicortical roots—organs which grow along the host branch and at intervals develop new haustoria. These haustoria are secondary in the sense that they are not outgrowths of the seedling's radicle. Usually these roots emerge from the base of the plant and grow up and down the host branch without regard for gravity. The many secondary contacts thus provided serve to increase the flow of nutrients to the parasite, but also anchor the mistletoe more securely. The latter function is dramatically shown where the original haustorium has broken away from the host, and the parasite is saved by its epicortical roots, which then develop massively. The same thing is true of seedlings that develop on host leaves: epicortical roots may yet establish haustorial contact on the adjacent branch before both leaf and seedling are dropped. Many American (Kuijt, 1964c), Australian (Hamilton and Barlow, 1963), and the majority of Indonesian Loranthaceae (Danser, 1931b), as well as a few African ones (Balle, 1964), are characterized by basal roots (fig. 2-3).

A number of mistletoes are able to regenerate leafy shoots from their epicortical roots. This phenomenon is rare in the American mistletoes; only one *Phrygilanthus* does so normally, and one *Oryctanthus* only occasionally (Kuijt, 1964a), possibly as a response to injury. But in some Old World species like *Ileostylus micranthus* of New Zealand a plant "walks" along the host branches with its roots, sprouting leafy shoots (or even inflorescences in *Dactyliophora* and *Amyema*) as it progresses. It is reminiscent of the behavior of *Gaiadendron*. This "mobile" parasitism gives the parasite great ecological flexibility. Most mistletoes are destined to live where they are first established (fig. 2-4). The ever-expanding crown of the host tree eventually overcomes and shades out an immobile plant, which often dies or is reduced to a vegetative existence. Vegetative reproduction as in *I. micranthus* gives the parasite a means of competing for light with the growing host canopy. The mistletoe can select, as it were, an advantageous station within the host crown. The same type of propagation allows *Bakerella grisea* from Madagascar to attain a length of 20 meters (Balle, 1964).

Another way of gaining the same ecological advantage is seen in the tropical American genus *Struthanthus* (Kuijt, 1964c). The shoots in *Struthanthus*, in a sense, achieve what the roots do in *Ileostylus micranthus*. All species of *Struthanthus* have rather long and trailing branches which may become entangled with surrounding vegetation. By means of epicortical roots emerging from these branches in strategic places the mistletoe anchors itself many times, making haustorial contacts every-

FIG. 2-3 *Phthirusa pyrifolia* from Costa Rica, grown in greenhouses of the University of British Columbia, Vancouver: *a*, epicortical roots growing from base of young plant extending to right—host is *Nerium oleander*; *b*, three individuals attacking *Codiaeum* sp., showing hypertrophy of hosts at primary haustoria (arrows); *c*, a mature secondary haustorium seen from side.

where. The species *S. costaricensis* has long and slender shoots of a spiral growth pattern which seem to grope for new hosts to conquer. The most advanced in this respect, however, is the almost ubiquitous *S. orbicularis*. On its attenuated young branches the stiffly recurved new leaves serve as grappling hooks (fig. 2-5). When a hooklike leaf rests on another branch, the petiole grows unequally in such a way as to twist itself around the captured organ. Soon afterward the same node forms epicortical roots which spiral the length of the host branch and make haustorial contact. That the "host" frequently turns out to be just another branch of the same individual seems to make no difference (fig. 2-6); indeed, this sort of self-parasitism is common in mistletoes with epicortical roots. (We have seen it in the anastomosing root system of *Phrygilanthus acutifolius*.) The almost unique parasitic strategy of *S. orbicularis* provides many haustorial contacts with the host, but also allows the mistletoe to grow on top of the host crown indefinitely. In many areas in Central America one often sees a great blanket of *S. orbicularis* weighing down a large portion of a tree, or reaching from tree to tree. Scarcely a host branch escapes attack in this inextricable mass of parasitic growth. Dr. D. Wiens has recently shown me a second species of *Struthanthus* from Ecuador showing a similar parasitism. The African *Globimetula braunii* has developed a *Struthanthus*-like behavior also; the secondary haustoria here either emerge directly from the mistletoe branch (as in *Cuscuta*), or the roots which give rise to them are exceedingly small (Balle, 1964).

It is easy to visualize the basal type of epicortical root as a primitive feature, a reminder of bygone days of terrestrial parasitism. In view of the distribution of Loranthaceae, it is significant that in continental Africa, one of the two geographical extremes of the family, only a handful of species have epicortical roots (Balle, 1964). A similar kind of scandent parasitism has evolved in a few paleotropical Santalaceae, otherwise a terrestrial family (see chap. 3).

In Viscacean mistletoes the primary absorptive system has undergone striking changes (see chap. 7). In this family only two genera are known to have epicortical roots. These two genera have in the past been regarded as primitive ones. The epicortical roots of *Antidaphne* are functional for only a limited period, after which they disappear (Kuijt, 1964c). An old plant shows scarcely a vestige of its juvenile root development. The related genus *Eremolepis* probably has a similarly evanescent root system (see fig. 7-32).

We are uniquely fortunate to have in living Loranthaceae a series which shows a complete gradient from terrestrial, root-parasitic habit to the immobile, shoot-parasitic habit (fig. 2-7). In all probability the evolution of the latter type has proceeded in the same direction. This existing series can be used in support of the idea of terrestrial rather than epiphytic origin of mistletoe parasitism. At the same time the epiphytic-parasitic dualism of *Gaiadendron* and possibly *Phrygilanthus acutifolius* demonstrates that the two ideas are by no means mutually exclusive. It is very likely that the ascent of the host tree from the root-parasitic habit was accomplished via some such versatile and paradoxical stage. From there on, a preference for a favorable position in the crown and the gradual disappearance of vegetative propagation brings us to the habits of *Phthirusa pyrifolia* and similar plants. The production of epicortical roots and/or haustoria other than from the base of the mistletoe (*Struthanthus*, *Globimetula*

FIG. 2-4. *Phoradendron* spp.: *a, P. californicum* on *Acacia* sp., Mohave Desert, California; *b*, an oak heavily attacked by *P. villosum*, near Jamestown, California; *c, P. undulatum*, San Pedro, Costa Rica.

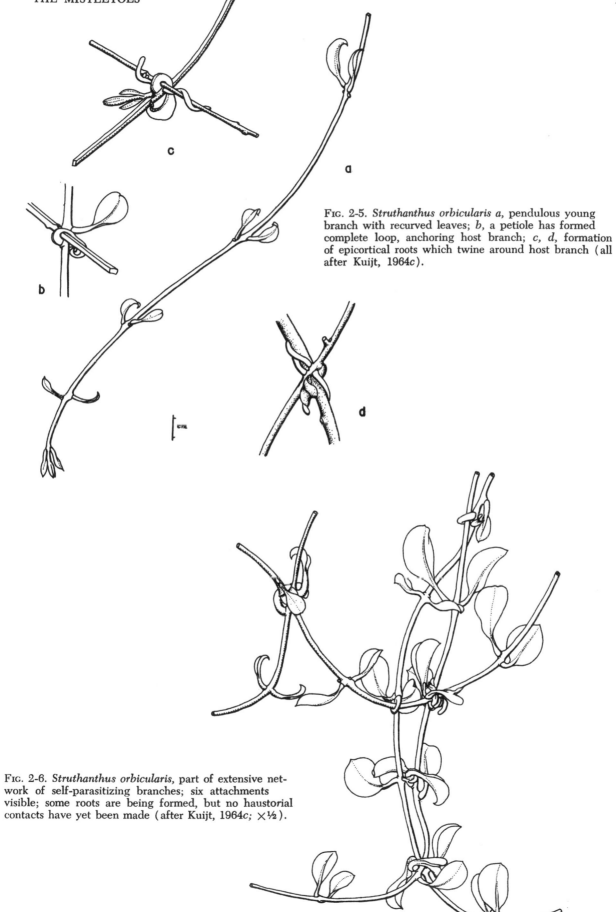

FIG. 2-5. *Struthanthus orbicularis a,* pendulous young branch with recurved leaves; *b,* a petiole has formed complete loop, anchoring host branch; *c, d,* formation of epicortical roots which twine around host branch (all after Kuijt, 1964*c*).

FIG. 2-6. *Struthanthus orbicularis,* part of extensive network of self-parasitizing branches; six attachments visible; some roots are being formed, but no haustorial contacts have yet been made (after Kuijt, 1964*c;* ×½).

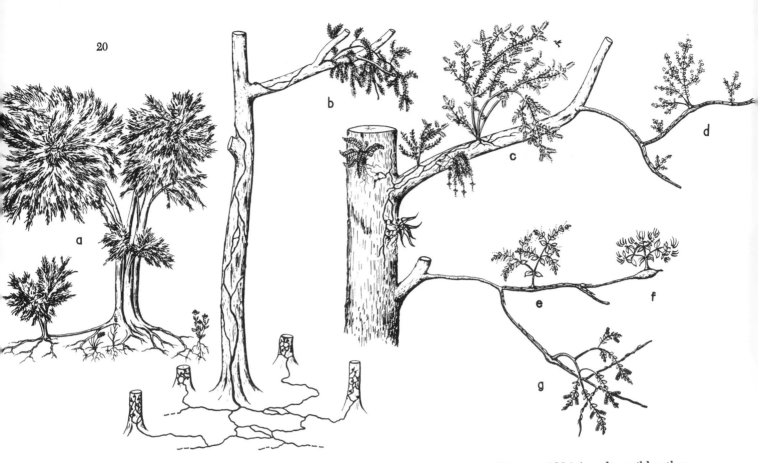

Fig. 2-7. Ascent of host tree as shown by present-day Loranthaceae: *a, Nuytsia floribunda* (after Herbert, 1918-1919); *b, Phrygilanthus acutifolius* (after Hoehne, 1931); *c, Gaiadendron punctatum* (after Kuijt, 1963*a*); *d, Ileostylus micranthus* and others; *e, Phthirusa, Oryctanthus*, etc.; *f, Psittacanthus* and many *Phrygilanthus*; *g, Struthanthus.*

braunii) probably represents an evolutionary sideline, a later addition to the parasitic arsenal. Loss of epicortical roots, perhaps through a stage of their evanescent existence, finally results in the completely stationary parasitism of *Psittacanthus* and mistletoes of similar habit which rely solely on the primary haustorium for anchorage and sustenance. In Viscaceae another development of striking interest has been the great expansion *within* the tissues of the host. In some of the most advanced members of this family, and in one or two Loranthaceae, vegetative reproduction has once more evolved, with the shoots and inflorescences arising from organs embedded within host tissues. This tendency finds its extreme expression in the dwarf mistletoes (*Arceuthobium*).

Other parts of the mistletoe plant, which have no direct relation to parasitism, show remarkably consistent patterns of branching. Opposite and decussate phyllotaxis is basic to the vegetative apparel of the mistletoe plant. The exceptions to this kind of leaf arrangement are rare. *Antidaphne, Eubrachion,* and *Eremolepis* have a spiral phyllotaxis. In some *Psittacanthus, Trithecanthera,* and sometimes in *Amy-*

ema aneityensis (Danser, 1934*a*) and possibly others, a verticillate phyllotaxis prevails; a solitary *Dendrophthora* is also trifoliate in organization (Kuijt, 1961*a*: see fig. 2-18,*a*). It is not uncommon to find the leaves of a pair separated along the stem, giving a very irregular appearance. With these exceptions in mind, we may still say that mistletoe leaves are usually arranged in decussate pairs. This puts a stamp of uniformity on many (especially Viscacean) mistletoes, whose appearance may be almost stylized in such plants as *Viscum album* (see fig. 1-5).

Even in profusely branched mistletoes, lateral branches, whether vegetative shoots or inflorescences, develop in an orderly manner: the first (primary) ones arise in the axil of the leaf; then the secondary laterals emerge from the axils of the minute prophylls at the base of the primary laterals. The prophylls of the secondary laterals, in turn, may bear a third generation of laterals, and so on. Laterals of the fourth order are visible in a diagram (fig. 2-8) based on a Costa Rican species of *Phoradendron*. With the aid of this diagram we can now appreciate the consistency of branching in the related *P. robustissimum* and others. Disappearance of the prophyllar organs in *Arceuthobium americanum* gives the impression of a simple verticillate or whorled branching pattern (Kuijt, 1960-1961).

The rigidity of this pattern is emphasized by the frequently forked habit of mistletoes. Both in Loranthaceae (*Tupeia, Gaiadendron,* some *Psittacanthus, Phrygilanthus* and *Oryctanthus*) and in Viscaceae (*Viscum album, V. indosinense*; Danser, 1938, fig.

2f), some species of *Phoradendron, Dendrophthora* (Kuijt, 1961c), *Ginalloa* (Danser, 1938a), and *Notothixos* (Danser, 1931b), inflorescences terminate the axis, with the axillary buds of the last leaves continuing vegetative growth. Technically, the resultant forked habit cannot be called a true dichotomy, but we can refer to it as dichasial branching. The shoot apex sometimes aborts immediately after forming its first pair of leaves (*Phoradendron robustissimum* and others; Kuijt, 1964d), a process repeated for every lateral vegetative branch.

A branching pattern unique in mistletoes is known for *Antidaphne*. Its small pistillate spikes are crowned by several minute fleshy leaves (Kuijt, 1964d). During the development of the fruit the spike elongates to several times its initial length, and the terminal leaves become small but otherwise normal foliage leaves. The regular system of axillary branching of other mistletoes is not here encountered.

In some species of *Arceuthobium* another, very distinctive branching pattern has evolved. The first lateral branch is followed by others directly below it, in descending order of size. *A. campylopodum* and

several of its American fellow species show this flabellate branching pattern. A similar serial branching is also known for lateral buds in many other woody plants where, however, the diminishing gradient is an ascending one (Goebel, 1932). The inflorescences of *Dendrophthora* and *Phoradendron* can be regarded as extreme expressions of the same serial tendency (Kuijt, 1959).

Leaf shape is remarkably uniform throughout the two mistletoe families and, indeed, the other Santalalean families. As in many predominantly tropical woody families, all Loranthaceae have simple leaves, without even marginal lobing. Some are fleshy, others thin, and there are minor variations in leaf apices, petioles, and pubescence. In *Struthanthus* and *Oryctanthus* there is a striking contrast between the at-

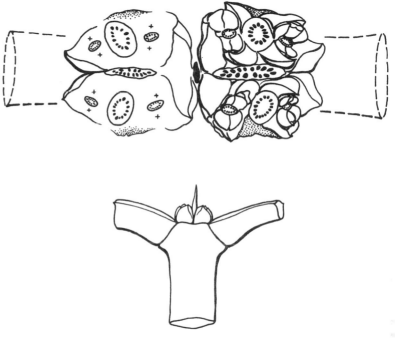

Fig. 2-8. (Right) *Phoradendron* sp., showing abortion of shoot apex (below) as origin of first branching dichotomy, and production of subsequent terminal deciduous spikes at same node (above); lateral inflorescence buds of fourth order (see +) visible (Kuijt No. 2465, Costa Rica; ca. ×2). Not related to *P. obliquum* from South America, as intimated in Kuijt (1964d).

Fig. 2-9. (Below) a, *Tupeia antarctica* from New Zealand, young plant (courtesy B. A. Fineran); b, *Struthanthus polystachyus* from Golfito, Costa Rica, showing juvenile and mature foliage of several young plants.

tenuate, pendulous juvenile leaves and the short and wide, fleshy and rigid mature foliage (fig. 2-9,*b*). A type of alternation of series of floriferous scale leaves with large leaves has developed in *Phoradendron crassifolium* (Kuijt, 1959) and in *Barathranthus axanthus*. Easier access by birds to fruit may have led to this sort of heterophylly. There are also a couple of deciduous mistletoes, one in Europe (*Loranthus europaeus*) and one in Africa ("*Loranthus*" *zeyheri;* Anonymous, 1964).

The reduction of foliage leaves to scale leaves in the mistletoes would seem to be related to photosynthetic activity and therefore to the degree of dependence on the host. The squamate habit is of general occurrence in many nonparasitic plants, particularly in arid regions. The squamate habit may be a consequence of radical changes in water economy, and the reduction of transference of the photosynthetic process may be a secondary result. Some leaf patterns represent halfway stations toward the squamate situation, especially in *Dendrophthora* (Kuijt, 1961*c*). But in this genus, as in *Phoradendron* and *Viscum*, large-leaved and completely squamate species exist side by side. *Phrygilanthus* contains one species with scalelike leaves, *P. aphyllus*. Three small Viscacean genera are exclusively squamate: *Arceuthobium, Korthalsella*, and *Eubrachion*. In Loranthaceae, in addition to *Phrygilanthus aphyllus*, the newly described and much reduced genus *Ixocactus* should be mentioned. The same evolutionary tendency is evident in Myzodendraceae and Santalaceae.

The genus *Lepidoceras* shows a peculiar semisquamate condition. All young branches are first enclosed by a mantle of scalelike, papery bracts. The bracts expand with the branch. It is as if the original scale leaves are raised by a foliar leaf which is formed in an intercalary fashion. The full-grown leaf carries a spinelike scale at its tip. Some of the scales appear to be deciduous, at least on the lower portions of inflorescences.

Leaf reduction does not necessarily mean an abandonment of photosynthesis. Chlorophyll is present in all mistletoes, although this pigment may be masked by others. *Arceuthobium* and *Korthalsella* have a good deal of chlorophyll, and so do leafless species of other genera, for example *Phrygilanthus aphyllus* (Follmann, 1963). Few plants are as brightly green as the squamate *Phoradendron* which parasitizes incense cedar (*Libocedrus decurrens)* in California and Oregon.

Independently, in several genera, the internodes have become greatly compressed and widened. A great deal of anatomical elaboration has taken place in these flattened stems, which in appearance and function resemble leaves (see chap. 9). Various degrees of this "transference of function" are seen in *Dendrophthora, Viscum, Phoradendron, Korthalsella, Ginalloa* (Danser, 1931*b*, 1934*a*), and *Ixocactus*. Stems with compressed internodes seem to undergo ontogenetic twisting, possibly allowing for maximum interception of light. In some herbarium material this orientation is an artifact due to the pressure of the plant press, but in certain *Korthalsella* and *Dendrophthora* species an early morphogenetic twist gives the appearance of a distichous phyllotaxy (Rutishauser, 1937; Kuijt, 1961*c*, under *D. opuntioides*). It is not impossible, instead, that the distichous arrangement is initiated in the apical meristem. In the latter two genera, however, decussate types prevail; in *K. geminata*, of distichous appearance, the spikes are again decussate.

FLOWERS AND FLOWER BIOLOGY

In Loranthaceae, there appears to be a single whorl of perianth members, as many as seven and eight in *Gaiadendron punctatum* and *Dactyliophora globiflora*, but most often six or five, and rarely four or three in others. Virtually all members of this family have bisexual flowers. In a few genera, unisexual flowers have evolved, as in *Tupeia, Loranthus* (*L. europaeus*), *Barathranthus* (Prakash, 1963), the entire genus *Struthanthus*, and possibly some *Phthirusa* (Krause, 1932; Rizzini, 1960) and a few others. Loranthacean species with unisexual flowers are always dioecious; monoecism is unknown. In *Struthanthus* and *Barathranthus* (Docters v. Leeuwen, 1954) dioecism appears to be of rather recent evolutionary development, since sterile stamens are always present in pistillate flowers. The latter fact and the great preponderance of bisexual flowers in Loranthaceae indicate the origin of dioecism from the perfect condition. Such an origin of dioecism appears to be common in angiosperms (Lewis, 1942), but is in distinct contrast to Viscaceae. Loranthacean stamens are usually well formed; although the anthers are nearly sessile in some American genera, in others on the same continent filaments are exceedingly long and have a versatile connection with the anther. The latter is four-chambered (or through confluence two-chambered) with a longitudinal dehiscence. Irregular sterile patches interrupt the continuity of the pollen mass in the elongated anthers of *Tapinostemma* (Johri and Prakash, 1965) and *Psittacanthus* (fig. 2-10,*d*). In the smallest-flowered member, *Ixocactus*, the position of the anthers is at two levels; in many other mistletoes with large flowers the same holds true but is more obvious. The structural variation of stamens is greatest in Africa (Balle, 1956).

With respect to morphology and embryology the simplicity of the Loranthacean flower is deceptive. The gynoecium, clearly inferior, could scarcely be simpler in its construction. The style, short and blunt in small-flowered species, may be improbably long, reaching a length of nearly a foot in *Aetanthus macranthus* (Hooker, 1848). Yet the stigma shows virtually no differentiation, and never is more than a small button. Within the apparently simple Loranthacean gynoecium, however, events take place which have no equal in any other flowering plants.

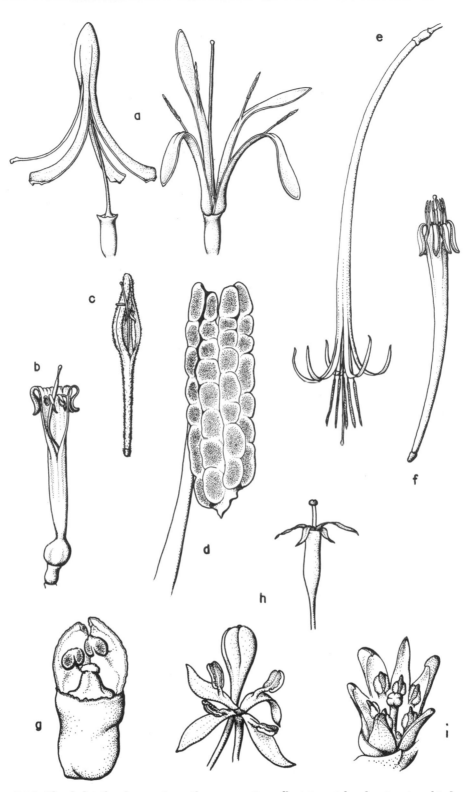

Fɪɢ. 2-10. Floral details of some Loranthaceae: *a, Peraxilla tetrapetala,* showing two kinds of anthesis (Dempster No. 3773, Australia, UBC; ×2); *b, Tapinanthus bangwensis* (after Balle and Hallé, Adansonia, 1961; × 1.5); *c, Socratina keraudreniana* (after Balle, Adansonia, 1964; magnification not known); *d, Psittacanthus schiedeanus,* anther showing transverse sterile septa in pollen sacs (Volcán Barba, Costa Rica; × 10); *e, Aetanthus macranthus* (after Hooker, 1848; ×½); *f, Macrosolen platyphyllus* (after Danser, 1931*b;* ×½); *g, Ixocactus hutchisonii,* two petals removed (after Kuijt, 1967*a;* × 20); *h, Tupeia antarctica,* male flower with bud, and female flower (Anderson No. 187, New Zealand; × 5); *i, Oryctanthus spicatus* (Kuijt No. 2390, Costa Rica, UBC; × 11).

Below the petals of Loranthacean mistletoes, and crowning the gynoecium, is an irregular rim of tissue which is variable in prominence. In such mistletoes as *Oryctanthus* the calyculus, as this rim is called, is difficult to demonstrate. In *Psittacanthus cupulifer* an enormous flaring calyculus seems to be present, but this turns out to be a sort of involucre formed around the ovary which is crowned by its usual insignificant calyculus. The morphological nature of this calyculus has been of concern to many botanists. Because of the absence of vascular tissue in even the larger calyculi, many students have regarded this structure as a mere gynoecial outgrowth (Engler and Krause, 1935). But in recent years some vascular tissue has been demonstrated in calyculi of *Nuytsia floribunda* (Narayana, 1958) and *Atkinsonia ligustrina* (Garg, 1958), two of the most primitive members of the family, and the matter seems conclusively settled. Venkata Rao (1964) thinks of the calyculus as a vestigial whorl of bracts, but that interpretation seems unnecessarily complex. To most workers the calyculus represents a calyx in various stages of phylogenetic disappearance. (This has interesting implications in fruit anatomy which are brought out below.) The apparent calyculus in *Viscum* (Schaeppi and Steindl, 1945) is perhaps no more than a slight constriction at the base of the petals, as the calyculus is otherwise known only in Loranthaceae. Another minor irregularity is the absence of a calyculus in the staminate flower of *Tupeia antarctica* (fig. 2-10,*h*); yet the pistillate flower is calyculate. It is not surprising that Engler and Krause (1935) placed *Tupeia* in Viscaceae instead. A calyculus is present in many Santalalean plants outside the mistletoe families, and may be a very striking feature.

The Loranthacean corolla is often a somewhat tubular organ, splitting into its component petals (figs. 2-10, 2-11, 2-12). It is generally actinomorphic. In African members, on the contrary, there is a strong tendency to zygomorphy. In such flowers the corolla splits deeply on one side and the flower has a marked similarity to a honeysuckle (*Lonicera*) flower. It is difficult to find a connection between zygomorphy and pollination here, as the normal pollinators for the colorful mistletoes of both the Old and New World are birds. These are the only zygomorphic flowers in the entire Santalalean order.

Nearly all Loranthacean flowers are associated in inflorescences of various types. This is not true of *Ixocactus* (Kuijt, 1967*a*), in which flowers occur singly—undoubtedly a very advanced condition.

The basic unit of the Loranthacean inflorescence is a triad or simple dichasium (fig. 2-13). Repetition along a specialized axis leads to the raceme, perhaps the commonest of inflorescences in the family. The raceme characterizes many *Struthanthus* species as an axillary inflorescence. In some *Psittacanthus* species the inflorescence tends to become terminal in position, and a forked branching habit results. *Struthanthus* and *Phthirusa* show various degrees of contraction of the stalks of the dichasium; the result is technically a spike, as all flowers are sessile along a single axis. In some mistletoes the triads are modified to consist of lateral or median flowers only. The former situation can be seen in *Psittacanthus allenii*, the latter in *Oryctanthus*; both occur in *Lysiana* (Barlow, 1963*a*). All the examples except the last are illustrated in Kuijt (1964*d*). Beyond these examples are many modifications of the types outlined. In some Loranthaceae flower production at the nodes is so profuse year after year that the nodes seem to have brilliantly colored balls of flowers, the leaves having been dropped (e.g., *P. allenii* and *Dactyliophora;* Danser, 1931*b*). *Phthirusa theobromae* has a number of racemes combined in a compound inflorescence (Eichler, 1868*a*). Very peculiar is *Tolypanthus* with its fused involucre surrounding several flowers (Engler and Krause, 1935). A capitulum reminding one of Compositae, that is, a series of clasp-

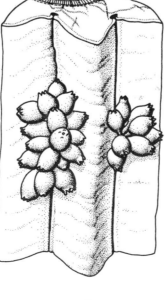

FIG. 2-11. *a, Viscum minimum* emerging from stem of *Euphorbia polygona*, with endophytic ramifications visible on cut surface of host stem (after Engler and Krause, 1908; × 1.5); *b, Psathyranthus amazonicus* (after Ule, 1907; × 1).

a

b

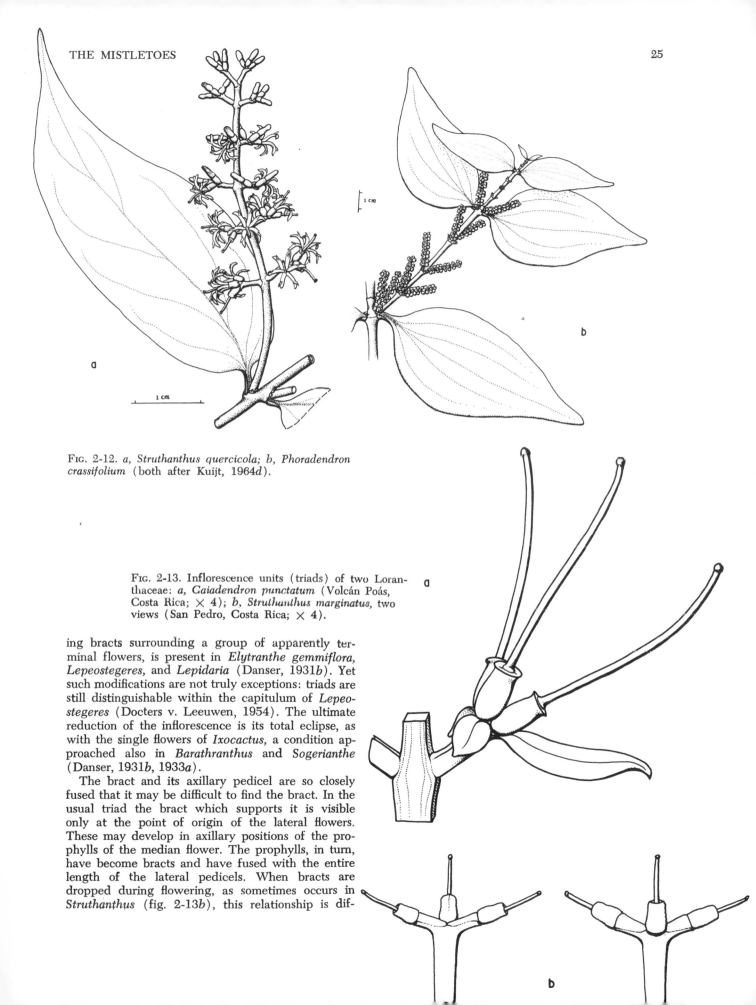

Fig. 2-12. *a, Struthanthus quercicola; b, Phoradendron crassifolium* (both after Kuijt, 1964*d*).

Fig. 2-13. Inflorescence units (triads) of two Loranthaceae: *a, Gaiadendron punctatum* (Volcán Poás, Costa Rica; × 4); *b, Struthanthus marginatus,* two views (San Pedro, Costa Rica; × 4).

ing bracts surrounding a group of apparently terminal flowers, is present in *Elytranthe gemmiflora, Lepeostegeres,* and *Lepidaria* (Danser, 1931*b*). Yet such modifications are not truly exceptions: triads are still distinguishable within the capitulum of *Lepeostegeres* (Docters v. Leeuwen, 1954). The ultimate reduction of the inflorescence is its total eclipse, as with the single flowers of *Ixocactus,* a condition approached also in *Barathranthus* and *Sogerianthe* (Danser, 1931*b*, 1933*a*).

The bract and its axillary pedicel are so closely fused that it may be difficult to find the bract. In the usual triad the bract which supports it is visible only at the point of origin of the lateral flowers. These may develop in axillary positions of the prophylls of the median flower. The prophylls, in turn, have become bracts and have fused with the entire length of the lateral pedicels. When bracts are dropped during flowering, as sometimes occurs in *Struthanthus* (fig. 2-13*b*), this relationship is difficult to demonstrate. Conversely, the situation is obvious in *Gaiadendron*, where each bract is a green foliar organ of considerable size (fig. 2-13,*a*). Foliar bracts are well developed in some species of *Diplatia* (Barlow, 1962) and *Psittacanthus*, but in the latter only for the median flower; the lateral ones are absent.

Viscacean flowers are invariably unisexual, in either a monoecious or a dioecious pattern. They are very insignificant structures. Some *Arceuthobium* species have bright reddish flowers, but in the rest of the family they are greenish or yellow, and very small. In many *Dendrophthoras* mature flowers are less than a millimeter in diameter. Naturally, we cannot expect much floral complexity on so small a scale. Even the number of perianth members is small, scarcely ever exceeding four in staminate and three in pistillate flowers; it may be difficult even to find the perianth in the latter. In *Arceuthobium* the perianth cannot be delimited from the ovary, and if it were not for two minute projections and their corresponding vascular strands we would be unable to distinguish any perianth members. The flowers of other genera are only slightly more elaborate. In the pistillate flower of *Antidaphne* (fig. 2-14,*b*) the two or three minute perianth members abscise during flowering and thus do not persist on the fruit. The stigma is a small knoblike protrusion in most genera, but in *Antidaphne* it is somewhat larger (Kuijt, 1964*d*). In general, therefore, pistillate Viscacean flowers show remarkably little variation.

It is in the staminate flowers that we find the most interesting evolutionary changes, at least one of which, in *Korthalsella*, may be unique in angiosperms. Several Viscacean genera seem to be based largely on androecial characters. A number of genera, presumably the most primitive, still have the standard angiosperm stamen, with two or four pollen sacs elevated above the supporting petal by a distinct filament (*Eremolepis, Antidaphne, Lepidoceras*). Reduction in *Phoradendron* and *Ginalloa* has resulted in a completely sessile anther of only two chambers (rarely one in *Ginalloa;* Rutishauser, 1937). In *Dendrophthora* even these two pollen sacs have become one, dehiscing by means of an irregular transverse slit. In *Arceuthobium* the archesporium is ringlike around a central sterile column, and a circular slit is said to develop, although somewhat irregularly (Cohen, 1968). *Korthalsella* is totally different: its three sessile anthers have joined to form a common structure. Although the six pollen sacs are still separate, they dehisce into a central cavity which opens through a small pore (Mekel, 1935; fig. 2-15,*d*). Especially odd is the androecium of *Viscum*, where even the anthers are scarcely distinct structures. In a cushion-like outgrowth of the petal a variable number of archesporial groups develops which liberate their pollen on the petal's surface (fig. 2-15,*b*). In *Notothixos* a short filament supports a broad anther transversely chambered; each chamber is said to open with a separate pore (Danser, 1931*b*). My

own observations on *Notothixos,* however, are quite in contrast to those of Danser. There is no filament at all. The anther, in fact, is not basically different from that of *Viscum*, except for its greater distinctness, and a transverse slit releasing pollen from about seven small circular pollen chambers (fig. 2-15,*c*).

It is not surprising that the vascular system of the staminate flower has also become progressively simplified. A separate vascular strand for the stamen is eliminated when the anther becomes sessile. The extreme in simplicity is seen in *Korthalsella* (fig. 2-15,*d*) and *Arceuthobium*, where petals are supplied by a single, nearly unbranched vascular strand.

Viscacean mistletoes are either monoecious or dioecious. Except for a single report of a bisexual flower in *Arceuthobium minutissimum* (Datta, 1951), bisexual flowers are unknown. Genera may be totally monoecious (*Ginalloa, Notothixos, Korthalsella, Eubrachion*) or dioecious (*Antidaphne, Arceuthobium, Lepidoceras*), or may contain species of both kinds (*Viscum, Dendrophthora, Phoradendron, Eremolepis*). Among the monoecious species are some interesting distribution patterns of the sexes (see fig. 2-19).

One of the striking facts about Viscacean inflorescences is that flowers are sessile or nearly so (fig. 2-16). A three-flowered spike of one terminal and two lateral flowers, as in many *Viscum* species, can perhaps be taken as the simplest condition (see fig. 1-5). A repetition of laterals, or a double series of three-flowered spikes along an inflorescence axis, leads to *Ginalloa*. A double series of three-flowered spikes, often on peduncles, along a common inflorescence axis is present in some species of *Ginalloa* (Danser, 1931*b*). In *Notothixos* racemes of those unit spikes occur, which here tend to have more than three flowers arranged in a fanlike structure (Danser, 1931*b*). Such fans are characteristic of several species of *Viscum* (see below).

In *Antidaphne, Eubrachion,* and *Eremolepis* the inflorescence is subtended by a number of scale leaves and is, particularly in the first genus, almost catkin-like. A peculiarity unique to *Antidaphne* is the postfloral expansion of the female spike (Kuijt, 1964*b*).

In some dwarf mistletoes, e.g., *Arceuthobium verticilliflorum* (Hawksworth and Wiens, 1965), two or more secondary flowers flank the axillary flower, without the benefit of prophylls. A veritable mass of such flowers, at least initially in neat series, is characteristic of *Korthalsella* (Mekel, 1935). If we now imagine the base of the internode to elongate while the flowers are initiated, thus separating them, we encounter one of the most unusual of angiosperm inflorescences, the spike of Phoradendreae.

The unique feature of the spike of *Phoradendron* and *Dendrophthora* is the production of flowers by means of an intercalary meristem (Kuijt, 1959). In the development of this inflorescence a series of flower-bearing internodes is telescoped (figs. 2-17, 2-18). In the axil of each bract of the fertile portion (the lower internode or internodes serve as a sterile

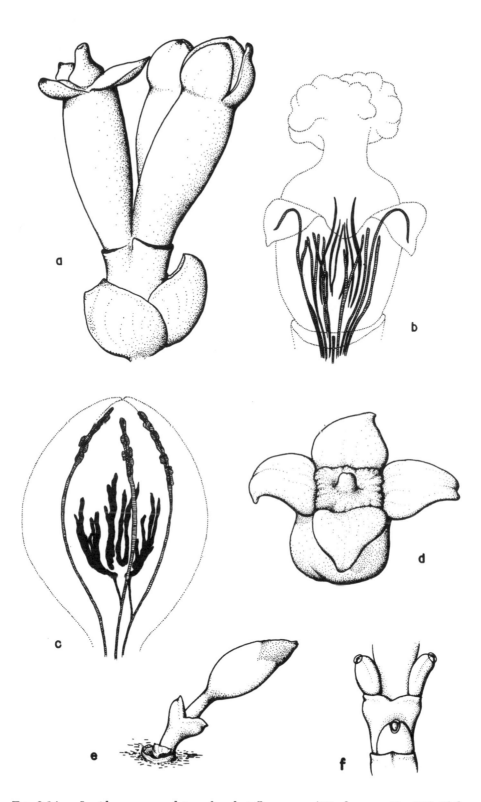

Fig. 2-14. *a*, *Lepidoceras punculatum*, female inflorescence (Werdermann No. 287, Chile, UC; × 30); *b*, *Antidaphne viscoidea*, cleared female flower (carpellary traces in black) (Kuijt No. 2433, Costa Rica; × 50); *c*, *Korthalsella* sp., same (Taylor No. 6235, Hawaii; × 50); *d*, *Eremolepis schottii*, female flower (Woytkowski No. 3417, UC; × 20); *e*, *Arceuthobium minutissimum*, entire shoot with female flower (Rodin No. 5695, Indian, UC; × 15), other shoots may have one or two lateral flowers; *f*, *A. campylopodum*, female flowers with receptive stigmatic droplets (Vancouver; × 10).

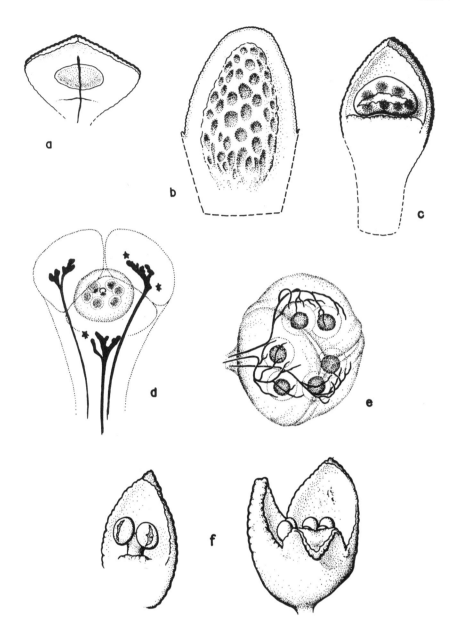

Fig. 2-15. Anther morphology of some Viscaceae: *a, Dendrophthora clavata* (Mexia No. 7604, Ecuador, UC; × 35); *b, Viscum album* (Canton-spark, Netherlands; × 10); *c, Notothixos cornifolius* (Clemens, Australia, UC; × 35); *d, Phoradendron pauciflorum* (after Kuijt, 1959; × 22); *e, Korthalsella* sp. (Taylor No. 6235, Hawaii; × 50); *f, Eubrachion ambiguum*, petal and flower (Source unknown; × 35).

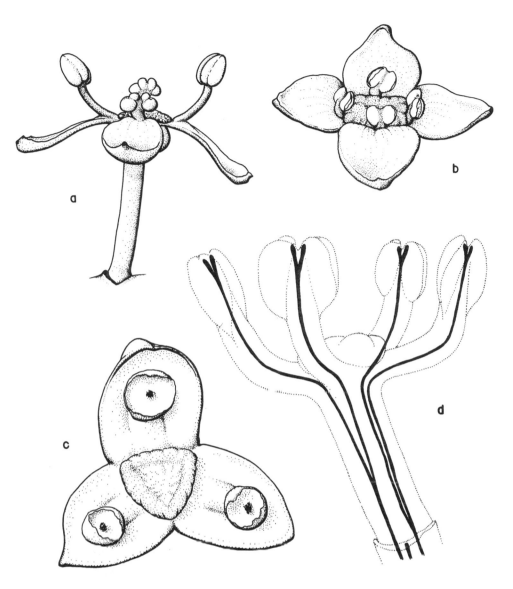

FIG. 2-16. Male flowers: *a, Lepidoceras squamifer* (Werdermann No. 315, Chile, UC; × 30); *b, Eremolepis schottii* (Woytkowski No. 34171, UC; × 20); *c, Arceuthobium campylopodum* (Vancouver; × 25); *d, Antidaphne viscoidea,* cleared to show vasculature (Kuijt No. 2433, Costa Rica; × 40).

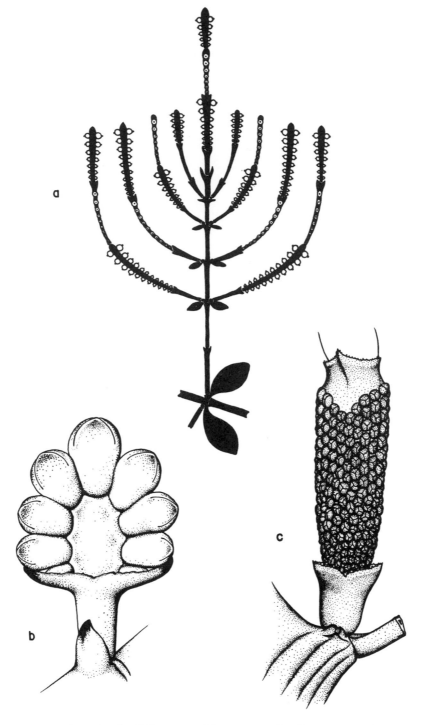

FIG. 2-17. *a, Dendrophthora paucifolia,* compound inflorescence (after Kuijt, 1961*c*; ×1); *b, Notothixos cornifolius,* inflorescence unit (Clemens, Australia, UC; × 15); *c, Phoradendron calyculatum,* basal fertile internode of spike (after Kuijt, 1959; × 6).

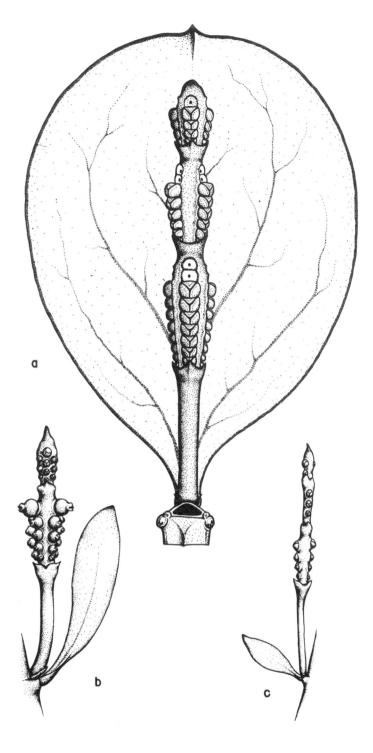

Fig. 2-18. Inflorescences in *Dendrophthora*: *a, D. ternata*, staminate
(× 2); *b, D. mesembryanthemifolia* (× 2); *c, D. ferruginea*
(× 2) (all after Kuijt, 1961*c*).

peduncle) more and more flowers are produced; these flowers form a longitudinal pattern because of the continuous lengthening of the internode. The youngest flowers are thus always lowest on the internode. Without such extension the spike would scarcely be distinguishable from that of *Korthalsella*.

Above each fertile bract the flowers assume one of several characteristic spatial patterns. The simplest pattern defines a large segment of *Dendrophthora*: it is a single longitudinal file directly above the bract (fig. 2-18,*c*). In more complicated types the apical flower is followed by three floral series or, when the median series drops out, by two. More lateral series may be added slightly lower. The two flower areas belonging to a single internode sometimes coalesce, and a jacket of flowers surrounds the stem of *Phoradendron calyculatum* (Kuijt, 1959; fig. 2-17*c*). When flowers are produced on so massive a scale the individual series tend to lose their identity rapidly. A flower area is always topped by a single apical flower. In one species of *Dendrophthora* (Kuijt, 1963*b*) a further novelty is the addition of two series per internode, one above each sinus formed by the supporting bracts. Since staminate plants add yet more series, the entire axis may again be covered with flowers. Spikes of Phoradendreae themselves may be arranged in elaborate compound inflorescences, often of beautiful symmetry (fig. 2-17,*a*).

In *Eubrachion* a single spike bears both types of flowers, the male below, the female higher (Engler and Krause, 1935). In the monoecious *Eremolepis* species the sexes are on separate spikes, scattered along the stem. *Ginalloa* has caused some controversy, perhaps because of the variation in the genus. Danser (1931*b*) describes abbreviated three-flowered dichasia (the terminal flower female, the lateral ones male) spaced along a common axis. Rutishauser (1937) found only single flowers along a raceme in

G. linearis, the basal flowers female, the distal ones male. In *Viscum*, sometimes the terminal flower of an inflorescence is male (*V. capitellatum, V. trilobatum*) and the lateral ones female; in most Indian mistletoes of this genus the reverse is true (Danser, 1941*a*; Rao, 1957). In either pattern more than two lateral flowers may develop, as in *V. capitellatum* and *V. orientale*, and especially *V. monoicum* and *V. heyneanum*, where fans of seven flowers are common.

The apical (or median) flower of *Korthalsella* is male; all others developing in the same axil are female (Mekel, 1935; Rutishauser, 1937). Some Phoradendreae are variable in their sex distribution but not irregular. *Dendrophthora clavata* may have unisexual spikes of both sexes, and bisexual spikes with the basal fertile internode entirely male, the distal one female (Kuijt, 1961*c*); *D. peruviana* and *D. flagelliformis* seem to show no regularity, and neither does *Phoradendron* cf. *obliquum*, where only one out of fifty flowers is male (Kuijt, 1964*d*). The most variable of all is *D. ambigua*, but its most common type (fig. 2-19,*h*) is comparable to the mixed spike of *D. clavata*. The basal fertile internode of *D. basiandra*, as its name implies, is entirely male; all distal ones are entirely female. Mixed internodes are present in *D. costaricensis*, where only the uppermost three flowers of the internode are male, and in *P. polygynum* and *P. platycaulon*, where each flower area has only one male flower, occupying an apical position.

It is surprising that from this mass of sometimes conflicting details we can formulate a general rule. When we seek a sequential pattern of male and female flowers within the Viscacean inflorescence, parallel to that of the sexual elements of a bisexual

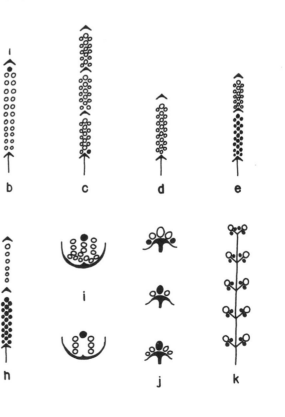

Fig. 2-19. Inflorescences of Viscaceae, diagrammatic representation of sex distribution (male flowers black; female ones, open circles): *a, Dendrophthora costaricensis*, spike; *b, Phoradendron polygynum, P. platycaulon*, etc., basal fertile internode; *c, Phoradendron*, spike (Kuijt, No. 19, 1964*d*); *d-f, Dendrophthora clavata*, inflorescence types; *g, D. peruviana*, spike; *h, D. ambigua*, one type of spike; *i, Korthalsella* (black crescent represents axillant bract); *j, Viscum*, spikes of various monoecious species; *k, Ginalloa* (according to Danser, 1931*b*); *l, G. linearis* (according to Rutishauser, 1937).

flower, we can obtain a large measure of satisfaction. The distal position of staminate flowers in several species of *Viscum* and possibly in *Ginalloa* would seem to violate this notion but most spikes of monoecious Phoradendreae can be harmonized with the floral sequence. In the unique development of these spikes the apical flowers are formed first, within a single internode; yet, when we consider the entire inflorescence, it is the *lowest* of the fertile internodes which is formed first. Thus we can generalize that male flowers or male internodes are initiated first and female ones later. Even *Eubrachion* and *Korthalsella* follow this pattern. The exceptions in Phoradendreae seem to be characterized by great irregularity (e.g., *D. peruviana, D. flagelliformis, P. cf. obliquum*).

Where, as in *D. clavata*, entirely male spikes occasionally develop in the prophyllar axils of female ones, there might be two successive growth periods. The situation in *D. ambigua* is even more confusing: here the two opposing flower areas of a single internode are sometimes opposite in sex as well.

In contrast to inflorescence and flower variation, the pollen of mistletoes shows relatively little structural variation. Loranthacean pollen is typically trilobate or triangular (fig. 2-20,*c d*), sometimes with very thin arms. Spherical pollen, however, is known from the family in *Tupeia* (Barlow 1964), *Oryctanthus* (fig. 2-20,*b*), and *Ixocactus* (fig. 2-20,*e*). Spherical pollen also characterizes Viscaceae, where the exine is often ornamented with spines. *Oryctanthus*

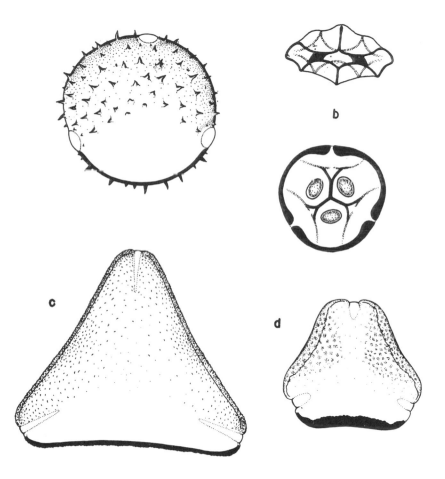

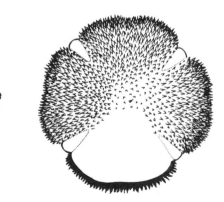

FIG. 2-20. Pollen grains of some mistletoes, from herbarium specimens (all × 700): *a, Lepidoceras squamifer* (Werdermann No. 315, Chile, UC); *b, Oryctanthus amplexicaulis* (Camp No. 3193, Ecuador, UC); *c, Psittacanthus calyculatus* (Mexia No. 867, Mexico, UC); *d, Atkinsonia ligustrina* (Maiden, Mount Tomah, Australia, UC); *e, Ixocactus hutchisonii* (after Kuijt, 1967*a*).

and *Ixocactus* show further exine modifications. The pollen of the latter is unique in Loranthaceae in being four-colpate—a condition not usual in the order Santalales except in *Aptandra, Ongokea,* and *Harmandia* (all in Olacaceae; Reed, 1955). The shape of *Oryctanthus* pollen is very complex and, surprisingly, finds a counterpart in that of the unrelated *Quinchamalium* of Santalaceae (see fig. 3-10,*b*).

Viscacean flowers are almost entirely insect-pollinated, as field observations attest. It is possible that wind pollination plays a minor role in *Viscum album* and *Arceuthobium.* Hymenoptera appear to be the main pollinators. In *Arceuthobium* a droplet exudes from the stigmatic region and serves as a collecting mechanism see (fig. 2-14,*f*). Withdrawal of the droplet brings the pollen in contact with the stigma (Jones and Gordon, 1965). Staminate Viscacean flowers are supplied with a glandular central disk which may also attract insects. In some genera, the pollen doubtless forms the main attraction for hymenopterous visitors, as in *Antidaphne* (Kuijt, 1964*d*). In a number of Loranthaceae, namely those with inconspicuous flowers, pollination is also, perhaps, exclusively by means of insects, as in *Struthanthus* (Kuijt, 1964*d*), *Tupeia* (Smart, 1952), *Barathranthus* (Docters v. Leeuwen, 1954), and *Loranthus europaeus* (Cammerloher, 1921).

About the year 1680, George Everhard Rumpf, a humble merchant of the Dutch East India Company, recorded in Amboyna that a restless species of bird *(Dicaeum vulneratum)* made a habit of visiting mistletoe flowers to drink their nectar. We know Rumpf (Rumphius) now as a towering pioneer of tropical biology, whose absorbing interest in the exotic world around him triumphed over his own misfortunes. Although he did not realize the pollinating service of these birds, his notes form the starting point of our present knowledge of one of the most fascinating aspects of mistletoes, and possibly of bird-pollination in general.

Perhaps all mistletoes with large and colorful flowers are generally bird-pollinated. That this is true in the Americas is becoming apparent (see fig. 2-1,*c*), although the more precise pollinating relations have developed in the tropics of the Old World. Much of the following information is based on the admirable work of Docters van Leeuwen (1954).

In the Americas hummingbirds are by far the most prominent pollinators; indeed, I have not found other birds mentioned in the literature, although they also may serve in this fashion. It has even been said that no other pollinators exist but hummingbirds (*Phrygilanthus tetrandrus*: Diem, 1950).

In the tropics of the Old World we find the mistletoe-pollinators par excellence in the family Dicaeidae, often called "flower-peckers" or "mistletoe birds." Although other birds may be of importance also (the Nectariniidae of the African region are typical visitors of mistletoe flowers), this discussion will be restricted largely to Dicaeidae. The family encompasses seven or eight genera and about fifty-five species. It is

related to the Nectariniidae; and another genus, *Zosterops,* is sometimes included in the family (Mayr and Amadon, 1947). The family is generally Indo-Malayan. The largest number of species occurs in New Guinea and the Philippines. The diet of this family falls into three distinct classes: nectar, berries, and insects and spiders. Interesting structural and functional modifications have evolved in the digestive tract as a result of this diet.

Early in the tropical daylight the first visitors, the smaller species of *Dicaeum,* seek out the flowering bushes of Indonesian mistletoes. In *Macrosolen cochinchinensis* (and the African *Tapinanthus kraussianus*) birds find no open flowers at that time of day. The oldest flowers, however, are under tension, apparently caused by a rapid transformation of starch to sugar in the adaxial tissues of the petals, which increases their turgidity. The bird squeezes the top of the flower in its bill or inserts its bill between the petals. The interlocking margins of the petals release their hold, and the flower springs open violently. This explosive anthesis may throw pollen onto the bird, as in *Tapinanthus,* where the style springs out of the way to prevent self-pollination. In another African mistletoe, *Erianthemum dregei,* the explosion breaks off the anthers, which fly into space, scattering their pollen as they go. No wonder the visitor may have great masses of yellow pollen on its head and breast. Preformed slits are available in the mature buds of some mistletoes (e.g., *T. kraussianus*). In Indonesia the early *Dicaeum* species are followed in regular succession by other birds, which then find the flowers opened for them. Many paleotropic mistletoes may, however, open their flowers without avian assistance: *Scurrula* flowers, for example, are open before sunrise. But in *M. cochinchinensis* and *T. kraussianus* the flowers have lost this independence and fall unopened when birds are absent. In the former the connate anthers, when the flower is dropped, rub along the stigma; it is possible that this may be an effective mechanism of self-pollination, by way of last resort, if cross-pollination fails. Such a mechanism is known from Compositae, and may be present in Orobanchaceae (see under *Orobanche uniflora*).

Dr. Calaway H. Dodson allows me to add the following observations which he made in January, 1963, at Rio Negro, along the road from Banos to Puyo, on the Amazonian slopes of Ecuador at an elevation of about 1,100 meters. Here a mistletoe grows, probably a species of *Psittacanthus,* whose orange-red flowers attract no fewer than nine species of humingbirds. The birds with long bills were seen to work only the open flowers, but the short-billed birds would thrust their beaks into the side of the mature bud and fly upward, thus "unzipping" the flower. Whether preformed slits or other adaptations of this sort exist we do not know.

With the exceptions noted earlier, the Loranthacean flower demonstrates the classical complement of adaptations to bird-pollination. (In the unusual large white flowers of *Elytranthe albida,* we might wonder about nocturnal pollinators such as bats.) Flower color ranges from white to orange and red, sometimes golden yellow, with trimmings of green, black, or purple, often in contrasting patterns. One of the most gorgeous sights in the tropics is a tree laden with the brilliant blaze of flowering mistletoe. The reference to fire here is no accident: vernacular names (e.g., "matches" and "fosforito") attest to this appearance, and it has even been suggested that Moses' "burning bush" (Exodus 3:2-4) could have been a waving bush of *Tapinostemma acaciae* (Smith, 1878). The flowers are virtually without scent, and lack the landing platforms of many entomophilous flowers. Great quantities of a rather viscous nectar are produced by a ring wall of tissue, near the base of the style; the corolla of *Lepeostegeres gemmiflorus* is more than half full, and that of *Helixanthera cylindrica* may overflow. Retention of nectar is aided by a capillary system of fine ridges on the inside of the corolla. The floral parts which are most likely to be damaged by the bird's beak often show structural reinforcements.

The other member of this partnership has undergone modifications which similarly increase its efficiency. The most striking one is seen in the tongues of some species of *Dicaeum* and *Anaimos.* That organ may have overlapping, upturned frayed edges, and may even be doubly forked, producing a capillary system excellently suited to the uptake of nectar (fig. 2-21,*b*). Knowledge of this intimate relationship, and of the dispersal mechanism, has thrown light on the geography of these birds, which are, not surprisingly, limited by treeless (= mistletoe-less) areas. Yet we should not underestimate the versatility of the birds, for at least one species of *Dicaeum* has been collected several times on Krakatoa, where no mistletoes were present. The bird apparently lives and reproduces there by means of other vegetation.

Many aspects of bird-pollination in mistletoes need further study. Danser (1931*b*) has suggested that mistletoe species might be competitive, one attracting more birds than another because of more attractive floral pattern and shape. When Docters van Leeuwen placed a flowering branch of *Macrosolen formosus* in his Bogor aviary, populated with a number of local flower-birds, a curious thing happened. The birds immediately noticed the flowers, which in this mistletoe are of an exotic and complex color pattern, but did not open any of them. As the birds were of local origin and the mistletoe did not occur in that area, the experiment may reveal flower preferences of importance for the mistletoe. Or must young birds be taught to open flowers, and thus remain limited to the flowers encountered in their youth? Docters van Leeuwen describes the jealousy of Nectariniidae (sunbirds) with regard to flowers, not only toward other birds but also toward insects. All competition is chased away from the source of nectar. I have observed something of the sort in *Gaiadendron* (Kuijt, 1963*a*). I also recall watching a vigorous though uneven battle between a hummingbird and a large black butterfly around the flowers of *Psittacanthus schiedeanus* on the slopes of Volcán Poás, Costa Rica. This might explain the fact that insects seem to avoid the bird-flowers. Even if the host tree's profuse inflorescences are alive with thousands of insects, they ignore the mistletoe flowers among them. Some birds break holes in the base of the corolla, from which they drink, by-passing the pollinating mechanism in the manner of a bumblebee biting a hole in the spur of *Linaria.* And there is the unsolved mystery of the flower of *Peraxilla tetrapetala* of New Zealand. The tips of its petals stick together even while they are separating from each other and the gynoecium (fig. 2-10,*a*; Laing and Blackwell, 1964). This results in a four-parted umbrellalike structure still containing the stamens and style, until the petals with their stamens fall to the ground separately. We cannot even guess at the meaning of this bizarre performance.

Aside from the springlike style of *Tapinanthus kraussianus*—perhaps the situation is similar in *Socratina* (Balle, 1964), where petal tips remain attached, style and stamens extending outward through separate slits below—there seem to be very few mechanisms preventing self-pollination. There are suggestions of protandry in *Dendrophthoe pentandra.* But several Indonesian mistletoes are known to produce fruit abundantly even if they are separated from their pollinators (Docters v. Leeuwen, 1954). Whether this is apogamy, or merely selfing, is an open question. Cleistogamy is believed to occur in some Loranthaceae, for example, *Helicanthes elastica* (Blakely, 1922-1928; Johri, Agrawal, and Garg, 1957).

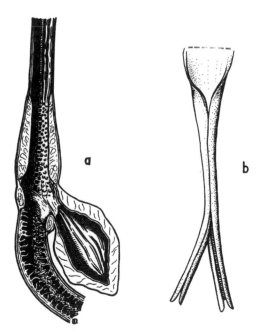

FIG. 2-21. *Dicaeum*: left, digestive canal of "mistletoe bird," *D. celebicum* (after Desselberger, 1931; ca. × 2.5); right, tongue of *D. trigonostigma,* dorsal view, ending in four equal-sized semitubular projections (after Gadow, 1890-1899; magnification unknown).

In a number of Loranthaceae (only bird-pollinated ones) the corolla has become bilaterally symmetrical when open. The occurrence of zygomorphy shows a geographic pattern which I cannot explain. In the Americas, without exception, Loranthacean flowers are radially symmetrical. Docters van Leeuwen informs us that in Indonesia only *Scurrula* shows zygomorphy. In contrast, in Africa zygomorphy is predominant (Balle, 1956): in this respect the pattern is "bipodal," with Africa and the Americas as the extremes.

The persistent notion that flower-pollinating birds are looking primarily for insects is nullified by the fact that some flowers depend on birds to open them. There can be no insects or spiders within an unopened flower.

The entire spectrum of pollination in Loranthaceae shows a remarkable correlation for which, again, I can provide no explanation. As we have seen, most Loranthaceae are bird-pollinated. All these flowers are bisexual. The reverse is not quite true, for the bisexual flowers of *Oryctanthus*, *Phthirusa*, and *Ixocactus* are almost surely entomophilous. Nevertheless, we can say that all bird-pollinated Loranthaceae have bisexual flowers. We can also say that, as far as is known, all dioecious Loranthaceae are insect-pollinated, and are inconspicuous in color. This is true of *Loranthus europaeus* (Cammerloher, 1921), *Tupeia antarctica* (Smart, 1952), *Barathranthus axanthus* (Docters v. Leeuwen, 1954), and *Struthanthus*. The mystery lies in the fact that no unisexual flowers are bird-pollinated. One might, indeed, wonder whether this is a general rule applying to other angiosperms as well. Certainly the classic examples of ornithophilous flowers are bisexual. Grant's (1950) paper on pollination does not explore this notion. Meanwhile, we are at a loss to find a reason for the apparent restriction of pollinating birds to bisexual flowers. But this impression may, after all, be a false one, based on the infrequency of unisexual flowers in angiosperms in general.

EMBRYOLOGY

The embryology of mistletoes is even more unusual than their flower biology. This chapter of mistletoe study also opened in an Indonesian setting when Melchior Treub, first director of the famous Botanical Gardens at Buitenzorg (Bogor), gave the first satisfactory account of mistletoe embryology. In recent decades, Maheshwari and his colleagues and students have studied many species of paleotropical Loranthaceae from this point of view; so we now have a reasonably accurate view of these bizarre phenomena (Maheshwari *et al.*, 1957). American mistletoes remain virtually unstudied in this respect. Since both the largest and the smallest of Loranthacean flowers occur in South America (*Aetanthus* and *Ixocactus*), many embryological novelties await dicovery there.

The gynoecium of mistletoes is exceedingly reduced. The inconspicuous central cavity contains not the ovules but a basal central body (the ovarian papilla or mamelon) of varying prominence (fig. 2-22). This organ is like a vestigial placenta. In *Lepeostegeres* the mamelon reaches far into the stylar canal. Instead of a mamelon, *Lysiana* has a central column, confluent with carpellary tissue at both ends, within which sporogenous cells develop. In the latter genus vertical partitions have produced four separate cavities. Traces of similar partitions can be found in a few other genera as well. But in at least five genera even the mamelon has disappeared. The sporogenous tissue, otherwise part of the mamelon, here arises from the cells immediately below the ovarian cavity. In none of the mistletoes is there anything we can call an ovule. In the strictest sense we can, therefore, not speak of *seeds* in Loranthaceae.

Fig. 2-22. Longisections of ovaries: *a*, *Lysiana*; *b*, *Lepeostegeres*; *c*, *Nuytsia*; *d*, *Arceuthobium*; *e*, *Taxillus* (all but *d* after Maheshwari *et al.*, 1957).

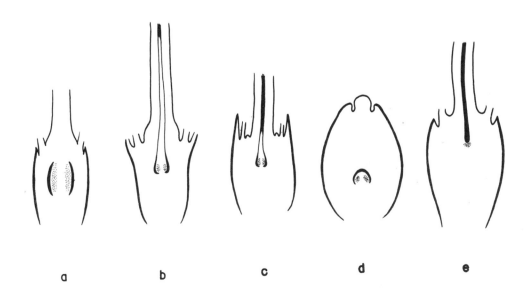

a   b   c   d   e

This is one of several places in the study of mistletoes where a rigid morphological terminology must be abandoned as inadequate. The dispersal units of mistletoes are seeds in function and evolutionary origin, and I shall continue to call them seeds.

Viscacean embryology is relatively simple and will serve as an introduction to Loranthacean embryology. In *Viscum, Korthalsella* (Rutishauser, 1935), *Arceuthobium* (Thoday and Johnson, 1930), *Ginalloa* (Rutishauser, 1937), and *Dendrophthora* (York, 1913), two spore mother cells originate in the mamelon. The resulting embryo sac is a simple one, consisting of four nuclei; its formation follows the Allium type. In *Korthalsella* the embryo sacs do not remain within the mamelon, but grow out basally and curve around into the gynoecial wall. The mature flower thus has two U-shaped embryo sacs. In *Ginalloa* (Rutishauser, 1937), *Arceuthobium* (Thoday and Johnson, 1930; Jones and Gordon, 1965), and *Viscum* (Schaeppi and Steindl, 1945) the embryo sac is straight. Five other genera of Viscaceae are unknown embryologically.

The "aggressive" behavior of the embryo sac in *Korthalsella* is surpassed by that in Loranthacean genera, in which the embryo sac digests its way into the style, and down into the floral base, where it meets an obstruction in the pad or cup of collenchyma. The upward extension of the female gametophyte into the style is most significant, for the height to which it rises seems to have some taxonomic constancy. In some species, only the base of the style is reached; in others the gametophyte extends to half or two-thirds of the total stylar length. Several instances have been recorded (Johri and Raj, 1965) of the embryo sac growing into the stigma, in *Helixanthera ligustrina* even reaching the stigmatic surface (fig. 2-23, *a, b*). A South African mistletoe goes even farther: its embryo sac, reaching the stigmatic summit, continues to grow in a recurved path, forming an inverted J (Johri and Raj, 1965). At the tip of this capillary embryo sac, the egg cell of the gametophyte fuses with the sperm nucleus. The female gametophyte, through its enormous linear growth, has in fact usurped the role of the pollen tube in traveling the length of the style.

After fertilization the young embryo will have to be pushed back, all the way, along the route of the embryo sac. Following an initial longitudinal division of the zygote—a rare occurrence in angiosperms—a number of transverse divisions results in a long, biseriate proembryo. Since several embryo sacs often develop to maturity, a veritable race now begins of several proembryos each following the same course, facilitated by elongation of the uppermost (suspensor) cells of the proembryos. Although the latecomers perish along the way, several reach the ovary. One proembryo soon dominates its rivals, and the latter gradually disintegrate. The victorious proembryo grows through the early cells of the endosperm and presses hard against the collenchymatous cup. Much later it retracts somewhat from the cup, allowing endosperm to envelop it at that pole.

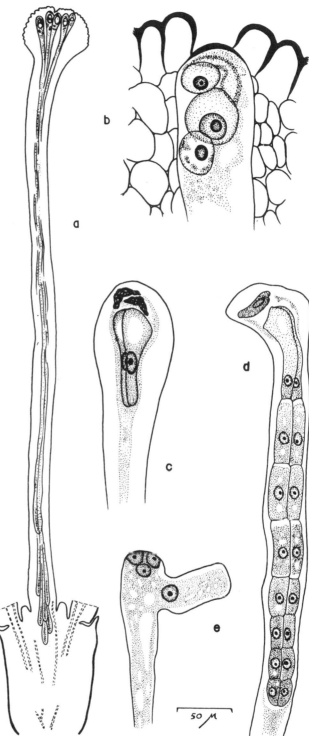

Fig. 2-23. *a, Helixanthera ligustrina*, longisection of gynoecium with mature embryo sacs (× 14); *b*, detail of embryo sac on far right in *a* (× 420); *c, Tolypanthus lagenifer*, two-celled proembryo (× 450); *d*, same, biseriate proembryo (× 450); *e, Nuytsia floribunda*, upper end of embryo sac with lateral caecum (all after Maheshwari *et al.*, 1957).

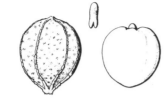

FIG. 2-24. (*Left*) *Lepidoceras punctulatum*, entire seed (left), endosperm with radicular tip (right), and embryo (Werdermann No. 287, Chile; × 14).

FIG. 2-25. (*Below*) *Psittacanthus schiedeanus*: transition zone between cotyledonary area of embryo and uppermost portion of massive multi-cellular suspensor; apical meristem of embryo visible between large cotyledons (Kuijt, 1967c; ca. × 34).

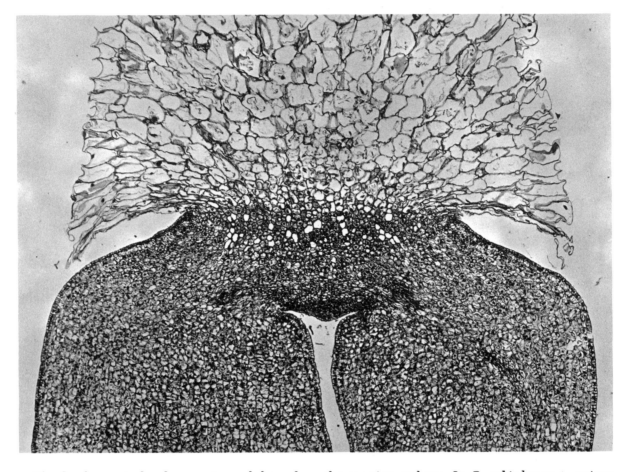

The development of endosperm is much less adequately known. The lower polar nucleus migrates upward to fuse with the upper polar nucleus and is eventually joined by a sperm nucleus. The primary endosperm nucleus thus formed travels to the lower end of the mature embryo sac. This appears to be another example of nuclear migration. It would not be surprising, considering the length of the embryo sac, if the fusion nucleus and second sperm occasionally failed to join. Perhaps the multiple endosperm of Loranthaceae guards against failure. The surprising thing is that the young endosperm is already at the goal before the first proembryo arrives.

In two mistletoe genera, endosperm is said to be lacking at maturity: *Lepidoceras* (Engler, 1894) of Viscaceae, and *Psittacanthus* of Loranthaceae. The former report is an error (fig. 2-24). No details are available for either, except that in *Psittacanthus*, endosperm is formed, but is subsequently absorbed by the growing embryo. In *P. schiedeanus* a unique situation has been discovered (Kuijt, 1967c). A great mass of cells in very young fruits forms a large central body, which appears to be endosperm-like. (fig. 2-25). The cotyledons which appear at one pole are not, however, truly emergent. The embryo actually takes its origin from a few small meristematic cells at the apex of this large body, which represents a combination of an incredibly massive suspensor and the proembryo. The embryonic pole of the suspensor shows a gradual histological transition, not only in central tissues but also in a discrete epidermis. Growth of the embryo gradually pushes back the suspensor until it is an insignificant, compressed, flaring cup around the radicular pole of the embryo. Loranthaceae therefore possess both the longest and the most massive suspensors in the angiosperms.

Except for *Psittacanthus*, then, the mature mistletoe embryo is surrounded by a mass of nutritive endosperm at its cotyledonary end. The endosperm sometimes shows prominent ridges. In *Gaiadendron* it extends in seven wings, each of which divides into two flanges. In Viscacaean endosperm—at least in *Viscum, Arceuthobium, Phoradendron, Dendrophthora,* and *Antidaphne*—much chlorophyll is present. This pigment has been demonstrated in endosperm of other angiosperms (Esau, 1960, p. 334), but not to the extent of Viscacean endosperm, which is obviously a seat of great photosynthetic activity. Structurally, the endosperm is further specialized in having a discrete epidermis, which in *Dendrophthora* lacks chlorophyll (York, 1913), as it does in sporophytic organs. But the two families cannot be clearly distinguished on this basis, for *Lysiana* also has green endosperm (Rigby, 1959).

The endosperm of Loranthaceae is unique in being compound: it is derived from several primary endosperm nuclei. Only one embryo survives, but several of the endosperms develop. They amalgamate so harmoniously that no sign of their multiple origin remains at maturity. In Viscaceae the endosperm is of simple origin. Small endosperm haustoria occur in *Arceuthobium* (Dixit, 1962; Jones and Gordon, 1965). *Korthalsella* would seem to have a similar development (Rutishauser, 1935). The occurrence of endosperm haustoria in these two genera of Viscaceae is intriguing in view of the elaborate haustorial processes in the endosperm of Santalaceae and Olacaceae (see chap. 3). No such organs are known from Loranthaceae.

The embryo dissected from mature mistletoe fruits is already dark green. It is normally dicotyledonous. In some of the squamate types, however, the cotyledons are scarcely differentiated. This is true of *Arceuthobium* (Kuijt, 1960) and of some *Korthalsella* (fig. 2-28,*b*), although Stevenson (1934) speaks of only a collar-like organ in New Zealand plants. Cotyledons in Viscaceae seem greatly reduced; the largest and most complex is known from *Antidaphne* (unpubl. inf.). In the majority of Old World Loranthaceae the two cotyledons are fused in a single conical organ. The two cotyledons are still separately endowed with vascular strands. Fusion is incomplete at the base of the cotyledons, leaving a slit through which the first true leaves will emerge. The cotyledons then form a compound absorptive organ, exhausting the endosperm and perishing with it. Much of this can be seen in *Loranthus europaeus* (fig. 2-26,*a*). In most neotropical Loranthaceae, and in the Elytranthinae of the Old World, the cotyledons are separate leaves which proceed from an absorptive and storage function to one of photosynthesis. Noteworthy in this regard are a number of polycotylous members of *Psittacanthus* (Kuijt, 1967*c*). The embryo of *P. schiedeanus* has up to a dozen awl-like fleshy cotyledons which make up nearly the entire seedling (fig. 2-26,*b*). This is presumed to be an advanced condition traceable to the more usual embryo.

What is known of the vascular skeleton of the Loranthacean embryo or young seedling shows each cotyledon to be supplied with two vascular traces (fig. 2-27). These traces reach into the cotyledonary tips, but along the way branch out into a complex network of veinlets, a network which is completely open but for an occasional anastomosis. In *Gaiadendron* the two traces unite above the middle of the cotyledon; this median vein reaches into the apex (fig. 2-27, *b*). In the region of the hypocotyl the first shoot traces take their origin, below which is found a transition to the root, or to the primary haustorium, depending on the species of mistletoe.

In Viscaceae the vascular structure of the young seedling is very simple (vascular differentiation may not have progressed very far in the seed). The simplest situation is met in *Korthalsella* (fig. 2-28,*b*) and *Arceuthobium* (Kuijt, 1960). A single central strand of xylem runs the length of the radicle to fork slightly just below the minute cotyledons. The seedling of the leafy *Phoradendron tomentosum* subsp. *macrophyllum* differs only in having longer separate cotyledon traces (Calvin, 1966). The *Antidaphne* cotyledon is unique in having three main veins which become somewhat frayed at the tip only, where a few anastomoses are present. The straight, unbranching part of each vein is accompanied by a heavy bundle of fibers, which represents another unique feature. I have not been able to trace the downward course of the vascular bundles, but presume that the median vein has arisen from basal branches of the lateral veins.

The radicular apex of the embryo is of particular interest. In the primitive Loranthacean trio (*Atkinsonia, Nuytsia,* and *Gaiadendron*) the root cap is not different from that of many autotrophic angiosperms (Kuijt, 1965*a*). But in other mistletoes of both subfamilies a radicular root cap is missing. A single discrete epidermal layer surrounds the entire embryo from pole to pole (fig. 2-29,*a*). It is as if the root apex had become shootlike, except that no foliar organs are formed. No wonder many morphologists have declared mistletoes to be rootless (Docters v. Leeuwen, 1954; Maheshwari, Johri, and Dixit, 1957). I see no reason, however, to deny mistletoes roots (see chap. 7).

There may be a functional meaning in this fundamental difference between primitive and advanced mistletoes. In primitive mistletoes a terminal haustorium is never formed. In advanced mistletoes the radicular apex becomes the primary haustorial organ, occupying a terminal position. A root cap in such mistletoes would serve no purpose; indeed, it would probably constitute a disadvantage, and consequently has been eliminated in the course of their evolution. Significantly, the apices of epicortical roots have root caps, but form lateral haustoria only.

The seedling of *Psittacanthus schiedeanus* can scarcely be accommodated by what has just been said about vascular structure and root apex. Each of its many cotyledons has a single trace which produces two laterals. Only the median trace reaches to the

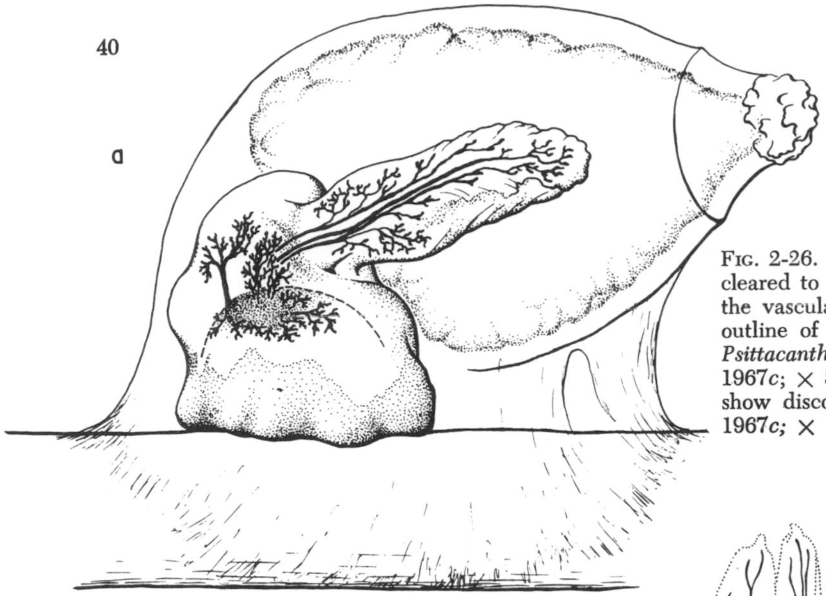

FIG. 2-26. *a, Loranthus europaeus,* germinating seedling cleared to show vascular system and fused cotyledons, the vasculature of only one of which is represented—outline of endosperm stippled (Vienna; × 15); *b, Psittacanthus schiedeanus,* seedling (after Kuijt, 1967*c*; × 5); *c,* same, unexpanded seedling cleared to show discontinuous vascular system (after Kuijt, 1967*c*; × 5).

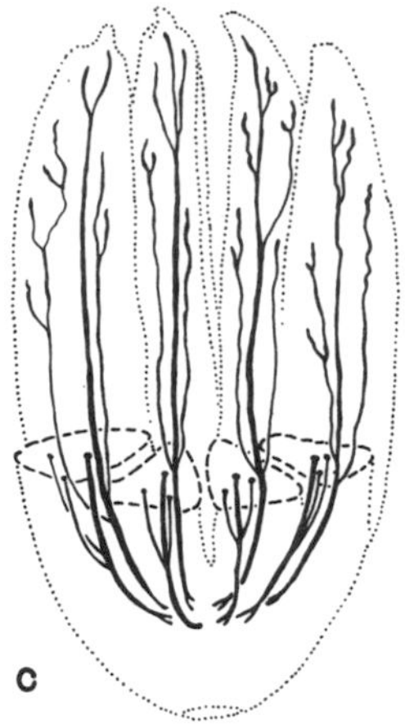

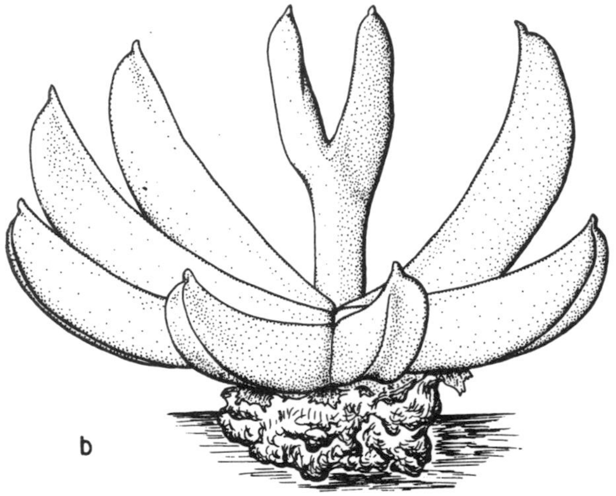

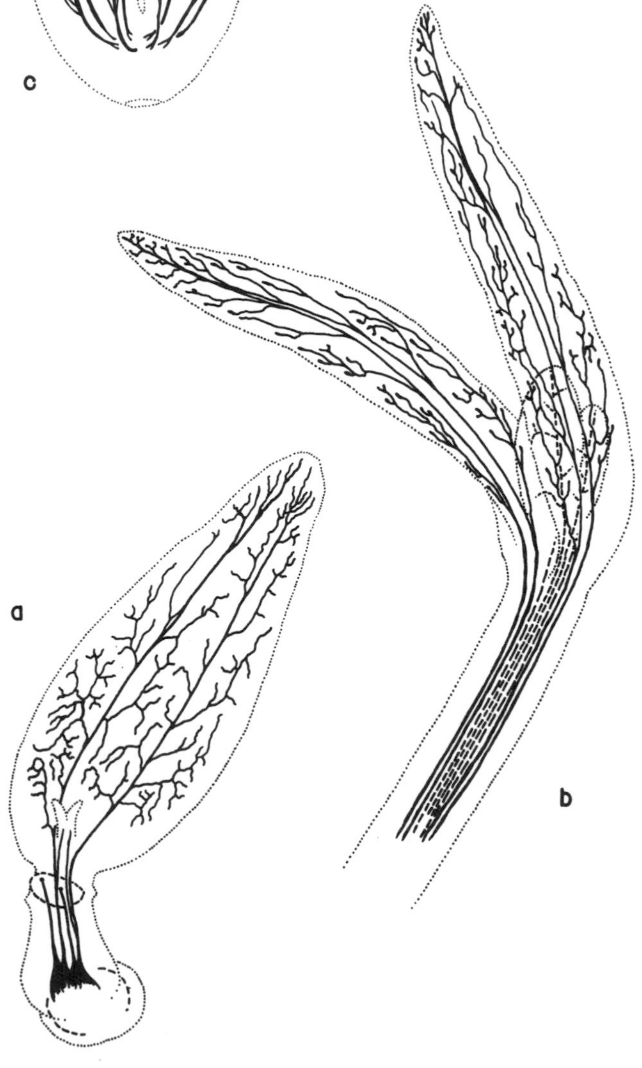

FIG 2-27. Cleared seedlings: *a, Phthirusa pyrifolia; b, Gaiadendron punctatum* (both from Costa Rica; × 7.5).

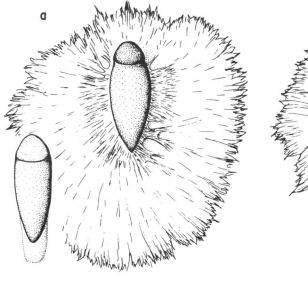

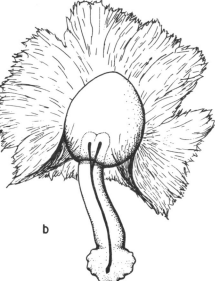

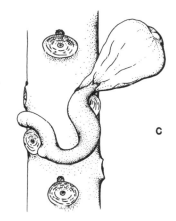

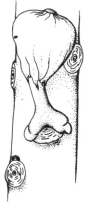

FIG. 2-28. *a, Arceuthobium campylopodum,* seeds immediately after ejection (left) and with viscin cells fully expanded, dried, and attached to substrate (right) (after Kuijt, 1960; × 10); *b, Korthalsella* sp., attached seed cleared to show vasculature (Taylor No. 6235, Hawaii; × 25); *c, Phoradendron densum,* seedlings with radicular lobes (California; × 10).

FIG. 2-29. *a, Arceuthobium americanum,* apical meristem of radicle (after Kuijt, 1960; × 150); *b, Phthirusa pyrifolia,* seed with dark green, somewhat flattened radicle emerging through translucent viscid collar; endosperm below (Costa Rica; × 5).

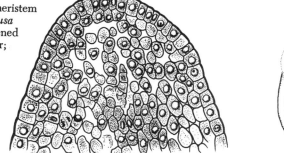

a   b

cotyledonary tip, but all three may ramify slightly (fig. 2-26,*c*). At least at the time of germination there is no continuity between the vasculatures of different cotyledons. Each strand runs down into the cotyledonary node but stops short of its fellows. How these strands are eventually linked up mutually and with the xylem of the growing epicotyl and haustorium, we do not know. Nor do we know how the haustorium develops from the radicular pole. The remains of the giant suspensor are still attached to the base of the embryo at maturity, and in the zone of attachment the discrete epidermal layer in most other mistletoes is lacking. Possibly a haustorial organ arises somewhat laterally. How little we really know about this beautiful genus of mistletoes!

FRUIT AND DISPERSAL

Although technical objections have been raised against such usage, the mistletoe fruit may be characterized as a one-seeded berry. It ranges from barely 2 mm to more than 14 mm in length. Its color variation rivals that of the flower. Everyone knows that the Christmas mistletoes of Europe (*Viscum album*) and North America (usually *Phoradendron serotinum*) have pearly white berries. Actually, fruit color in *V. album* ranges through various tints of yellow and orange to bright red in the eastern parts of its natural range (Danser, 1941 *a*). *Struthanthus* fruits may be dark blue, dull purple, brilliant red, or orange (Kuijt, 1964*d*); *Scurrula atropurpurea*, dark yellow; *Elytranthe albida*, black; and so forth. In *Phthirusa pyrifolia* I have observed a complex series of color changes including purple, orange-red, and yellow which slowly descend the berry from its tip. The surface of the fruit is smooth except in a couple of unusual species of *Viscum* which have warty fruits (Rao, 1957). The pedicels of maturing *Lepeostegeres* fruits elongate and carry them out of the involucre, thus catching the eye of fruit-eating birds. Possibly the same explanation applies to a similar phenomenon in *Struthanthus costaricensis* (Kuijt, 1964*d*), where only the lateral fruits of a triad become elevated.

From the bird's point of view, of course, it is the flesh of the berry which is attractive. It is covered by hard, rindlike tissues which include a strongly cutinized epidermis. In *Gaiadendron* more than two-thirds of the fruit consists of flesh (fig. 2-30,*a*), in *Dendrophthora costaricensis* perhaps half (fig. 2-31,*a*), and in *Antidaphne viscoidea* slightly less. But many mistletoe fruits, quite unexpectedly, seem to have scarcely any flesh. In *Struthanthus orbicularis* some fleshy tissue can be seen, but in *S. quercicola*, *Phthirusa pyrifolia*, and *Phoradendron californicum* the rind seems to abut directly upon the viscid tissue (fig. 2-30, 2-31). Perhaps the most likely explanation is that the viscid tissues are partly nutritive, thus combining two major functions. Or would it be too far-fetched to suspect that the birds are duped through a likeness of the mistletoe fruit to a truly nu-

tritive one belonging to another local plant? The variety of colors on mistletoe fruit makes one wonder if color preferences might not play an evolutionary role here, as they might in bird-pollination. Bewildering possibilities are suggested by the situations where the same species of birds live off both mistletoe nectar and mistletoe fruits, as do Dicaeidae. Accurate and systematic field observations of the most elementary kind could add much to our understanding of this many-faceted partnership.

The variable position of vascular tissue in the mistletoe fruit has given rise to misconceptions. In virtually every major systematic work on mistletoes one of the differences between the two mistletoe families is said to be the position of the viscid layer (Danser, 1931*b*; Maheshwari *et al.*, 1957; Dixit, 1962; Barlow, 1964). In Loranthacean fruits the viscid zone is said to be situated outside the vascular bundles; in Viscacean fruits, within them. This notion, as far as I am aware, was introduced by Danser (1931*b*) in his classical monograph on Indonesian mistletoes, and has subsequently gained much credence. It was cited, for example, as an instance of transference of function from one tissue to another (Corner, 1958). On the basis of the same criterion, it has been suggested that the bird-dispersal mechanism in the two subfamilies may be of different evolutionary origin (Barlow, 1964).

All these arguments fail to take full account of the Loranthacean calyculus, which must be regarded as a reduced calyx. Where, in relation to the viscid zone, would the vascular bundles of the calyx have run? Narayana's (1958) work on *Atkinsonia*, where the calyculus has a remnant of vasculature, leaves little doubt. The viscid zone in the Loranthacean progenitors must have been flanked by corolla traces on the inside and by calyx traces on the outside. The viscid zone was within the calyx bundles, as it is in present-day Viscaceae. The inconspicuous bundles in Loranthacean flowers supply the corolla and the androecium; the more central, slender bundles run into the style. The central "basket" of bundles (*Korthalsella*, fig. 2-14*c*) of pistillate Viscacean flowers belongs to the gynoecium only. A search for vascular remnants of a second perianth whorl is futile, for these flowers are monochlamydeous in contrast to the dichlamydeous Loranthacean flower. There has been no transfer of function of viscid tissue, and not necessarily an independent origin of habit; it is the same tissue, in the same position, except that the calycular vasculature has vanished in Loranthaceae.

The viscid tissue is variable in amount and position, and this variation has certain consequences at the time of germination. In many mistletoe fruits there is a great bulge or ring of the tissue at the radicular pole of the seed. In *Phthirusa pyrifolia* a seed extracted from the fruit has such a viscid ring through which the dark green radicle is apparent (fig. 2-29,*b*). Yet in the fruit this layer runs to the very base of the flower. Both color differences and cell alignments indicate a single continuous tissue

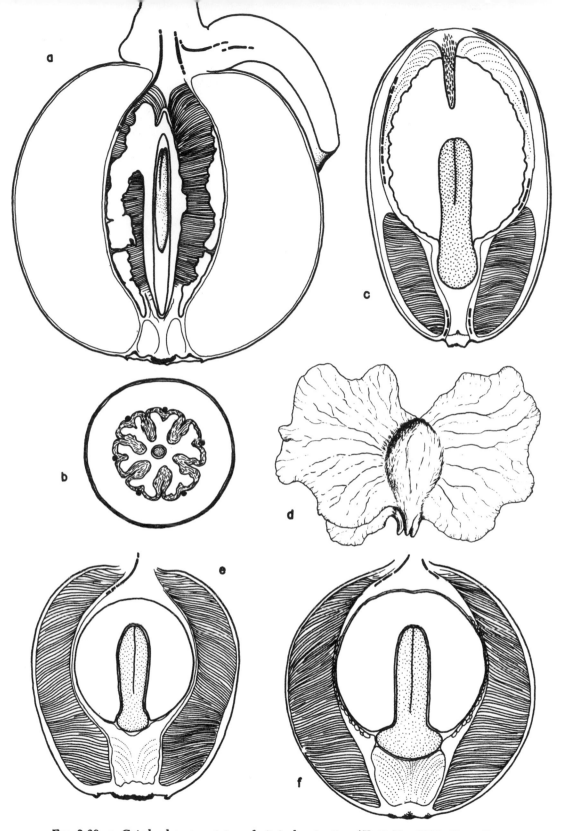

Fig. 2-30. *a, Gaiadendron punctatum,* fruit in longisection (Kuijt No. 2396, Costa Rica, UBC; × 5); *b,* same, transection of fruit showing.dichotomizing flanges of endosperm (white) (× 8.5); *c, Struthanthus orbicularis,* longisection of fruit (Costa Rica; × 7.5); *d, Nuytsia floribunda,* dry, winged fruit (after Blakely, 1922-1928; × 3); *e, Phthirusa pyrifolia,* longisection of somewhat immature fruit (Kuijt No. 2419, Costa Rica, UBC; × 10); *f, Struthanthus quercicola,* longisection of fruit (Kuijt No. 1535, Costa Rica, UC; × 10).

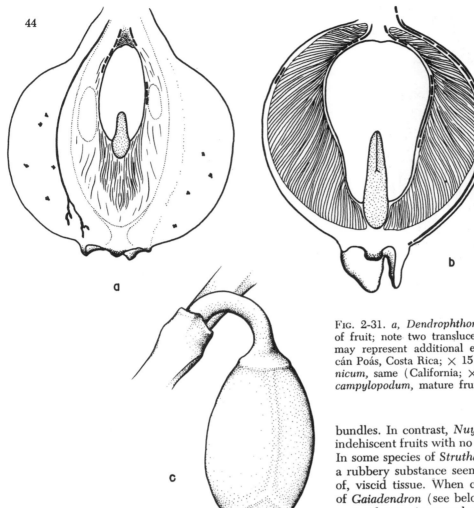

FIG. 2-31. *a, Dendrophthora costaricensis,* longisection of fruit; note two translucent elliptical bodies, which may represent additional embryos or endosperm (Volcán Poás, Costa Rica; × 15); *b, Phoradendron californicum,* same (California; × 15); *c, Arceuthobium campylopodum,* mature fruit (Vancouver, B.C.; × 10).

which is fleshy (yellow) at the base and viscid (opaque white) at the tip of the fruit. This "radicular" viscid tissue occurs also in *Antidaphne, Struthanthus quercicola* (fig. 2-30,f), *S. orbicularis* (fig. 2-30,c), *Dendrophthora costaricensis* (fig. 2-31,a), *Arceuthobium* (Kuijt, 1960), *Korthalsella* (Docters v. Leeuwen, 1954), and many others. In contrast, *Lepeostegeres* and *Macrosolen* fruits develop a mass of viscid tissue at the opposite pole, near the base of the fruit (Docters v. Leeuwen, 1954). This means, of course, that the initial growth of the former type is directed toward the host branch; that of the latter, away from it. In many of these plants, however, the somewhat elongated fruit usually adheres along its length. This is true of many small-seeded Viscaceae. Even so, the contraction, through desiccation, of the viscid "tail" may direct the radicle obliquely toward the substrate. Many Loranthacean seeds also adhere along their length, as in *Helixanthera, Dendrophthoe, Scurrula,* and *Barathranthus* (Docters v. Leeuwen, 1954).

In *Dendrophthora* (fig. 2-31,a) a mass of flesh is sometimes developed outside the main vascular

bundles. In contrast, *Nuytsia* fruits are dry, winged, indehiscent fruits with no flesh at all (Barlow, 1964). In some species of *Struthanthus* (e.g., *S. orbicularis*) a rubbery substance seems to aid, or take the place of, viscid tissue. When considering the germination of *Gaiadendron* (see below) it is not surprising that a viscid zone is scarcely recognizable, as this tissue represents a specific adaptation to germination on smaller host branches.

The viscid tissue of mistletoe berries provided the birdlime of long ago. Although the preparation and use of this substance in most of Europe does not seem to have persisted beyond the early decades of our century, many recipes survive (Tubeuf, 1923). The berries were often soaked and boiled in water, and were then crushed and kneaded intensively by hand. After rinsing, the lime can be kept for an extended period of time. The final product was placed on sticks near some food (mistletoe berries?). The birds soon became entangled in the strong viscous threads. The irony of a bird being caught by the fruits of a plant whose dispersal he carries out did not escape the Ancients, who already used birdlime. The saying *Turdus ipse sibi cacat malum* (The thrush prepares his own misfortune) is ascribed to the Roman Plautus (ca. A.D. 254). The practice of birdliming was common even at the time of Theophrastus (371-286 B.C.), and is known also from West Africa (Dalziel, 1937). *Viscum album* was not the only mistletoe used for birdliming. In many countries in Europe, *Loranthus europaeus* was used for the same purpose, and the African use is based on still other mistletoes. The same material was used to catch flies.

Not very well known is the fact that birdlime may be manufactured from other plants such as *Lantana* roots or *Ilex* bark, the latter in Japan.

In nature an agent or mechanism of dispersal is required to take the seed from its fruit and place it in a position where it will germinate. The dispersal of mistletoe seeds has unexpected facets of great biological interest.

It has been known for centuries that mistletoe fruits are eaten by birds, which, by voiding them, disperse the seeds. Yet, more than a thousand years after Theophrastus, students of nature found this remarkable cooperation between bird and mistletoe difficult to believe. Even the outstanding botanist Jean Bauhin, a contemporary of Rumphius (second half of the seventeenth century) denied the origin of mistletoe from bird droppings (Docters v. Leeuwen, 1954). He maintained that mistletoes develop from the tree itself, much as warts emerge from the human body. Here, Bauhin was merely voicing the general opinion of his time; this particular notion survived at least until the early nineteenth century (cf. Tubeuf, 1923, p. 609).

Ironically, when early Western botanists accepted the "information" of the Ancients, both were frequently wrong; and where they doubted the Ancients, or elaborated on what they had said, the Ancients were often closer to the truth. The belief that mistletoe seeds were not viable unless they had passed through the bird's alimentary canal was current in Ancient Rome (Tubeuf, 1923, p. 609), but even today it is a common misconception about these plants. Not until the second half of the eighteenth century was it shown that mistletoe seeds may germinate without avian assistance.

The association of seeds with bird excrement may have produced the Germanic name of mistletoe. The correspondence of the German words for manure and mistletoe ("der Mist," "die Mistel," respectively) is too close to be accidental. Other names exist also, usually with reference to use or folklore (e.g., the Dutch "vogellijm," birdlime, and "maretak," connected with evil nocturnal spirits).

Ornithological studies of our day have shown that in other ways truth, indeed, outdoes fiction. Important discoveries have revealed the extent of avian adaptation, both in structure and behavior, to a diet of mistletoe berries. This seems to have been explored only in Dicaeidae (Desselberger, 1931) and Euphonia (*Tanagra;* Wetmore, 1914).

Many (perhaps most) of the birds which eat mistletoe berries demonstrate no particular modifications with regard to this diet. Mistletoe berries are no more than an incidental part of the diet of waxwings, many thrushes, and some other birds. In the same category is the use of mistletoe berries by martens (Tubeuf, 1923) and by the fruit bats of Réunion, Mascarene Islands (Allen, 1962). Tubeuf's (1923) extensive summary of bird dispersal in European mistletoes shows that nearly every fruit-eating bird there eats the berries. Even the European mistletoe thrush, *Turdus viscivorus* L., while famous for its preference, is a rather unspecialized frugivorous bird. The silky flycatcher, *Phainopepla (Ptilogonys) niteus,* seems to be very closely associated with the desert mistletoe, *Phoradendron californicum* in North America. But from the bird's occurrence on Santa Catalina Island (Crouch, 1943) it appears that this relation is not an obligate one, for the island has no mistletoe of any sort (Millspaugh and Nuttall, 1923).

The most fascinating conditions exist where an obligate or nearly obligate interdependence of bird and mistletoe has evolved. Such conditions are known so far in two bird groups, both in the order Passeriformes, but naturalists in tropical areas would be well advised to watch for other instances. The first known case is that of the American Euphonia, more particularly *Tanagra musica sclateri* from Puerto Rico (Wetmore, 1914).

The stomach of Passeriformes generally includes a smaller, constricted and glandular portion, the proventriculus, connected with the esophagus; and a larger, muscular ventriculus (gizzard) leading to the pyloric sphincter. The stomach of Euphonia is fundamentally different in several ways. The proventriculus is unusually large; the ventriculus, on the contrary, is a thin transparent zone passing directly into the dilated beginning of the small intestine. There is neither a pyloric nor a cardiac constriction.

The diet of these birds is believed to be mistletoe fruits exclusively. The best way to find them is to station oneself near a fruiting mistletoe bush. In various islands of the Lesser Antilles this obvious association has given Euphonia the name of "mistletoe bird." The seeds pass through the digestive tract in almost the same condition as when swallowed. The time required for this passage has not been recorded, but we may be certain that it is short. No insects have been recovered from the stomach.

The second known case is the paleotropical flowerpecker or mistletoe bird, *Dicaeum.* The organization of the digestive tract, especially of *D. celebicum,* reveals a unique combination of specialization and versatility (Desselberger, 1931). A functional and structural duality has evolved in the digestive apparatus, paralleling the two main types of food used. The glandular proventriculus is adapted to the digestion of fruit; the gizzard, to that of spiders and insects (fig. 2-21,*a*). The gizzard is a lateral muscular organ which opens into the base of the proventriculus, just above the pyloric sphincter, by means of an aperture too small for mistletoe fruits to enter. The remarkable fact is that, when animal food is taken, the sphincter closes while the gizzard aperture extends, thus forcing the food into the gizzard. The undigestible chitinous remnants are thrown up as pellets; the resulting liquid materials are now allowed to leave the gizzard and pass through the sphincter into the absorptive portion of the alimentary canal. When fruits are eaten the sphincter remains lax, permitting a direct passage from the glandular proventriculus to the bowel; the gizzard is not involved. This unique division of labor is not known from other Dicaeidian genera.

There are conflicting reports on the precise way in which birds handle mistletoe berries. But we should not expect uniform avian behavior here. The occasional consumers of mistletoe berries are likely to be careless and clumsy. Perhaps some of these wipe a fruit off their bill now and then. This is emphatically denied as a mechanism of dispersal in the more specialized mistletoe birds. Many birds swallow the entire berries. I have observed this with a thrush and a silky flycatcher feeding on Gaiadendron fruits in Costa Rica; Sutton (1951) reports it for a cedar waxwing, several kinds of Euphonia, and a silky flycatcher. The Puerto Rican Euphonia, strangely, seems to discard the rind and swallow only the seed (Wetmore, 1914). In Dicaeum t. trochileum and D. sanguinolentum the fruits are placed carefully in the mandibles and squeezed; the seed then pops out and is swallowed. The rind is discarded, but not until the pulp is pressed out and swallowed (Docters v. Leeuwen, 1954). According to a recent observer, a Ceylonese species of Dicaeum is interested only in the pulp; the seed sometimes, perhaps accidentally, slips into the mouth and is swallowed (Weeraratna, 1960). Normal dissemination is said to be by means of wiping off seeds on branches. Considering the variety of mistletoe fruits and birds we can scarcely be surprised at behavioral differences.

The slippery and viscous seed, for all its attraction to Dicaeum, is an awkward, often bulky item to void. After ingesting several berries the bird usually quiets down and remains stationary for a while (Ali, 1931), and members of Euphonia and silky flycatchers do likewise (Sutton, 1951). While voiding the seeds the bird (Dicaeum) executes a hilarious maneuver (translated from the Dutch, v. Heurn, 1922): "When defecating the bird suddenly squats visibly, and simultaneously shifts its position with astonishing agility along the twig while its body rocks rapidly to and fro, thus covering a distance of 20 - 30 cm." In other words, the bird actually pastes the seeds to the branch. Docters v. Leeuwen (1954) has confirmed this extraordinary behavior.

The rapid passage through the highly modified digestive tracts of some birds is of great advantage to the mistletoe. It ensures, in return, future crops of berries for the bird's offspring. There are, of course, many birds which possess a stomach not so accommodating (Kerner v. Marilaun, 1896-1898). In the gizzard of fowl or pigeon the extended muscular action, frequently aided by sand and pebbles, is too much for the mistletoe seed, which is quickly ground to a pulp. Rapidity of passage would seem to reflect greater adaptive intimacy between bird and mistletoe. Seeds of Viscum album spend about thirty minutes within the body of the mistletoe thrush (Tubeuf, 1923, p. 614); Ali (1931) reports a period of no more than four minutes in Dicaeum for a bird eating fruits of Dendrophthoe falcata. Docters v. Leeuwen (1954) gives a time of about twelve minutes for Dicaeum sanguinolentum.

A short digestive period does not necessarily prohibit mistletoe dispersal over considerable distances. A partly digested mistletoe seed is obviously a rather nasty item to get rid of, and requires a specialized maneuver if the voiding is to be done properly, at least in Dicaeum. Consider a bird being overcome by this need while in the air! There is no stopping the seed, and it might even become entangled in the bird's feathers. In this fashion seeds could be carried much farther than a rapid digestive passage would indicate.

Not all mistletoes depend on birds for dispersal. The fruits of Nuytsia floribunda are dry, and we know nothing about their dissemination. But far more interesting are the explosive fruits of some Viscacean mistletoes. This phenomenon has been documented in India for one species of Korthalsella (Sahni, 1933), but we do not have information on other species. The fruit of Viscum congolense is said to dehisce by means of an equatorial transverse slit (Balle and Hallé, 1961), which would seem to imply forceful ejection.

The entire fruit of the dwarf mistletoe (Arceuthobium) is about four mm long, and of an elongated ovoid shape. It has developed from a minute pistillate flower which is remarkable in its degree of reduction and simplicity. Shortly before the fruit matures its pedicel recurves so that the tip of the fruit points directly downward (fig. 2-32,b). The seed, situated near the base of the fruit, has an abundance of chlorophyllous endosperm with a minute, undifferentiated embryo. The viscid tissue forms a prominent tail on the embryonal (i.e., distal) pole of the seed. Considerable pressure builds up at maturity within the fruit, but the way in which this is accomplished and the tissues involved are unknown. The outer layer of the fruit is very tough and possibly elastic, thus containing the mounting pressure.

At the base of the fruit a transverse zone of weakness eventually gives way, and the bullet-shaped seed is shot more or less vertically upward. The violence of this explosion is astonishing. Recent photographic studies (Hinds et al., 1963; Hinds and Hawksworth, 1965) are illuminating, for in nature the process is too rapid for the human eye. With the aid of photography it has been possible to calculate initial velocities of about 24 meters per second for several species. Seeds in one species are expelled to an average distance of 5 meters, but sometimes nearly 15 meters. These distances are measured to a place on the same level as the point of discharge. In nature, a horizontal position of the fruit rarely occurs, and thus horizontal spread cannot occur as rapidly as these figures imply. Balancing this qualification is the fact that the effective horizontal distance in nature may be greater than that measured in these reports, because of the elevated position of many dwarf mistletoe plants. Moreover, an undisturbed fruit in nature may reach a higher pressure. At any rate, this is an astonishing feat for a seed of less than three mm in length, and implies a tremendous pressure in the mature fruit.

It is no wonder that one may experience a stinging sensation when hit in the face by a dwarf mistletoe seed, as the initial velocity is formidable. The pressure that activates expulsion is believed to develop in the viscid tissue, but there is no real evidence to support this notion; the surrounding parenchyma cells may also be involved. Photographic studies reveal that a large amount of liquid material is expelled with the seed which may be derived from jetlike action of the mucilaginous cells. The seed experiences a good deal of vertical tumbling after the first few centimeters. By this time, however, its velocity is not too great for immediate adherence to an object in its path, although in earlier stages of its flight it may easily ricochet. The entire surface of the seed is extremely sticky. During the first rain, the tail of viscid cells lengthens greatly, and its cells

FIG. 2-32. *Arceuthobium* spp., *a, A. oxycedri* on *Juniperus* sp., Mount Olympus, Greece (courtesy Prof. Dr. W. Rauh); *b, A. campylopodum* on *Pinus monticola,* Slocan Lake, B.C.; plant bears both female flowers and mature berries, so oriented that seed will be ejected upward (after Kuijt, 1955; courtesy Canada Dept. Forestry, Calgary); *c, A. campylopodum,* young male plant on *Tsuga heterophylla,* Vancouver, B.C.; *d, A. pusillum* on branch of broom of *Picea glauca,* near Hudson Bay, Saskatchewan (after Kuijt, 1961-1962).

FIG. 2-33. *a, Arceuthobium campylopodum* seed immediately after having been ejected and intercepted by a leaf of *Abies* sp. (courtesy Dr. R. F. Scharpf, after Scharpf, 1963); *b,* similar seed after viscin cells have expanded following rain (also after Scharpf, 1963); *c, Viscum album* berries, Cantonspark, Baarn, Netherlands.

spread in all directions (fig. 2-33,*b*). When dry, this thin film holds to the substrate tenaciously, and the stage is set for germination. Further rains will not affect the viscid tissue's hold, as its hygroscopic qualities disappear.

A fresh seed of dwarf mistletoe soaked by rain is very slippery. This feature may be of importance in adaptation (Roth, 1959). Many of the expelled seeds adhere to the erect leaves of the host, especially if the latter is a pine. Hawksworth (1965) cites an example of more than 90 percent of all the seeds first adhering to pine needles. These seeds, when moistened, slip down along the needles until they reach the host branch. Thus their final emplacement leaves them with a much greater chance of success. Seeds may slip to the lower surface of the host branch, which provides some protecton from desiccation. That the same thing happens in many other mistletoes is indicated by the frequent position of mature plants on the underside of a host branch, signifying a positional change from the place where birds voided them.

The unique feature of *Arceuthobium* dispersal, therefore, is that the genus has emancipated itself from birds. Although birds have been seen to eat the fruits (Kuijt, 1955), this by itself is an unreliable criterion. There is no doubt that dwarf mistletoes can expand their territories efficiently without avian assistance. Thus to the pathology of this genus is added a feature that is all too familiar to the North American forester. Dwarf mistletoe tends to occur in foci which expand slowly but inexorably within the forest, leaving behind a ravaged area of broomed and malformed trees (Hinds and Hawksworth, 1965). Mistletoes which depend on birds do not show advancing frontiers of this sort, and occur erratically or in patterns determined by the habits of birds.

GERMINATION

In mistletoe germination the radicle grows out from the seedling toward the host, and produces a terminal, disklike swelling. From the interior of this disk—reminiscent of the appressorium of rust fungi—the intrusive organ enters the host.

Under what conditions does the mistletoe seed germinate, and how does it manage to grow toward its host? Various suggestions have been made regarding substances, produced by the host, which are required to set germination in motion. This is not, actually, an unreasonable idea, as something of the sort happens in the germination of certain root parasites, as in *Orobanche* and *Striga.* But in the mistletoes and their allies there is no evidence for such host influences. Many field observations have been made of mistletoe seeds germinating successfully on stones, wires, and other inanimate objects. Whatever else influences germination, the host is not directly involved.

There has been much controversy over the function of the viscid tissue in germination, which has been ably summarized in a monograph on *Viscum album* (Tubeuf, 1923). A number of workers are of the opinion that the viscin inhibits germination, either through an inhibitor substance or through its hygroscopic powers. In *V. album* there are really two viscid layers: an outer, apparently digestible, cellulosic slime layer; and an inner, indigestible, pectic layer. In *Loranthus europaeus,* in contrast, the viscid tissue is of the pectic type only. In fact, it is essential here that we do not restrict our scope to one or two mistletoes. We might wonder, for example, how the viscid tissue of *Arceuthobium* and many *Phoradendron* species can inhibit germination, as nothing remains but a taut, dry membrane soon after emplacement of the seed. Nevertheless, we must be prepared to find such inhibition in some mistletoes, probably only in species of the temperate zones, where it may be related to the dormancy of seeds (e.g., about six months in *V. album*). Tropical *Viscum* species apparently have no dormancy whatever (Wiesner, 1894).

In contrast to the sluggish development during

and after germination of temperate-zone Loranthaceae is that of tropical ones (Wiesner, 1894; Docters v. Leeuwen, 1954). I have personally followed the development of *V. album* and *Phthirusa pyrifolia* in Vancouver. The former is of European origin (Baarn, Netherlands); the latter, Costa Rican (San Pedro). The *Viscum* seeds were planted on *Sorbus*, also the host of the mother plants. Germination was visible four months afterward. Two years later, although successfully established, the mistletoes had not grown significantly in size and had not yet formed their first foliage leaves. Contrast this with *Phthirusa*: within two weeks after emplacement the seedling stands erect on its terminal disk, its two large cotyledons spread apart and separate from the collapsed endosperm. Even under the cloudy skies of Vancouver, in a greenhouse, some plants had flowered and borne ripe fruit before two years had elapsed. After a similar period, no foliage leaves were visible in *V. album*. Even in the highly advanced *Arceuthobium*, also of the temperate zone, a life cycle of three to five years seems minimal (Kuijt, 1961a).

Of the known germination requirements, water is perhaps the most important. Whether the endosperm and embryo really need water is not always clear, but certainly the viscid cells must have it in order to function properly. Mistletoe seeds of arid regions, as in some species of *Arceuthobium* and *Phoradendron*, probably do not need to absorb water from their surroundings while germinating. Considering the absence of a true seed coat, perhaps desiccation is the greatest danger.

Some mistletoes do not germinate in darkness; others germinate equally well in light and darkness. Although *V. album* belongs in the first group, germination continues in darkness once the process has been set in motion (Tubeuf, 1923). *Loranthus europaeus* germinates readily in darkness (Wiesner, 1897), and I can report the same for *Phthirusa pyrifolia*. Since the endosperm of *V. album* is chlorophyllous, while that of the other two is not, could it be that in Viscaceae the light stimulus which triggers germination is received in the endosperm, and that plastid pigments function as light receptors? If inhibition by darkness could be demonstrated in other Viscaceae, this possibility would become stronger.

How does the radicle manage to grow toward the host, no matter on what side of the host branch it is situated? Again, the behavior of the *Viscum* seedling provides some clues (Tubeuf, 1923). The radicle's ability to find the host appears to be due to a combination of negative phototropism and negative geotropism. The former is dominant under conditions of illumination, forcing the mistletoe to grow to the darkest area of its environment, that is, its substrate. But below a certain light intensity the underlying negative geotropism makes itself felt. In nature this usually corresponds to a position on the lower side of a host branch, where the radicle thus will grow up toward the substrate. Any negative phototropism remaining in such situations reinforces growth in the right direction.

In many mistletoes the radicle is so short that it never emerges from its membranous shell of dried viscin, but grows directly to the host and forms the haustorial disk. This is visible in *Loranthus europaeus* (fig. 2-26,a), and is very common in paleotropical species. Growth of the radicle is also at a minimum in some Viscaceae, for example *Phoradendron tomentosum* subsp. *macrophyllum*. In others, such as *P. californicum*, *Arceuthobium*, and *Phrygilanthus aphyllus* (Reiche, 1904), phenomenally long radicles may develop; in the last-named species, a length of six to eight cm is not uncommon. It is not surprising that such radicles are very thin, almost hairlike.

The terminal disklike swelling is rather weakly developed in Viscaceae, although *Antidaphne* is an exception (Kuijt, 1965b). In most Viscaceae the physical connection with the immobile seed may compensate for the fact that little adhesion can be expected from the haustorial disk. In this family the seedling generally does not break contact with the seed to become erect until haustorial penetration has made its hold secure. This is not true of the Elytranthinae of the Old World or of most New World Loranthaceae. I was able to demonstrate this clearly for *Phthirusa pyrifolia*. After allowing the seedling to germinate completely on a glass plate, I carefully removed the erect little plant, after moistening it thoroughly. It was then made to adhere with its haustorial disk to a young branch of *Citrus*. Since that time the mistletoe has developed normally. I suspect that this could not have been accomplished in *Arceuthobium*, *Phoradendron*, or *Viscum*, where the disk is much less of an adhesive organ.

Hardly any of the above applies to the primitive Loranthacean trio *Atkinsonia*, *Nuytsia*, and *Gaiadendron*. The first two are exclusively terrestrial; the third is either epiphytic or terrestrial, depending on its circumstances (Kuijt, 1963a). Their germination pattern is like that of many nonparasitic flowering plants. The radicle grows out of the seed to penetrate the soil, soon after which the hypocotyl lifts the seed from the ground. The cotyledons dissociate themselves from the withered endosperm, and the root prepares to form its first laterals. Any root may form a haustorium, but always as a lateral organ. Indeed, the striking developmental difference between these three genera and other mistletoes is their total lack of a primarily haustorium. Their early development is in no way different from that of the more primitive families Santalaceae, Olacaceae, and Opiliaceae. *Gaiadendron* is, furthermore, unique in its formation of a tuber, probably a water-storage organ. How long such mistletoes can remain independent of a host is an interesting question; I have hazarded a guess of at least six months for *Gaiadendron*, but *Nuytsia* seedlings have been grown without the benefit of hosts for at least a year (Main, 1947).

In truly epiphytic mistletoes the Elytranthinae seem to represent the next evolutionary step. A long

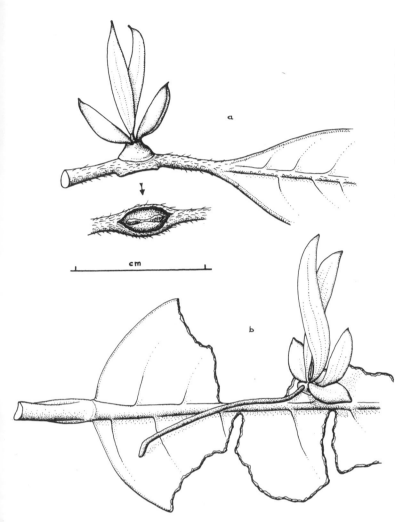

FIG. 2-34. *Oryctanthus occidentalis*, seedlings on leaves: *a*, haustorium has penetrated entire petiole (see arrow); *b*, seedling has produced first epicortical root (after Kuijt, 1964*c*).

ate. The latter is an enlarged haustorial disk, covered with hairlike projections perhaps functioning as aids to attachment, and by means of this disk the entry of the host is effected. Essentially the same events take place in *Arceuthobium* (Kuijt, 1960) and possibly occasionally in *Phoradendron californicum*. The shoot apex aborts, and all subsequent shoots trace their origin, via the haustorial system, from the radicular apex.

This marks the beginning of the endophytic existence of the mistletoe. The intrusive organ will establish itself securely within the tissues of the host, and channel the latter's resources to the growing shoot system of the parasite. The haustorial organ, with its great complexity, variable structure, and overriding biological importance warrants detailed discussion. Since haustorial structure and action in Loranthaceae is similar to that in related Santalalean families, the absorptive organ will be discussed in chapter 7.

THE MUTUAL AFFINITIES OF MISTLETOES[2]

Notwithstanding a slight overlap of characters, the two families are quite distinct. In fact, they are more clearly separate than Olacaceae are from Santalaceae. To guard against submerging the actual differences in a nomenclatural formula, I have assembled the important contrasting points, freely adapted from Barlow (1964) and Dixit (1962). Outstanding exceptions are indicated in parentheses.

LORANTHACEAE	VISCACEAE
1. Calyculus present; flowers dichlamydeous (*Tupeia*, male)	1. Calyculus absent; flowers monochlamydeous (*Viscum?*)
2. Polygonum-type embryo sac	2. Allium-type embryo sac
3. Many embryo sacs per flower	3. Two embryo sacs per flower
4. Cleavage of zygote vertical	4. Cleavage of zygote transverse
5. Pollen ca. trilobate (*Oryctanthus, Tupeia, Ixocactus*)	5. Pollen ca. spherical
6. Suspensor very long	6. Suspensor very short or absent
7. Flowers rather large, brightly colored (*Barathranthus, Struthanthus, Ixocactus, Tupeia, Phthirusa, Oryctanthus, Loranthus*)	7. Flowers very small, not brightly colored
8. Endosperm compound	8. Endosperm simple
9. Endosperm achlorophyllous (*Lysiana*)	9. Endosperm chlorophyllous
10. Flowers bisexual, or plants dioecious; monoecism absent	10. Plants monoecious or dioecious; bisexual flowers absent
11. Petals four or more	11. Petals four or less
12. No nodal constrictions	12. Nodal constrictions (*Antidaphne, Eremolepis, Eubrachion, Lepidoceras*)

radicle or hypocotyl grows out to meet the host, forming the usual adhesive disk at the place of contact, and lifting the remnants of the seed. When the endosperm is shed it becomes clear that the cotyledons are the first functional foliage leaves. The same can be said of virtually all American Loranthaceae (fig. 2-34); *Phrygilanthus aphyllus* is the only known exception (Reiche, 1904).

The most advanced germination pattern is doubtless that of a majority of Old World Loranthaceae (fig. 2-26,*a*). The cotyledons, frequently fused to form a conical absorptive organ, remain in the endosperm and never become photosynthetic. A narrow slit at their base allows the first foliage leaf to emerge.

The cotyledons of *P. aphyllus* (Reiche, 1904) also remain in the endosperm, but this is a separate development. The radicle, having lengthened considerably to reach the surface of the host, serves to transfer food materials from the endosperm to the clavate radicular apex. Once this has been accomplished all but the swollen portion begins to deterior-

[2]See footnote 1, this chapter.

For generic affinities within each family, I refer the reader to Barlow (1962), whose tentative classification is reproduced below.

VISCACEAE
  Tribe Eremolepidae v. Tiegh.
    Subtribe Eremolepidinae Engl. (*Eremolepis, Antidaphne, Eubrachion*)
    Subtribe Lepidoceratinae Engl. (*Lepidoceras*)
  Tribe Phoradendreae v. Tiegh.
    Subtribe Ginalloinae (*Ginalloa*)
    Subtribe Korthalsellinae Engl. (*Korthalsella*)
    Subtribe Phoradendrinae Engl. (*Phoradendron, Dendrophthora*)
  Tribe Arceuthobieae v. Tiegh. (*Arceuthobium*)
  Tribe Visceae v. Tiegh. (*Viscum, Notothixos*)
LORANTHACEAE
  Tribe Elytrantheae Danser
    Subtribe Gaiadendrinae Engl.
    Subtribe Elytranthinae Engl.
  Tribe Nuytsieae v. Tiegh.
  Tribe Lorantheae Engl.
    Subtribe Loranthinae Engl.
    Subtribe Psittacanthinae Engl.

It is easier to criticize a system than to replace it with a better one. Nevertheless, I should 'like to comment on several items in Barlow's outline. First, on the basis of what is known of Viscacean chromosomes (Barlow, 1963b) and inflorescences, I feel that *Phoradendron, Dendrophthora, Korthalsella,* and *Arceuthobium* may easily be combined in a single tribe, notwithstanding their androecial divergence and the advanced parasitism of *Arceuthobium*. Second, it would be more appropriate to place *Ginalloa* and *Notothixos* nearer *Viscum*, as indicated again by chromosome numbers and by inflorescence similarities. Third, in Loranthaceae the separation of *Nuytsia* from *Atkinsonia* and *Gaiadendron* at the tribal level seems highly artificial; I refer the reader to a recent paper on *Gaiadendron* (Kuijt, 1965a). Fourth, I agree completely with Barlow's opinion and that of MacBride (1937) as to the unreliability of *Psittacanthus* as a natural taxon, and therefore of subtribe Psittacanthinae Engl. as a natural group. Support for some of these notions is supplied by cytotaxonomy.

In paleotropic mistletoe taxonomy the name of Danser looms very large. In a series of papers, mostly in the 1930's, he was able to lay the foundations of a rational organization of at least Loranthaceae of the Old World. The mass of subsequent embryological work has tended to agree with Danser's views to a remarkable extent. This is not to say that changes will not be made. In Africa, much sound work has been accomplished by Balle and co-workers. For Loranthaceae the "dark continent" is South America.

I summarize my own tentative views as follows. *Phrygilanthus* in the Americas is an artificial entity nearly all species of which can be incorporated in either *Gaiadendron* or *Psittacanthus*. The former genus accommodates the species of *Phrygilanthus* which resemble it in general habit and parasitism

(e.g., *P. acutifolius*). The foliar bracts in *G. punctatum* and others should not be used as a generic criterion; in some other genera foliar bracts are also present in at least some species (*Psittacanthus* and *Diplatia*; Barlow, 1962). Added to *Psittacanthus* may be the species with large red or orange flowers and often fleshy leaves (e.g., *Phrygilanthus heterophyllus, P. sonorae, P. aphyllus, P. flagellaris*). *Psittacanthus* becomes a genus with variable endosperm development, and includes *Aetanthus*. We may thus discern in American Loranthaceae four groups totaling seven genera:[3]

  I. *Gaiadendron*, incl. *Phrygilanthus*, p.p.
     Struthanthus
 II. *Phthirusa*, incl. *Dendropemon*
     Oryctanthus
III. *Ixocactus*
 IV. *Psittacanthus*, incl. *Aetanthus* and *Phrygilanthus*, p.p., *Psathyranthus*

Although this arrangement does some violence to the overlap between *Phthirusa* and *Struthanthus*, both these genera and *Oryctanthus* (and possibly *Ixocactus*) may be traceable to a single offshoot from a *Gaiadendron*-like mistletoe population. *Oryctanthus, Ixocactus,* and most of *Phthirusa* have retained bisexual flowers; *Struthanthus* has switched to dioecism completely. The origin of dioecism in a few species of *Phthirusa* may be quite independent from that in *Struthanthus*.

Tropical America, for all its wealth of species, does not constitute as great a center of mistletoes as has previously been thought. This impression is probably due to my general taxonomic attitude, which is in contrast to the somewhat more restricted generic concepts of Danser, Balle, and Barlow. But even if we include *Dendropemon, Phrygilanthus,* and *Aetanthus*, there are still no more than ten Loranthacean genera in the Western Hemisphere. The Indonesian and African regions have more than twenty genera each, four of which are found in both regions (Danser, 1931a; Balle, 1954, 1964). At the generic level, therefore, each of these regions is twice as rich in Loranthaceae as is America. Even Australia has eleven genera (Barlow, 1962), although in terms of species numbers the neotropical region is superior to Australia.

The cytotaxonomy of mistletoes has not achieved a level at which it can make unequivocal contributions to systematics. Nevertheless, tentative conclusions can be drawn from work by Barlow (1963b), Wiens (1964a, 1967), and earlier workers. In Loranthaceae the coincidence of primitive germination patterns with a chromosome number of n = 12 in *Gaiadendron, Nuytsia, Atkinsonia,* and Elytranthinae indicates that this may be the basic haploid number. Other more advanced genera have numbers as low

---

[3]See the discussion of Santalalean taxonomy at the end of chapter 3 for a number of items of relevance to mistletoe taxonomy.

as n = 8, which Barlow accepts as evidence of aneuploid reduction. At least two species of *Phrygilanthus* from America, however, also have n = 12. Others from the same genus in America have n = 8, suggesting again that *Phrygilanthus* is an artificial entity. In Viscaceae the situation is less clear, but the basic number there may be n = 14, adding some support to the familial separation.

Actually, there seems to be little variation in chromosome numbers for families of this size. Genera are constant in this respect, with one or two exceptions, as in *Viscum,* with haploid numbers of 10, 11, 12, and 20. It seems that the first three numbers may correspond to Danser's (1941*a*) divisions of *Viscum. Dendrophthora* appears to have an increasing aneuploid series of n = 14, 15, and 17. Not a single case of polyploidy is known for the larger of the families, Loranthaceae. In Viscaceae I know of only four: two in *Viscum* (Barlow, 1963*b*) and two in *Phoradendron* (Wiens, 1964*a*; the second is a recent unpublished count by the same author).

Since chromosomes differ little in size within a genome, karyotype work does not hold much prom-ise. In Loranthaceae, metacentric or submetacentric chromosomes prevail; this family shows karyotype symmetry in Stebbins' (1950) sense. The usual effect of chromosome reduction, highly telocentric chromosomes, is not in evidence in Loranthaceae.

Many mistletoes have chromosomes which are among the largest in the plant kingdom. It is interesting to note, incidentally, that the presumedly primitive features inherent in the germination and chromosome numbers of *Nuytsia, Atkinsonia,* and *Gaiadendron* are correlated with relatively small chromosomes (Barlow, 1962; Wiens, 1967). Covas and Schnack (1946) claim the record for all angiosperms in *Psittacanthus,* but *Viscum cruciatum* and some Australian genera have chromosomes of the same magnitude (Coutinho, 1957; Barlow, 1962). More comparative data have been added by Wiens (1964*a*).

The vast majority of mistletoes have not yet been studied chromosomally. No one knows what complexities await us in this regard, or what light cytotaxonomy may throw upon the systematics and evolution of the mistletoes.

3

Sandalwoods and their Relatives

THE remaining three families in Santalales — Santalaceae, Olacaceae, and Myzodendraceae—are rather closely related to one another. Balanophoraceae are usually incorporated in *Santalales,* but the grounds for this assignment are weak; the arrangement would seem to survive only through the extreme reduction of Balanophoraceae, severely limiting the number of comparative criteria available. Cronquist (1965) even suggests that Rafflesiaceae, Hydnoraceae, and Mitrastemonaceae (here included in Rafflesiaceae) are more at home in Santalales. I do not believe that many botanists will take this notion seriously. The latter three families, and Balanophoraceae, all holoparasites whereas Santalales *sensu stricto* contain only semiparasites, are treated in chapter 5.

Two additional families, Grubbiaceae and Octoknemaceae, have been members of Santalales in the past. Grubbiaceae is now known to be a nonparasitic family not related to the present order (Gutzwiller, pers. comm.). Thorne (pers. comm.) regards Grubbiaceae as a family belonging in his Pittosporales. Octoknemaceae contain two genera, *Octoknema* and *Okoubaka,* both of which have now been placed in existing families: *Octoknema* in Olacaceae (Fagerlind, 1948a), and *Okoubaka* in Santalaceae; so Octoknemaceae has ceased to exist (Stauffer, 1957).

SANTALACEAE

"Sandalwood" is a slightly misleading name for this family of parasites; most members are not arborescent, but are small perennial herbs or inconspicuous shrubs (fig. 3-1). Nevertheless, the true sandalwood (*Santalum album*) and its look-alikes have been highly important trees in some parts of the world, even to the extent of being responsible for the rise in power of a dynasty (see chap. 1), and we are therefore justified in using this vernacular name.

The family is a large one. A reasonable estimate might be 250 species (Hieronymus, 1889; Rendle, 1925) distributed over more than thirty genera. When surveying the entire family, one is impressed by the large number of small genera. About twenty genera have four or fewer species each; of these, ten genera are monotypic. In keeping with other Santalalean families (except Myzodendraceae) many sandalwoods are tropical plants. In comparison with Loranthaceae, however, there is a greater shift to subtropics and even temperate zones. *Thesium* in Europe reaches into southern Sweden; *Geocaulon* grows even in Canada's Yukon Territory and in

Alaska; *Nanodea* and *Arjona* inhabit Patagonia. The family is reminiscent of the mistletoes also in that, except for five anomalous entities, Santalacean genera are restricted to either the Old World or the New World. The exceptions are interesting. *Buckleya* consists of two species, *B. quadriala* in Japan and *B. distichophylla* in the eastern United States. Another inhabitant of the latter area is *Pyrularia pubera,* whose only congener is *P. edulis* from the Himalayas. A very unusual geographic disjunction is seen in the monotypic genus *Comandra,* with three subspecies in North America; the only other subspecies is limited to the region of the lower Danube (Piehl, 1965c). The largest genus of the family, *Thesium,* belongs to the Old World but has two Brazilian members. These two species have recently been placed in a new genus (Hendrych, 1963) on what appears to be rather weak evidence. No single explanation will fit all these unusual distribution patterns.

Habit

The leaves of Santalaceae are always simple, and not much variation is evident in their shape. The illustrations of *Buckleya* (fig. 3-4,*b*), *Cladomyza* (fig. 3-2,*b*), and *Acanthosyris* (fig. 3-2,*d*) summarize leaf shape in the family. In a number of genera, however, leaves are reduced to leaf scales. In *Cladomyza* (fig. 3-2,*b*) groups of expanded leaves alternate with branch portions which are squamate. This arrangement is reminiscent of the mistletoe *Phoradendron crassifolium* (Kuijt, 1959) and is perhaps related to the degree of accessibility of pollinators to flowers, or birds to the fruits. In both *Cladomyza* and *Dendromyza,* however, shoot dimorphism may in reality be related to parasitism and not to reproduction, for Danser (1940c) maintains that the elongated, twining, haustoria-producing branches are squamate, whereas the short, nontwining branches are leafy. In *Anthobolus, Leptomeria, Omphacomeria, Choretrum, Phacellaria,* and *Daenikera* all species are squamate: this condition predominates also in *Exocarpos.* In the latter genus, *E. latifolus* and several others have rather large leaves, but in yet others expanded leaves are limited to the more vegetative parts of the plant. Fig. 3-4,*a* shows the situation in *E. gaudichaudii* from Hawaii. The percurrent shoots bear foliage leaves; the profusely branched inflorescences have deciduous scale leaves only. The same is said to be true of *E. pseudocasuarina* Guill. (Stauffer, 1959). Whatever the shape of the leaf, the latter's arrangement along the stem is nearly always alternate; the opposite phyl-

FIG. 3-1. *Exocarpos bidwillii,* New Zealand (courtesy B. A. Fineran, after Fineran, 1963b).

lotaxis of *Buckleya, Okoubaka, Colpoön, Rhoiacarpos, Eucarya,* and most of *Santalum* is exceptional. Many leathery leaves are found in evergreen tropical and subtropical members; the leaves of temperate genera such as *Pyrularia* and *Buckleya* are thin and deciduous. The peculiar, nearly rhombic leaves of *Jodina* from Argentina and vicinity bear three sharp spines each (fig. 3-2,*a*), but the spines of *Acanthosyris* borne on long shoots are short shoots reduced to stout spines (fig. 3-2,*d*).

In *Osyris,* and in the species with opposite phyllotaxis, flowers or inflorescences are terminal, and the lateral branches below carry on vegetative growth in the next growing season. The dichasial branching habit typical of many mistletoes, therefore, finds a close parallel in these sandalwoods. In *Buckleya* even the branches which do not flower abort terminally (fig. 3-4,*b*). The same effect is achieved in the broomlike members of *Anthobolus, Exocarpos, Choretrum,* and *Leptomeria,* nearly all inhabitants of arid regions. In these plants, as in the "Scotch broom" (*Cytisus scoparius*) which they closely resemble, branches are determinate and do not form terminal buds. Abortion of the apex results in sympodial growth.

In *Exocarpos* we find various degrees of phylloclade formation. Such determinate branches, as in the familiar Christmas cactus, are much flattened, and the resultant phylloclade is often mistaken for a leaf. It betrays its stemlike nature by its axillary insertion and by the fact that it, in turn, may bear flowers or inflorescences (fig. 3-2,*c*). In photosynthetic function, of course, it resembles a leaf. (See chap. 9 for parallel instances from other squamate parasites where stems have become modified to function as photosynthetic organs.)

Except for the phylloclade formation in *Exocarpos* and the shoots of *Cladomyza* and *Dendromyza,* the only other instance known to me of dimorphism of woody shoots is in *Acanthosyris falcata* from tropical South America (fig. 3-2,*d*). This tree bears both short and long shoots, restricting flowers to the former. In two other genera in Santalales we find a similar division of function (*Schoepfia* and *Ximenia,* both in Olacaceae).

In stature we encounter as much variation in Santalaceae as can be expected in any family. The majority of sandalwoods are root parasites (actually, parasitism remains to be confirmed in many genera; see table 1). This is true even of the trees in *Santalum* which may reach a height of 13 meters (Engler and Volkens, 1897), for shrubs such as *Exocarpos* (fig. 3-3,*a*) and *Buckleya,* and for small perennial herbs of the genera *Arjona, Quinchamalium, Nanodea, Comandra, Geocaulon,* and most species of *Thesium.* The smallest members of the family are found in *Nanodea* (fig. 3-7,*b*) and *Quinchamalium,* some species of which flower when no more than 5 cm high. *Thesium* and *Quinchamalium* have a frequently suffrutescent habit, producing many flowering shoots from the crown of the taproot.

The roots of some terrestrial shrubs in the family have considerable regenerative power. In Australia this ability seems to bear an ecological relationship to the frequency of fires. Local species of *Santalum* and *Exocarpos* regenerate shoots from even distant parts of their root systems after injury to the mother plant (Coleman, 1934). In such shoots there is a temporary return to juvenile foliage, a phenomenon seen in many other woody plants. It has been recorded (Coleman, 1934) that some of these roots can regenerate individuals nearly twenty feet away from the mother plant. This wide reach gives an indication of the number of hosts which may be attacked even by a tree standing by itself. *Pyrularia*

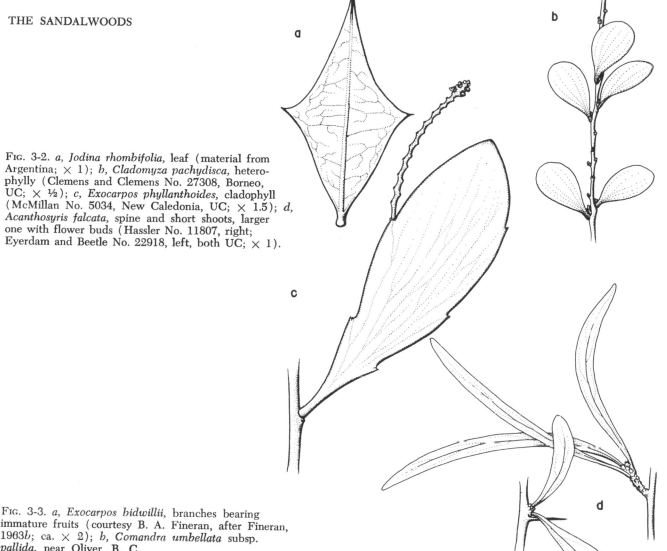

FIG. 3-2. *a, Jodina rhombifolia,* leaf (material from Argentina; × 1); *b, Cladomyza pachydisca,* heterophylly (Clemens and Clemens No. 27308, Borneo, UC; × ½); *c, Exocarpos phyllanthoides,* cladophyll (McMillan No. 5034, New Caledonia, UC; × 1.5); *d, Acanthosyris falcata,* spine and short shoots, larger one with flower buds (Hassler No. 11807, right; Eyerdam and Beetle No. 22918, left, both UC; × 1).

FIG. 3-3. *a, Exocarpos bidwillii,* branches bearing immature fruits (courtesy B. A. Fineran, after Fineran, 1963*b;* ca. × 2); *b, Comandra umbellata* subsp. *pallida,* near Oliver, B. C.

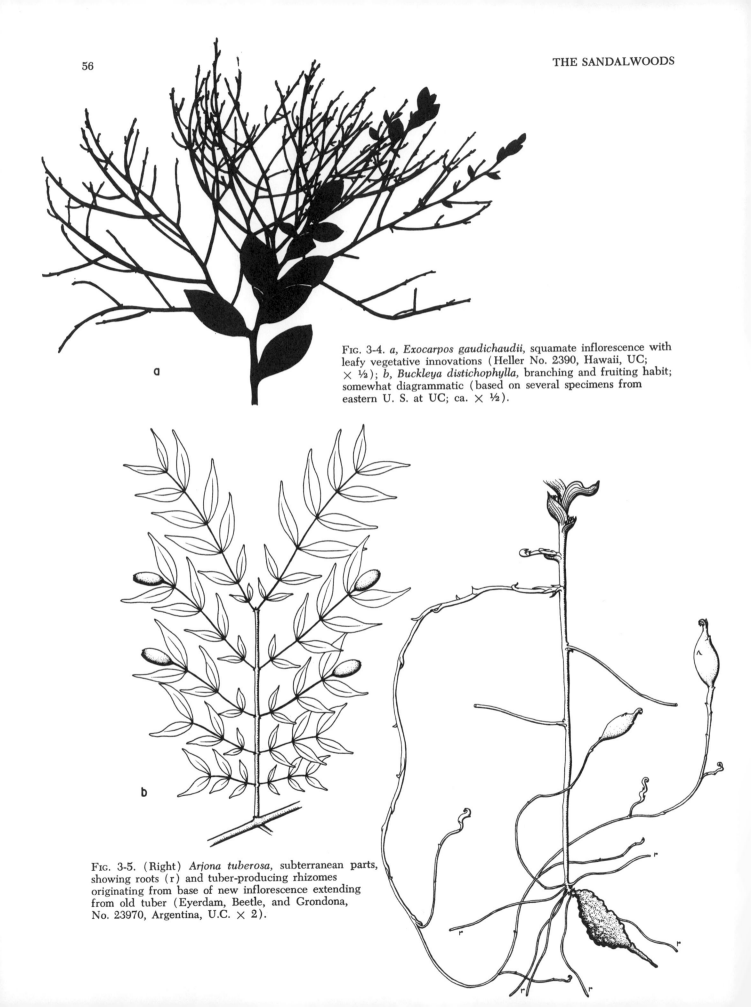

FIG. 3-4. *a, Exocarpos gaudichaudii,* squamate inflorescence with leafy vegetative innovations (Heller No. 2390, Hawaii, UC; × ½); *b, Buckleya distichophylla,* branching and fruiting habit; somewhat diagrammatic (based on several specimens from eastern U. S. at UC; ca. × ½).

FIG. 3-5. (Right) *Arjona tuberosa,* subterranean parts, showing roots (r) and tuber-producing rhizomes originating from base of new inflorescence extending from old tuber (Eyerdam, Beetle, and Grondona, No. 23970, Argentina, U.C. × 2).

also has the power of vegetative reproduction (Claus, 1955).

In a separate category is the type of vegetative reproduction apparently limited to *Arjona* and *Comandra* (figs. 3-5, 3-6). These plants are rhizomatous (as are also *Geocaulon, Nanodea,* and *Nestronia*). *Comandra* rhizomes are rather thick, pale organs which traverse the soil in various directions. Those of *Arjona* are very slender and not always easy to distinguish from roots. In both, scale leaves are distributed along the rhizomes, and the apex of the rhizome is conspicuously recurved. Lateral rhizomes are occasionally seen. A peculiarity noted especially in *Comandra* is the origin of roots directly above the axillary buds on the rhizome (fig. 3-6,*b*). No other roots have been seen in this genus. During the growing season the distal por-

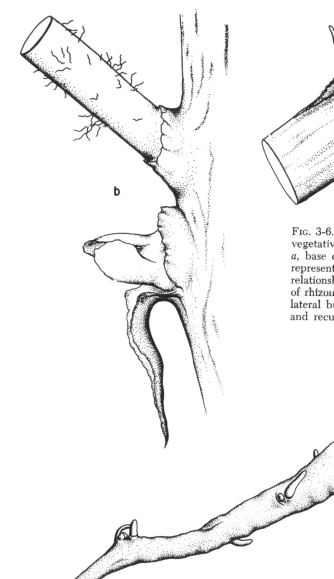

FIG. 3-6. *Comandra umbellata* subsp. *pallida*, details of vegetative reproduction (material from Oliver, B.C., May): *a*, base of young inflorescence, with basal buds probably representing future vegetative shoots (× 10); *b*, spatial relationship of lateral root (cut) and lateral bud in axil of rhizome scale (× 25); *c*, swollen tip of rhizome with lateral buds and roots in axillary positions, and recurved rhizome apex (× 2.5).

tions of the rhizomes become fleshy storage organs from which growth will be resumed after the unfavorable season has passed. The tuberous organ is rather slender in *Comandra*, but in *Arjona tuberosa* the tubers are large enough (1 to 2 cm) to be gathered and eaten by Patagonian inhabitants. The apex of the *Arjona* tuber normally seems to develop into an aerial shoot during the next growing season, initiating a number of new rhizomes in the area between the tuber and the first foliage leaves. In *Comandra*, roots are by far the most abundant at the "crown" (Piehl, 1965c); in *Arjona* they may be limited to the same region, just above the tuber of the previous season. The tubers are exhausted and die at the end of the season, and, since the aerial shoot which has issued from it also dies, mutual contact between the "sibling" tubers is severed. In other species of *Arjona* long-lived storage rhizomes, sometimes fleshy and remarkably like those of *Comandra*, take the place of tubers (Skottsberg, 1916).

Another feature related to branching in the two genera also indicates a much closer taxonomic affinity than has been supposed: in major works, *Arjona* and *Comandra* are usually placed far apart (Pilger, 1935; Smith and Smith, 1943). From the base or upper part of each flowering stem emerge a number of unbranched, completely vegetative shoots which seem to have a purely vegetative function (fig. 3-3,b). The same dimorphism occurs in *Quinchamalium*, but not in other perennial Santalaceae.

Especially interesting are those sandalwoods which, like most mistletoes, are stem parasites. A number of paleotropical genera have achieved this habit: *Cladomyza, Dendromyza, Dendrotrophe, Dufrenoya, Hylomyza,* and *Phacellaria* (fig. 3-7,a). The first five genera, judging from field observations, are rather chaotic scandent shrubs on the branches of trees in the Indo-Malayan region. *Dendrotrophe* may also be terrestrial (Danser, 1940c). When Lam (1945) saw some of these parasites in the mountains of New Guinea he was reminded of the tangle characteristic of *Cassytha* and *Cuscuta*. This habit has resulted in some ambiguity as to their parasitism. The parasitism of *Dendrotrophe*, in fact, has not yet been demonstrated. It is probably safe to assume, however, that all Santalaceae — especially those growing on tree branches — are parasitic, but it is unfortunate that these plants have not been studied in greater detail. Both primary and secondary haustoria probably develop at least in some species. The former is described as a "swollen base at insertion" for *Cladomyza cuneata;* the latter can be seen on some herbarium specimens, and are mentioned in Danser's posthumous paper on these plants (1955). Secondary haustoria here are very small structures issuing directly from the twining and climbing stems of the parasite. Their structure and origin are unknown.

The most advanced genus with regard to parasitism is *Phacellaria*, which ranges from southern China to the Malay Peninsula (Danser, 1939). This small genus of greenish, squamate parasites is remarkable chiefly because of its endophytic development, reminiscent of some of the more advanced Viscaceae. *Phacellaria rigidula* and *P. malayana* are said to emerge in fascicled manner from the swollen nodes of their hosts. Danser's photographs confirm this emergence pattern. Other species have single shoots from thinner host twigs. We do not know what implications these patterns hold for the distribution of the endophyte. A considerable amount of vegetative reproduction from within the host is certain. But what about the puzzling restriction to host nodes? Is the endophyte continuous between successive host nodes? If so, why are the nodes favored for shoot production? Another possibility lies in the mode of dispersal, about which nothing is known. Just as the seeds of *Arceuthobium* may slide down pine needles, and thus seem to favor the host stem near a fascicle as a court of entry, *Phacellaria* seeds might also be deposited on host leaves, perhaps by fruit-eating birds, slide down during the next rainstorm, and eventually germinate in the nodal area. If this is true, the endophyte would extend only a little way into the host tissues in each direction. The swollen nodes of the host also suggest such a restriction. *Phacellaria* seems to parasitize only other stem-parasitic Santalaceae and mistletoes.

Even in *Exocarpos*, usually considered a typical terrestrial root parasite, the tendency for stem parasitism is present. Coleman (1934) reports an example of *E. cupressiformis* successfully attached to the stem of *Eucalyptus* in Australia. A species from New Guinea, *E. pullei*, is able to parasitize either roots or stems (Lam, 1945). It would be simple to dismiss these instances as chance occurrences. The ability to germinate in the stem crevices of the host and to penetrate bark, however, suggests a parasitic versatility which may easily lead to a more permanent stem-parasitic habit. Perhaps we are witnessing here a stage in the slow evolutionary changes similar to those of other stem parasites in the past.

Flowers and Inflorescences

There is a rather high degree of uniformity throughout the flowers and inflorescences of Santalaceae. The typical sandalwood flower is a perfect one, not very conspicuous in color and size. Some of the stem-parasitic forms such as *Dendromyza* and *Cladomyza*, and the terrestrial *Quinchamalium* from South America, have white, yellow, or reddish flowers, but most flowers blend with the surrounding foliage. The only pollinators referred to in the literature are insects.

The flowers are always radially symmetrical. They have three to five perianth members, usually separate, but they may be connate basally to form a campanulate or even tubular corolla, as in several herbaceous genera. There is no trace of a calyx or even a calyculus, except perhaps in the female flower of *Buckleya* (fig. 3-8,b), where four foliar bracts seem to be fused with the gynoecium to

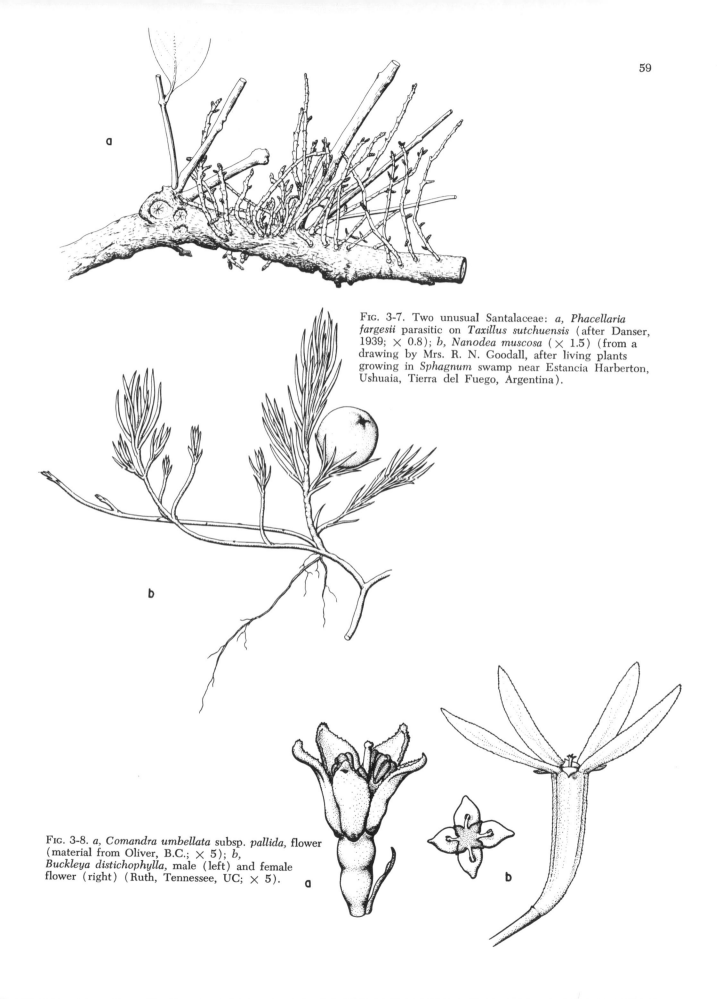

FIG. 3-7. Two unusual Santalaceae: *a, Phacellaria fargesii* parasitic on *Taxillus sutchuensis* (after Danser, 1939; × 0.8); *b, Nanodea muscosa* (× 1.5) (from a drawing by Mrs. R. N. Goodall, after living plants growing in *Sphagnum* swamp near Estancia Harberton, Ushuaia, Tierra del Fuego, Argentina).

FIG. 3-8. *a, Comandra umbellata* subsp. *pallida*, flower (material from Oliver, B.C.; × 5); *b, Buckleya distichophylla*, male (left) and female flower (right) (Ruth, Tennessee, UC; × 5).

extend above in a winglike fashion. In *B. joan* (fig. 3-11,*d*) these wings persist and even enlarge in fruit, but can scarcely be an adaptation to wind dispersal, for which the fruit is too heavy.

In *Quinchamalium* a conspicuous toothed calyculus appears to be present, but this structure turns out to be the fusion of several bracts, and would seem to be akin to the still separate bracts of *Thesium* (see fig. 3-11,*c*). In *Quinchamalium* this compound structure becomes a sclerenchymatous envelope surrounding the ovary or fruit but not coalescing with it. One stamen is associated with each perianth member and is sometimes fused with it. In the center of the flower is a very simple style and stigma, the latter often with several lobes. The most striking feature of the flower is the disk, or central cushion, reminiscent of Viscacean flowers. This disk shows prominent lobes alternating with those of the perianth (e.g., *Buckleya*, fig. 3-8,*b*).

In *Santalum* and others the "disk" gives the impression of a whorl of floral elements, fused with one another and the "perianth" to form a short floral tube. In fact, one might wonder whether the "perianth" is not the calyx, within which the corolla exhibits various degrees of reduction. The ovary is rather variable in position: in Antholboleae it is superior, but in most, like *Thesium* and affiliated genera, it is inferior. In *Jodina* and *Santalum* the position is intermediate.

In some sandalwoods such as *Buckleya* and *Osyris*, flowers occur singly. The inflorescences of others are never very elaborate. Groupings of flowers in small axillary groups, on short spikes or racemes sometimes tending to be capitulate, or in terminal or axillary corymbose groups, are virtually the only types of inflorescence. Worthy of mention are the elongated, often branched spikes of *Exocarpos latifolius* (fig. 3-9,*d*) which may produce several hundred sessile, starlike flowers. A further modification of the spike is seen in *Phacellaria* (see below).

The staminal apparatus is also rather simple. The stamens are sometimes nearly sessile (*Choretrum*), but are normally provided with a distinct filament. The four pollen sacs (two in *Phacellaria*) dehisce longitudinally, but in *Choretrum* the pollen sacs make up sectors of the anther, with each sector tearing open individually. Something of the sort seems to occur in *Exocarpos* also (Hieronymus, 1889). The filament is frequently accompanied by a tuft of hairs on the adaxial side of the opposing perianth member. The function of these hairs is unknown.

The majority of sandalwoods have perfect flowers, but at least half a dozen genera are dioecious: *Cladomyza, Dendromyza, Osyris, Pyrularia, Okoubaka, Exocarpos, Buckleya,* and *Thesidium.* Two clear-cut instances of heterostyly, accompanied by pollen dimorphism, are known from *Arjona* (Skotts-

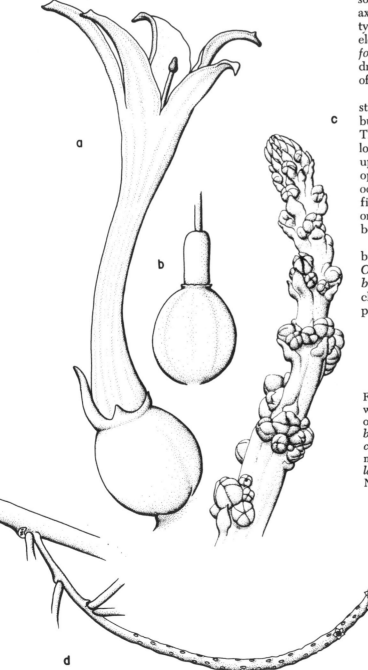

c

a

b

d

F᙭ɢ. 3-9. *a, Quinchamalium majus,* flower complete with sclerenchymatous bracteal cup surrounding ovary (Eyerdam No. 10281, Chile, UC; × 10); *b,* same, ovary and pedestal-like base of style (× 10); *c. Phacellaria rigidula,* portion of inflorescence with male flowers (after Danser, 1939; × 7); *d, Exocarpos latifolius,* portion of compound spike (Bur. Sci. No. 44630, UC; × 3).

berg, 1915, 1916). Heterostyly is exceedingly rare in Santalales; the only other instance that has come to my attention is that of *Schoepfia chrysophylloides* of Olacaceae (Fawcett and Rendle, 1914).

The most remarkable genus of all with respect to sex distribution is *Phacellaria* (Danser, 1939). The genus has spikes like *Exocarpos,* except that many flowers may spring from a single axil (fig. 3-9,*c*), where a number of bracteoles exist. Many adventitious flowers may emerge, presumably endogenously, from nearby areas of the internodes, together forming irregular cushions of buds, flowers, and bracteoles. The parallel to the inflorescence of certain Viscaceae is made even more convincing by the tendency of late intercalary expansion to separate the flowers, thus lengthening the internodes. The remarkable fact is that in *Phacellaria* there is not only monoecism and dioecism but perfect flowers as well: *P. malayana* is said to develop one female flower per axil first, usually followed by several male flowers; *P. fargesii* has perfect flowers only. In *P. rigidula,* Danser speaks of male and female flowers on different stems of the same plant. All three types of flowers may grow on a single individual *P. gracilis,* whereas *P. compressa* has a nearly dioecious regime. For *P. rigidula* the possibility cannot be excluded that the individuals which Danser mentions are in reality colonies consisting of both male and female individuals. Since *Phacellaria* fruits are almost certainly bird-dispersed, a number of seeds might frequently be voided together and become established as a mixed colony.

The work of Swamy (1949) and Straka (1966) has shown that sandalwood pollen is tricolpate with a smooth or slightly granular exine. Three furrows often exist without a definite pore. In *Quinchamalium* a pollen grain has evolved which is nearly indistinguishable from that of the mistletoe genus *Oryctanthus* (compare figs. 3-10,*b* and 2-20,*b*). As Swamy interprets the pollen of *Quinchamalium* differently, further study is necessary.

Nothing but entomophily is known for sandalwood flowers. Insect visits have sometimes been observed (Piehl, 1965*c*), and are indicated also by the nectar-producing disk at the base of the flower (Pilger, 1935) and by the fragrance of some, for example *Jodina* (Hieronymus, 1889). Not all sandalwood flowers are inconspicuously colored. No bird-pollination has been reported in the family. It is not impossible, however, that *Quinchamalium* is ornithophilous. The flowers of this genus are tubular, grouped in crowded spikes which are usually well above the foliage. Red, orange, and yellow are the dominant colors, extending even to the adjacent stem and leaves, somewhat in the manner of *Castilleja.* In *Q. majus* a peculiar pedestal-like body supports the style, and may represent a nectar-producing organ, possibly an elongated disk comparable to those of other genera (fig. 3-9,*b*). It is rather strange that Agarwal (1962) mentions no such body in *Q. chilense;* it seems to be better developed in older flowers, and might be related to fruit dispersal.

Taken together, these features would seem to indicate ornithophily, probably by hummingbirds. It would be interesting to have field observations on this genus as well as *Arjona,* which shares some of these features with *Quinchamalium.*

Embryology

The Santalacean gynoecium has a single ovarian cavity although the upper and lower portions may show lobed contours. From the bottom of this cavity a (usually) contorted stalk projects upward, with a number of ovules in a subterminal position (fig. 3-10,*a*). In some genera (e.g., *Santalum*) the column projects beyond the ovules, somewhat like the attenuated mamelon of the terrestrial mistletoe *Nuytsia floribunda* (fig. 2-22,*c*). The projection is

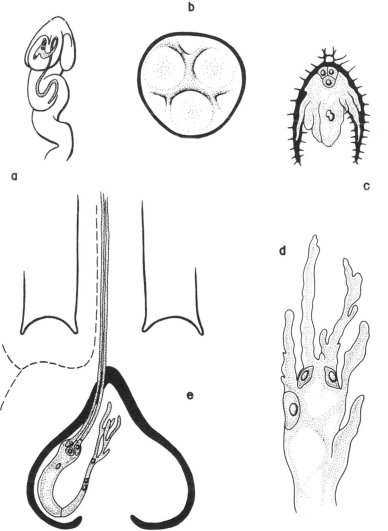

FIG. 3-10. *a, Comandra umbellata,* placental stalk bearing three ovules, with embryo sac of one shown (after Ram, 1957; ca. × 100); *b, Quinchamalium excrescens,* pollen grain in equatorial view (Wagenknecht, No. 18104, Chile, UC; × 75); *c,* and *d, Exocarpos menziesii,* embryo sacs (after Fagerlind, 1959; × 100); *e, Quinchamalium chilense,* longi-section of ovary, showing one mature embryo sac (after Agarwal, 1962; × 100).

formed at about the time of meiosis in *Exocarpos* (Fagerlind, 1959). The ovules are greatly reduced; in *Thesium* and *Comandra* an integument is still discernible, but in *Mida* and *Santalum* it is scarcely separable from the nucellus (Johri and Bhatnagar, 1960). *Phacellaria* has an ovarian papilla similar to that of many mistletoes (Stauffer and Hürlimann, 1957). The orientation of the ovule is variable in the family.

The embryo sac of sandalwoods, as in all other Santalales but Viscaceae and *Olax wightiana* (Olacaceae), is of the Polygonum type. The solitary exception in the family is *Buckleya* (Raj, 1963), which, like the exceptions above, has an Allium-type embryo sac. The Santalacean embryo sac frequently invades gynoecial tissues. Perhaps the most striking in this regard is *Quinchamalium chilense* Lam. (Agarwal, 1962). The pole occupied by the egg apparatus enlarges in a clavate fashion (fig. 3-10,*e*).

The two synergids, even before fertilization, grow out of the embryo sac and form slender tubelike processes. These structures, which are interpreted as haustorial in function, emerge from the placental body, traverse the ovarian cavity, and grow into the style for about one-third of its length, following a vascular strand. The antipodal portion of the embryo sac, meanwhile, is sealed off by a transverse wall. The tip of this antipodal chamber now also elongates and penetrates the apex of the placental body, where it branches several times. When, after fertilization, the lower four endosperm cells form one tubular haustorium each, these structures simply retrace the course of the antipodal haustorium into the placental tip. We thus have three separate sets of haustorial structures in the gynoecium of *Quinchamalium*.

(This is not comparable to the behavior of the Loranthacean embryo sac, for there the synergids are passive.)

With respect to synergid and antipodal haustoria, *Quinchamalium* appears to be unusual, if not unique. A more representative embryology is known from *Comandra* (Ram, 1957). Here the embryo sac itself forms a lateral caecum just below the egg apparatus (fig. 3-10,*a*). This extraordinary structure, which is found also in some Loranthaceae, now grows into and follows the twisted placental column for some distance. After fertilization a set of endosperm haustoria such as those in *Quinchamalium* develop and follow the path of the caecum. Subsequently, the endosperm consumes not only the major part of the ovules and placental column but also much of the parenchymatous endocarp. For *Santalum* it has been reported (Paliwal, 1956) that the endosperm of several ovules may coalesce to form the compound endosperm also typical of Loranthaceae. During germination, of course, the endosperm in its turn is consumed by the embryo.

Perhaps the closest resemblance to the Loranthacean embryo sac is seen in *Exocarpos* (Fagerlind, 1959; Ram, 1959). The embryo sac appears to absorb the epidermal placental cells, and at the same time forms a number of finger-like processes downward. Eventually the entire ovarian cavity is filled with the octopus-like embryo sac. The addition of embryo sac extensions upward, sometimes well into the style, completes the embryo sac.

Fruits, Dispersal, and Germination

The same semantic objection could be raised in Santalaceae and in both mistletoe families to the use of the word "seed," for integuments are rarely recognizable in the ovule. However this may be, the fruit of sandalwoods contains only one seed, often surrounded by a sclerenchyma layer to form a nutlike structure (fig. 3-11). A rather thin fleshy layer in turn surrounds the nut, and is covered by a (frequently) brightly colored ectocarp. There have been few reports on dispersal of seeds, but various rodents and nut-eating birds are possible agents (see Piehl, 1965*c*). In *Thesium alpinum* the floral pedicel becomes fleshy in fruit and may serve as an elaiosome in ant-dispersal (Ulbrich, 1907). In *Thesidium* the red "pseudodrupe" contrasts vividly with a fleshy basal ring of white tissue (Marloth, 1913). In the tribe Anthoboleae the floral pedicel just below the fruit proper enlarges and becomes very fleshy (Coleman, 1934; Stauffer, 1959). This pedicel and the nut above it are often of a bright and contrasting color pattern, as in *Exocarpos strictus* (white and purple-brown) and *E. cupressiformis* (red and green), and it is not surprising that *Exocarpos* fruits are part of the diet of the Australian emu (Stauffer, 1959). It is one of the few native Australian fruits suitable for human consumption: *E. cupressiformis* is often called "cherry tree." Many early writers on Australian natural history refer to these amazing "cherries with the stone outside." It is not known whether the nuts survive intestinal action in a viable state.

The advanced condition of the fruit of the habitual stem parasites among Santalaceae bespeaks adaptation to dispersal by birds and to germination on host branches, and it would be interesting to know the extent of parallelism with mistletoes. Unfortunately, only fragments of information are available. It seems reasonably certain that brightly colored, fleshy, one-seeded fruits predominate. This is stated to be true of *Cladomyza* and *Dendromyza* (Danser, 1955). *Dendromyza reinwardtiana* has green and red fruits; *D. papuana* and *D. salomonia* and *Cladomyza ledermannii*, red fruits; and *C. cucullata*, yellow fruits. In *Phacellaria* the drupelike fruit has a fleshy outer layer which is probably green (Danser, 1939).

Especially interesting are the various modifications of the pericarp (Danser, 1940, 1955). In *Dendromyza reinwardtiana* the exocarp differentiates a number of strong, unbranched, longitudinal fibers which are attached to the smooth seed at its apex (fig. 3-13,*a*). This bundle of recurved bristles is likely to remain intact during intestinal action. In *Cladomyza robustior* similar fibers are present, but small radial fibers connecting mesocarp and exocarp produce a pilose surface on the seed. Many of the small fibers are attached to the longitudinal ones, giving the effect of long, lax feathers inserted at one pole of the seed (fig. 3-13,*b*). Seeds of *Hylomyza* and *Phacellaria* are clothed in a dense mass of membranous fibers of the mesocarp. These elaborate modifications of the ovarian wall probably serve to attach the germinating seed to its host. Indeed, one wonders if the membranous fibers of *Hylomyza* and *Phacellaria* might not be comparable to the viscid tissue in mistletoe seeds, which is also formed from the mesocarp. The feather-like fibers in *Cladomyza*, in turn, are very reminiscent of the attachment fibers of the seed of *Myzodendron*. Field observations on living plants of these genera would not only enrich our understanding of Santalaceae, but would also throw interesting sidelights on the extent of evolutionary parallelism in stem parasites in general. This is equally true of the germination of these plants, since a rather specialized behavior is required of the radicle to contact the host. Might not *Phacellaria* produce a terminal primary haustorium, as stem parasitic mistletoes do? And surely here, also, positive geotropism has been subordinated to negative phototropism.

In terrestrial sandalwoods, as far as they have been observed, germination is either of the hypogeous (Piehl, 1965c, for *Comandra*) or the epigeous type. The epigeous germination sequence of *Exocarpos aphyllus* is reproduced in figure 3-14. The primary root grows directly into the soil, and the elongating hypocotyl slowly lifts the seed above the soil. Eventually the two cotyledons enlarge also and push the entire seedling out of the seed coat, which shows three longitudinal ruptures. After holding the empty seed coat on the cotyledonary tips briefly, this is dropped and the seedling stands erect and free. Meanwhile, lateral roots have been formed, and these may already have established the first haustorial contacts. The precarious and never-ending search for new host roots has begun, and the suc-

cessful completion of the life cycle of the parasite probably depends mainly on the structure and efficiency of the haustorial organ (see chap. 7).

The genera of Santalaceae whose parasitism has been demonstrated are recorded in table 1. It may safely be assumed that all members of the family are parasites; the same opinion has recently been expressed by Piehl (1965c).

Table 1

DOCUMENTED PARASITISM IN SANTALACEAE

GENUS	DOCUMENTATION
Arjona	Illin No. 27, Argentina (UC)
Buckleya	Kusano, 1902; Piehl, 1965b
Cladomyza	Danser, 1940c
Comandra	Piehl, 1965c
Daenikera	Hürlimann and Stauffer, 1957
Dendromyza	Danser, 1940c
Exocarpos	Fineran, 1963b-f, 1965a, 1965b; Philipson, 1959
Geocaulon	Moss, 1926
Hylomyza	Danser, 1940c
Mida	Philipson, 1959
Nanodea	Skottsberg, 1913a
Nestronia	Melvin, 1956
Osyris	Ferrarini, 1950
Phacellaria	Danser, 1939
Quinchamalium	Ruiz and Roig, 1958
Santalum	Barber, 1907a, 1907b, 1908
Thesium	Mitten, 1847

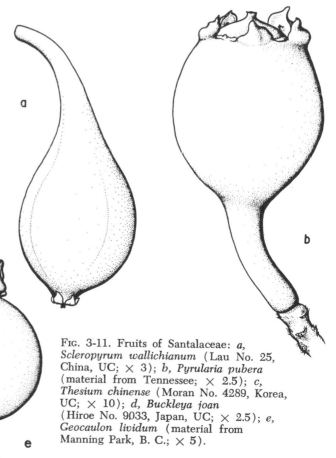

FIG. 3-11. Fruits of Santalaceae: *a, Scleropyrum wallichianum* (Lau No. 25, China, UC; × 3); *b, Pyrularia pubera* (material from Tennessee; × 2.5); *c, Thesium chinense* (Moran No. 4289, Korea, UC; × 10); *d, Buckleya joan* (Hiroe No. 9033, Japan, UC; × 2.5); *e, Geocaulon lividum* (material from Manning Park, B. C.; × 5).

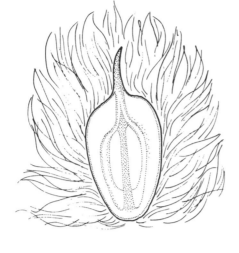

a

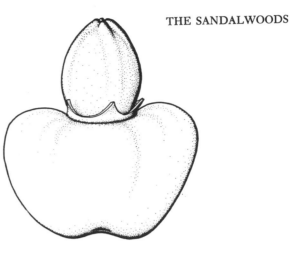

b

Fig. 3-12. *a*, *Hylomyza sphaerocarpa*, longisection of seed with membranous fibers of mesocarp (after Danser, 1955; × 6); *b*, *Exocarpos strictus*, "cherry with the stone outside"; stone (above) purplish black, fleshy pedicel (below) translucent white (after Stauffer, 1959; ×6).

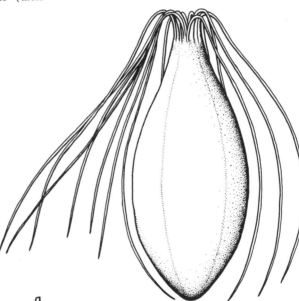

a

Fig. 3-13. *a*, *Dendromyza reinwardtiana*, endocarp of ripe fruit showing wirelike fibers of exocarp (after Danser, 1940*c*; × 10); *b*, *Cladymyza robustior*, same, exocarp fibers feather-like (after Danser, 1940*c*; × 10). The soft exocarp has been removed in both specimens.

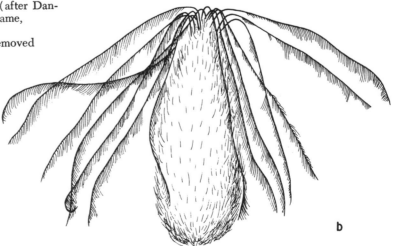

b

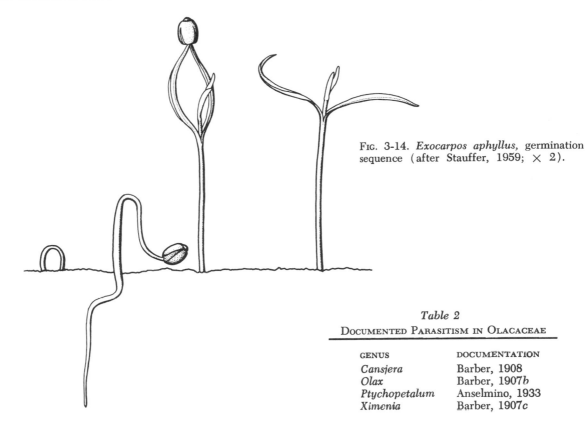

FIG. 3-14. *Exocarpos aphyllus*, germination sequence (after Stauffer, 1959; × 2).

Table 2
DOCUMENTED PARASITISM IN OLACACEAE

GENUS	DOCUMENTATION
Cansjera	Barber, 1908
Olax	Barber, 1907*b*
Ptychopetalum	Anselmino, 1933
Ximenia	Barber, 1907*c*

OLACACEAE

The Olacaceae, in Engler's (1889) sense, represent the great terra incognita of the order. The changing taxonomic fortunes of these plants make a concise treatment of them impossible, and I can present only a brief survey. The following treatment should not be interpreted as support of the one-family status: indeed, more than the usual Olacaceae — Opiliaceae division will probably be required to make this difficult complex intelligible. For the present I am limiting the discussion to the genera discussed in Engler (1889).

Part of the confusion in this region of Santalales arises from our ignorance with respect to the occurrence of parasitism. A number of genera are known to be root parasites (see table 2), but they constitute only a fraction of the genera here included. If we here reach the periphery, or perhaps the source, of the Santalalean parasitic habit, it is likely that many species are completely autotrophic. I have recently established the fact that mature fruiting plants of *Heisteria longipes* at Monte Verde, Costa Rica, lack haustoria. It is well, however, to keep in mind that the list of known parasites in the order is continually expanding. Only recently, the last mistletoe presumed by some to be independent was shown to be parasitic after all (Kuijt, 1963*a*). In Santalaceae, also, the prospect looms of an exclusively parasitic family.

When we consider the complexities of the *Cansjera* haustorium (Barber, 1908), it would seem difficult to conceive of a truly related plant showing no parasitism at all. In any case, we may be sure that more genera of Olacaceae will turn out to be parasites. Such an assemblage of plants of vague mutual relationships is difficult to characterize. The forty genera which have been accommodated in Olacaceae inhabit the tropics almost exclusively; but a few, like *Schoepfia* and *Ximenia*, penetrate subtropical areas. The generic geographical distinctness of the Western and Eastern Hemispheres so striking in the three previous families is not evident in Olacaceae. There is, for example, considerable correspondence between South American and African genera: *Heisteria, Schoepfia, Ximenia, Ptychopetalum,* and *Aptandra* are found in both. The tribe Couleae contains three closely related genera, one in South America (*Minquartia*), one in Africa (*Coula*), and another in tropical Asia (*Ochanostachys*) (Stauffer, 1961*d*). *Ximenia americana*, belying its name, appears to be truly pantropical (Nevlin, 1960).

Habit

All are terrestrial, woody plants. Some are sun-loving shrubs like *Schoepfia, Ximenia,* and the lianas, but there are also small trees growing in shady habitats. *Ximenia americana* seems to prefer sandy areas just behind beaches in tropical and subtropical regions

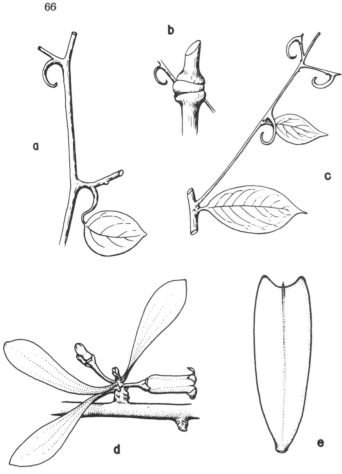

Fig. 3-15. *a, Anacolosa uncifera,* persistent recurved petioles (after Louis and Boutique, 1947; × ½); *b, Olax* cf. *scandens,* constriction of supporting stem by parasite's woody tendril (after Treub, 1883); *c, Olax* cf. *scandens,* branch with hooklike tendrils (after Fagerlind, 1940); *d, Schoepfia californica,* short shoot with flowers (J. H. Thomas No. 8208, Baja California, UC; × 2.5); *e, Olax retusa,* leaf (Clemens, Queensland, UC: × 5).

throughout the world, including many isolated islands. The scandent habit is fully developed in a number of Olacaceae. *Cansjera rheedii* is a shrub which may either be erect or climb on existing supports. The specific names of *Olax scandens* and *Erythropalum scandens* indicate a similar behavior. *Rhopalopilia, Anacolosa,* and *Opilia celtidifolia* are also said to be lianas. There are many medium-sized trees in the family, for example in *Opilia* and *Chaunochiton.* Trees of up to nearly 20 meters in height exist in *Agonandra, Cathedra, Minquartia,* and possibly in *Scorodocarpus* and *Ochanostachys.*

Olacaceae, in common with other Santalales, have simple leaves only, which are rarely deciduous (*Worcesterianthus*). There seem to be no squamate types in the family. The only example of heterophylly I have discovered is the alternation of scale leaves with groups of foliage leaves along the stems of *Heisteria longipes.* The range of leaf size is enormous. *Olax phyllanthi* from Western Australia has leaves less than 1 cm in length, while those of *Champereia platyphylla* from the Philippines often reach 11 x 25 cm.

In the genera *Anacolosa, Erythropalum,* and *Olax* there are certain adaptations to a scandent life. The recurved spines which allow the long branches of *A. uncifera* to anchor themselves to the surrounding vegetation are modified petioles (fig. 3-15,*a*). The

leaf blade is eventually dropped. *Erythropalum* has developed axillary tendrils which appear to be thigmotropic as they coil themselves around any object within reach (see fig. 3-17). These tendrils are branchlike in nature: not only are they axillary in position, but also they occasionally produce a short lateral branch. In *Olax* a thigmotropic, axillary branch spine differentiates which curls itself actively around a captured object (fig. 3-15,*b* and *c*) (Treub, 1883; Fagerlind, 1940). This results in a very tight clasp on the support which, if a living branch, often swells considerably. Although graft fusions occur where another branch of the same individual is thus "attacked," no physiological contact with other species seems to be formed in this fashion. In fact, at maturity the hook is a rigid, dead clamp.

From the illustration of *Olax* (fig. 3-15,*c*) another aspect of vegetative diversity is visible: the lateral branch bearing hooks is a branch specialized for this purpose. It is usually determinate and nearly leafless—a collection of three or four grappling hooks. Three different types of branches can therefore be recognized in *Olax*: normal branches, hook-bearing laterals, and the hooks themselves. The division into short and long shoots which occurs also in *Acanthosyris* of Santalaceae has evolved in *Schoepfia* and *Ximenia* (fig. 3-15,*d*; 3-16,*b*). In the latter, the short shoots are frequently inserted on one of the straight, rigid lateral spines, which are modified branches and make this a very inhospitable shrub. A number of Olacaceae have regular supernumary branching (fig. 3-16,*c*).

Inflorescences and Flowers

The inflorescences of the family show little diversity. A few genera have single flowers or few-flowered racemes or umbels, as in *Schoepfia, Ximenia, Worcesterianthus, Olax,* and others. Compound racemes or spikes are also present in *Coula* and *Ongokea,* but are rather rare. The very slender compound dichasium of *Erythropalum* (fig. 3-17) may be unique in the family and, in fact, makes one suspect the affinities of the genus with the rest of the family.

The occurrence of scalelike leaves or bracts in

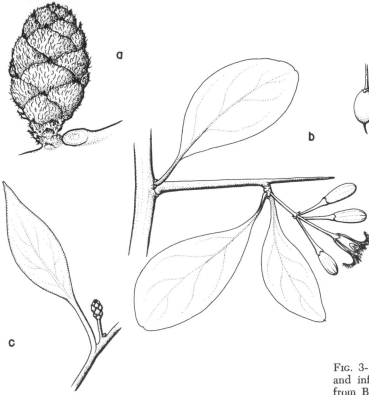

FIG. 3-16. *a, Opilia celtidifolia,* unexpanded spike (Tanner No. 1866, Tanganyika, UC; × 5); *b, Ximenia americana,* habit of spines, short shoot, and inflorescence, and nearly mature fruit (based on material from Florida; × 1.5); *c, Agonandra racemosa,* young leaf and unexpanded spike (Jones No. 27330, Mexico, UC; × 2).

FIG. 3-17. *Erythropalum scandens,* leaf, tendril, fruit, and inflorescence (after several specimens from Borneo at UC; × 1.5).

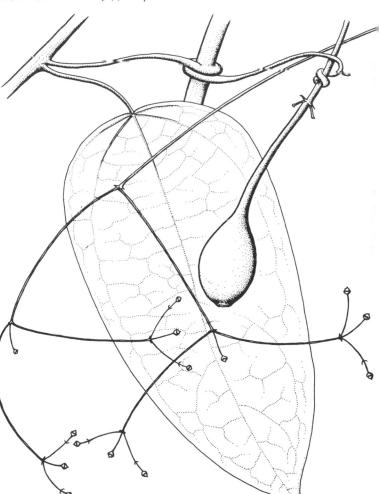

association with the inflorescence is rather common. It is most evident in the genera often separated as Opiliaceae (fig. 3-16,*a*) where the bracts on the inflorescence fall away as soon as the axis and floral pedicels elongate. The inflorescences are thus rather catkin-like. Only a few other examples of such deciduous bracts are known, in Myzodendraceae and the mistletoe *Antidaphne.*

The Olacacean flower is generally similar to the Santalacean flower. Dioecism is very rare, and is known only from some species of *Aptandra,* in *Agonandra, Octoknema, Worcesterianthus,* and *Harmandia.* The single report of heterostyly in *Schoepfia chrysophylloides* requires substantiation (Fawcett and Rendle, 1914). Superior ovaries are in the majority, but that this is not a uniform character may be seen in *Rhopalopilia, Octoknema,* and *Strombosiopsis.* Even within a single genus, *Olax,* both conditions of the gynoecium may be present (Agarwal, 1963). Within the ovary, partition is variable. Commonly a placental column stands erect in the ovarian cavity and bears several more or less pendent ovules with one, two, or no integuments. In *Agonandra* reduction has led to a virtually undifferentiated ovarian papilla or mamelon, as in Loranthaceae (Fagerlind, 1948*a*). Several genera of Olacaceae have a complete placental column attached at the upper end to the ovarian wall. According to illustrations of the following workers, such columns are found in *Anacolosa* (Hutchinson, 1959), *Chaunochiton* (Hooker, 1867), *Octoknema* (Mildbread, 1935), *Strombosiopsis* and *Ximenia* (Sleumer, 1935), and *Phanerodiscus* (Cavaco, 1954). A complete placental column is, on the one hand, reminiscent of the gynoecium of the mistletoe genus *Lysiana* (see fig. 2-22,*a*); on the other hand, it may

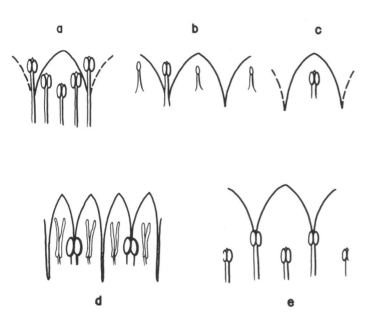

Fig. 3-18. Stamens in Olacaceae: *a, Coula edulis;*
b, Olax viridis; c, Schoepfia; d, Liriosma; e, Heisteria
(*a* and *b*, after Louis and Léonard, 1948; *c* and *e*, after
Sleumer, 1935; *d*, after Monachino, 1961).

The only studies dealing with Olacacean pollen,
as far as I am aware, are those of Reed (1955) and
Straka (1966). It is apparent from this work that
pollen of Olacaceae is generally very similar to the
prevailing Santalacean type.

Pollination in this family has received even less
attention than in the sandalwoods. Certain species
have very pleasant floral perfumes, and the inconspicuous flowers would suggest entomophily. It is
difficult, however, to conceive of pollinating insects
for the flowers of *Chaunochiton loranthoides.* The
size and appearance of its yellowish flower, the association of many flowers in terminal racemes
(Hooker, 1867), and the peculiar terminal anthers
strongly suggest the inflorescence of *Psittacanthus*
of Loranthaceae. It would be interesting to know
whether this species, like the mistletoes it resembles,
is bird-pollinated as even Porsch (1929) suggested.

Many Olacacean flowers have structures inter-
posed between the perianth and the androecium—
structures which have been very difficult to interpret
morphologically. Very common are staminode-like
organs alternating with petals, as in *Agonandra,*
Rhopalopilia, and *Opilia* (fig. 3-19,*f*). In *Agonandra*
lacera these organs bear irregular tubercular pro-
liferations (Toledo, 1952). When we think of the
tendency toward staminodial formation in other
members of the family, there seems little doubt that
these organs are reduced stamens. It is true that
these staminodia are without vasculation; but, since
even the stamens are devoid of xylem (at least in
Opilia amentacea), this criterion is meaningless.

In *Aptandra* we may find an irregularly lobed
rim somewhat like staminodia, but outside the androecium (fig. 3-19,*d*). This rim grows out to a disklike pedestal supporting the fruit (fig. 3-23,*a*). It
is not clear whether this structure functions as an
aril, and what its morphological nature is. In *Ery-*
thropalum there is a similar rim of tissue, this time
not involved in fruit formation (Sleumer, 1942).

The corolla of Olacaceae is very simple. It is almost invariably four- or five-parted, and sometimes
shows a little fusion at the base to form a cuplike
structure. It is not very conspicuous, except in
Chaunochiton loranthoides, where the yellowish-
white petals are about 8 cm long (fig. 3-20). The
inner surface of the corolla in some genera bears a
dense mass of white or yellowish-brown hairs, as in
Scorodocarpus, Heisteria, Ptychopetalum, Ximenia,
and *Minquartia.* In *Chaunochiton* the throat of the
perianth tube is filled with such hairs, and *Octo-*
knema has dense tufts of hairs on the outer surface
of the petals.

have significance in the search for Olacacean ancestry.

Stamens differ greatly in number and position,
but not in structure, in the family (fig. 3-18). In
Coula the maximum number seems to be reached:
three or four stamens for each petal, and an additional stamen between each two adjacent petals. A
certain amount of variation is present within the
species of all Couleae (Stauffer, 1961*d*). A more
common pattern is that illustrated for *Heisteria,* with
stamens opposing as well as alternating with petals.
In *Olax* some of these stamens are absent, and others
become merely staminodial, all without a regular
pattern. In the simplest situation, characteristic of
most Santalales, one stamen opposes each petal, as
in *Schoepfia.* It seems fairly clear that these do not
represent several individual staminal whorls, some
of which fail to materialize: rather, a single whorl
is variable in the number of its stamens.

In the three genera constituting Aptandreae (*Ap-*
tandra, Ongokea, and *Harmandia*) a strikingly different androecium has evolved (fig. 3-19,*b* and *d*).
The four stamens are completely fused, forming a
staminal tube surrounding the entire ovary. The
eight terminal pollen sacs open by means of flaps
which bend down against the staminal tube in a
regular, graceful manner. The tip of the style only
just emerges from the surrounding ring of pollen
sacs. *Harmandia* has unisexual flowers, and the pollen sacs of the male flower have fused, closing off
the central opening completely.

Notwithstanding this unusual trio of genera,
Olacacean stamens are not significantly different
from those of Santalaceae. The only other deviation
is the remarkable anther of *Chaunochiton* (fig. 3-20),
which opens like a minute, four-parted yellow flower, at the end of an exceedingly long and slender
filament.

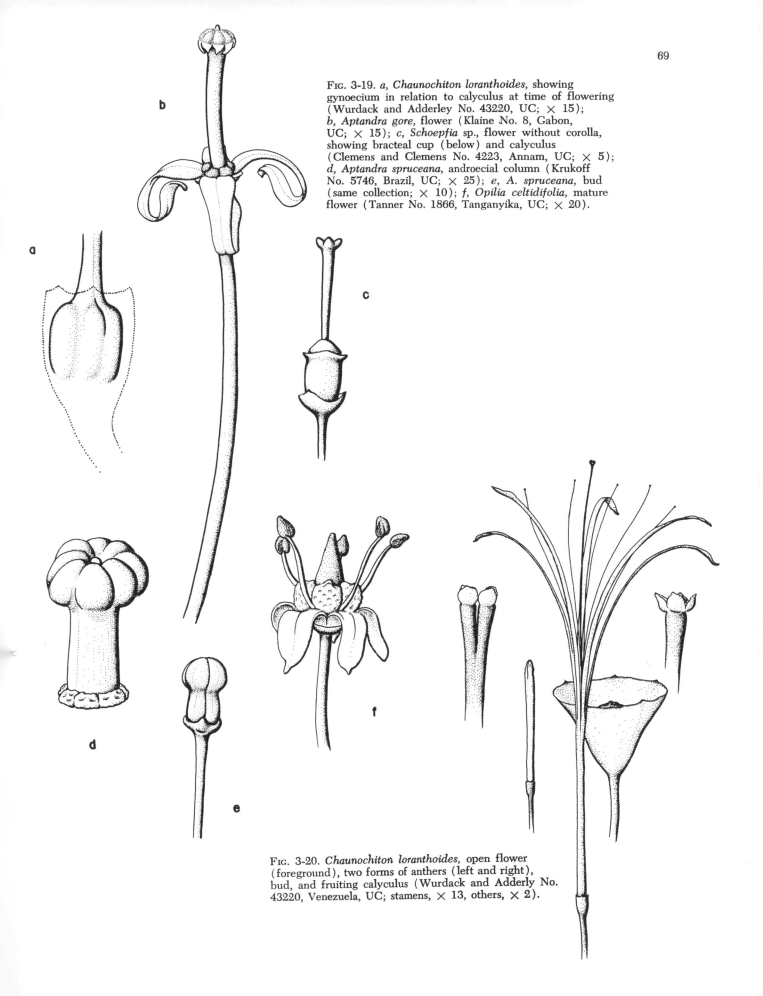

Fig. 3-19. *a, Chaunochiton loranthoides,* showing gynoecium in relation to calyculus at time of flowering (Wurdack and Adderley No. 43220, UC; × 15); *b, Aptandra gore,* flower (Klaine No. 8, Gabon, UC; × 15); *c, Schoepfia* sp., flower without corolla, showing bracteal cup (below) and calyculus (Clemens and Clemens No. 4223, Annam, UC; × 5); *d, Aptandra spruceana,* androecial column (Krukoff No. 5746, Brazil, UC; × 25); *e, A. spruceana,* bud (same collection; × 10); *f, Opilia celtidifolia,* mature flower (Tanner No. 1866, Tanganyika, UC; × 20).

Fig. 3-20. *Chaunochiton loranthoides,* open flower (foreground), two forms of anthers (left and right), bud, and fruiting calyculus (Wurdack and Adderly No. 43220, Venezuela, UC; stamens, × 13, others, × 2).

A calyx or calyculus is present, sometimes as poorly differentiated as in Loranthaceae, at other times extremely large. The flower of Schoepfia (fig. 3-15,d) shows a greatly reduced calyx, and in this regard is no different from the Loranthaceae. Much the same is true of Opilia (fig. 3-19,f), Rhopalopilia, Agonandra, and others. In Octoknema, as in Tupeia of Loranthaceae, a reduced calyx is visible in the female, but is absent in the male flower.

In one group of Olacacean genera, however, the inconspicuousness of the floral calyx is misleading. The flower of Aptandra (fig. 3-19,b) is supported by a small calyx of four or five lobes. After fertilization this minute calyx resumes growth until it has formed a giant, flaring cup extending far beyond the fruit (fig. 3-23,a). This enormous calyx often reaches 5 cm in diameter and has a very intricate, leaflike vasculation. Similar fruiting calyces are met in Chaunochiton (fig. 3-20), Olax (fig. 3-23,b), Heisteria (fig. 3-23,d), Erythropalum (fig. 3-17), Cathedra, Liriosma, Ongokea, and Harmandia. In Heisteria the calyx becomes a recurved disk or a five-armed star, but in Olax, Erythropalum, and Ongokea it forms a tight envelope around the fruit, at maturity tearing open into several parts in the last two genera. These are by far the most striking calyces in Santalales, and supply important evidence for the interpretation of the calyculus of Loranthaceae.

Embryology

The embryology of Olacaceae has, unfortunately, not had the detailed study of the mistletoes and sandalwoods. Nevertheless, the few studies relating to this family hint at some of the same extraordinary events which occur in the other families, with a few extra ones added. The ovular structure is highly variable (Reed, 1955; Kapil and Vasil, 1963). Two distinct integuments can be recognized in some (Coula, Heisteria, and some related genera); in others, only one (Strombosia, Octoknema, and others); still others show no discernible differentiation between nucellus and integuments (Opilia, Schoepfia, and others). Some species of Olax have a clear integument; others do not (Agarwal, 1963). The reduced, mamelon-like condition of the Agonandra placenta has been mentioned (Fagerlind, 1948a). It has been supposed (Reed, 1955) that there is a correlation between the degree of integumental reduction and an increasingly parasitic habit in the family, but this suggestion, plausible as it may be in the general context of Santalales and other parasites, lacks the necessary factual basis. Only one member of Olacaceae (Heisteria longipes) is known to be autotrophic with some degree of certainty.

The heterogeneity of Olacaceae is visible also in the behavior of the embryo sac and the endosperm. In Strombosia and Olax stricta the embryo sac remains within the ovule and shows no sign of growth at the chalazal end (Agarwal, 1961, 1963). In contrast, the embryo sac of Opilia amentacea produces

an outgrowth which at maturity appears like a straight, unbranched tube extending through placental parenchyma toward the floral base (Swamy and Rao, 1963). This haustorial tube may be comparable to the caecum known in some Santalaceae and Loranthaceae. The primary endosperm nucleus moves into position at the top of this tube, where fertilization takes place. The first division of the resulting primary endosperm nucleus provides a cross wall separating the micropylar from the chalazal compartments. Further cell divisions are limited to the former, eventually crushing all cells of the ovule. The haustorial tube, which may now be considered an endosperm haustorium, resumes its downward growth, even penetrating the pith of the floral pedicel. This remarkable cell never divides.

A similar lateral caecum, and the same initial transverse partition, is reported from Olax scandens, O. wightiana, and O. stricta (Agarwal, 1963) (fig. 3-22,a). The initial endosperm division here takes place just below the lateral caecum, which at this time extends into the funiculus. The chalazal chamber thus cut off from the V-shaped remainder of the embryo sac grows vigorously downward; the nucleus divides twice to form a tetranucleate haustorium. It branches profusely when reaching the base of the ovary, consuming the surrounding tissues and even invading the pith of the pedicel. This ramified haustorial system can still be seen in the fruit. Meanwhile, the remainder of the embryo sac, including the caecum, proceeds to form the endosperm proper. The caecum also initiates a diffuse system of haustorial tubules which grow into the funiculus. The part of the embryo sac containing the egg apparatus in Olax scandens and O. wightiana emerges from the ovule and grows through the ovarian cavity to the base of the stylar canal.

The early stages of endosperm development in Cansjera are nearly the same as those of Opilia (Swamy, 1960). Again a simple caecum extends downward, and the upper part becomes enveloped in the endosperm proper, which takes its origin from the micropylar chamber. The caecum, however, continues to grow down to the base of the gynoecium, destroying the axial system of vasculature (there is a separate peripheral system). A spindle-shaped bulge develops in the middle of the caecum, and there the single nucleus remains. The free end of the caecum eventually reaches the extreme base of the ovary and proliferates into two to four major branches. Each branch continues to grow, but forms a number of short tubular processes. The resulting system, coralloid in appearance, digests its way through the axial vascular system, showing some preference for xylem parenchyma but proceeding to phloem later. Meanwhile the endosperm proper has become a massive tissue, forming a thick jacket around this primary endosperm haustorium. The endospermal cells adjacent to the haustorium now develop similar haustoria, growing as a bundle of secondary haustoria around the primary one, and proliferating in the same way. Wherever lateral

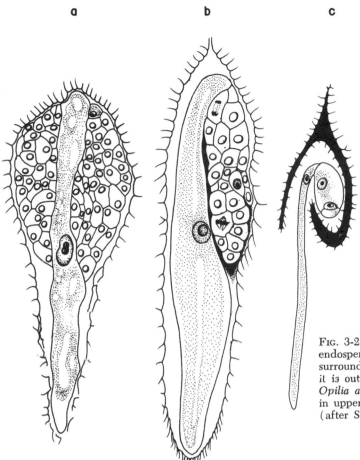

FIG. 3-21. *a* and *b*, *Cansjera rheedii*, two views of primary endosperm haustorium; in *b*, nucleus of zygote, surrounded by endosperm, has been shaded, but in *a* it is out of view (after Swamy, 1960; ca. × 500); *c*, *Opilia amentacea*, mature embryo sac with haustorial tube in upper portion of which rests primary endosperm nucleus (after Swamy and Rao, 1963; × 63).

haustorial walls abut, fusions take place, resulting in a highly complex system of interconnected tubules which supplies the growing endosperm and embryo. A transverse cuplike mass of sclereids limits the downward spread of the compound haustorium.

From an embryological point of view, Olacaceae are as fascinating and extraordinary as many of the other Santalales. No doubt some unusual discoveries await study in the genera which have not yet received this kind of attention.

Fruit and Germination

The drupelike or nutlike fruits of some Olacaceae are the largest in this alliance of families. *Scorodocarpus* may well be the largest, with a diameter of about 4 cm (fig. 3-24,*a*). Long stalks are developed on some (*Anacolosa*, fig. 3-23,*c*; *Erythropalum*, fig. 3-17). We find spherical fruits (*Heisteria*, fig. 3-23,*d*; *Agonandra*, fig. 3-24,*e*) and various ellipsoid and ovate fruits (*Anacolosa*, fig. 3-23,*c*, *Champereia*, fig. 3-24,*b*). Mention has been made of the arilloid base of the fruit of *Aptandra* and the fruiting calyx of this and a few other genera. Some Olacacean fruits are edible, but none constitutes a significant source

of food for human consumption. The presumption is that they are eaten and dispersed by animals, but observations are lacking.

Germination is of the hypogeous (*Ximenia*) or epigeous type (*Ongokea* and *Strombosia*). In some genera the cotyledons remain in contact with the copious oily endosperm, and are dropped with the empty seed. In some *Strombosia* species, however, the cotyledons emerge and become expanded photosynthetic organs (Heckel, 1901). *Ximenia* germination is very peculiar in that a number of awl-shaped scale leaves are formed above the cotyledons, before regular foliage leaves appear (fig. 3-24,*d*). Just above each cotyledon, one of these structures grows down into the groove formed by the cotyledonary petioles to reach nearly the middle of the seed (Heckel, 1899). These unique structures are puzzling both in function and in morphology. Could they be auxiliary absorptive organs with respect to the endosperm? Their close adherence to the cotyledons and somewhat branched vasculature would suggest such a function, although their slenderness would cast doubt upon it. They are believed to be foliar in nature. Their superaxillary position, how-

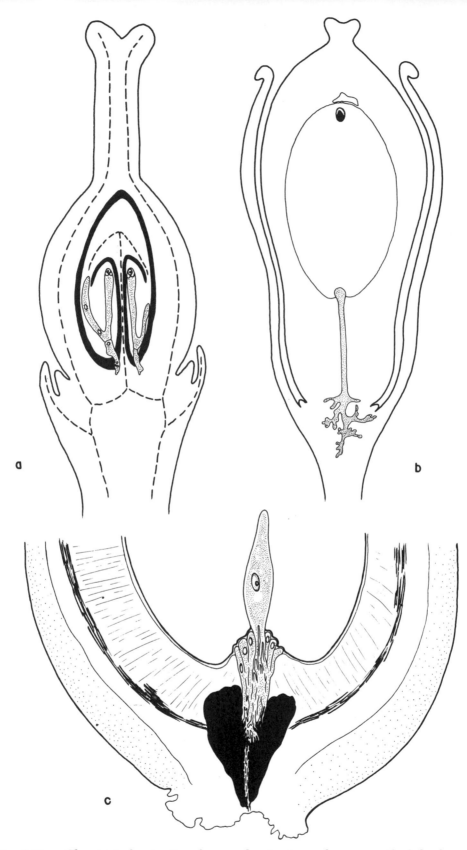

FIG. 3-22. *a, Olax stricta,* longisection of ovary, showing two embryo sacs with chalazal haustoria extending into receptacle (after Agarwal, 1963; × 45); *b, O. stricta,* showing endosperm haustorium reaching from base of endosperm into receptacle of young fruit (after Agarwal, 1963; × 13); *c, Cansjera rheedii,* base of young fruit, showing primary and secondary endosperm haustoria (adapted from Swamy, 1960; × 64).

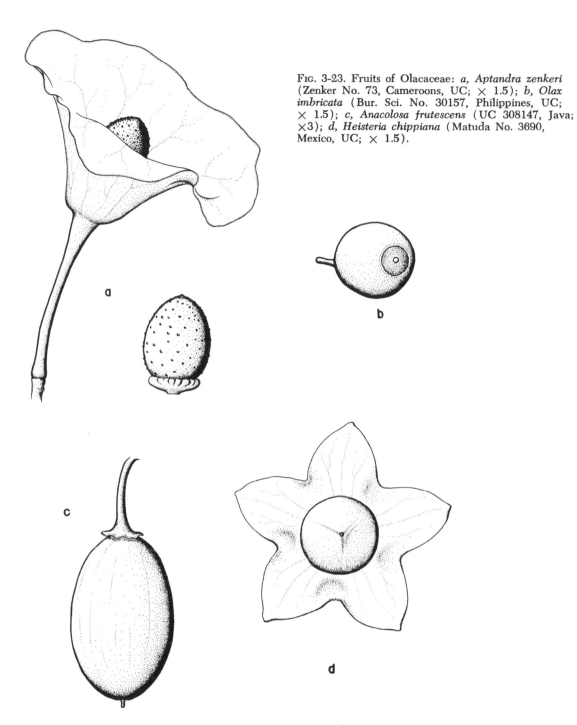

FIG. 3-23. Fruits of Olacaceae: *a, Aptandra zenkeri* (Zenker No. 73, Cameroons, UC; × 1.5); *b, Olax imbricata* (Bur. Sci. No. 30157, Philippines, UC; × 1.5); *c, Anacolosa frutescens* (UC 308147, Java; ×3); *d, Heisteria chippiana* (Matuda No. 3690, Mexico, UC; × 1.5).

ever, and the occurrence of corresponding acicular structures in the axils of the first foliage leaves render such a notion somewhat dubious. In *Ongokea* there are some scale leaves at first, but nothing like the two spurlike organs of *Ximenia*.

Variation is great also in the number of cotyledons (Reed, 1955). The three above-mentioned genera have the normal pair as typical of dicotyledons, but other genera have two to four cotyledons, or even as many as six in *Octoknema*.

If haustoria develop, they do so on the lateral roots of the seedling as in other terrestrial Santalales. The presumed primitive position of Olacaceae within the order might lead one to expect rather simple haustoria, but nothing could be farther from the truth. The haustoria of *Cansjera* are the most complex known, not only in Santalales but also in parasites in general (see chap. 7). We cannot, of course, extrapolate to other Olacaceae from *Cansjera*, but, if other members are found to be parasitic, it seems safe to predict great structural complexity in their haustorial organs.

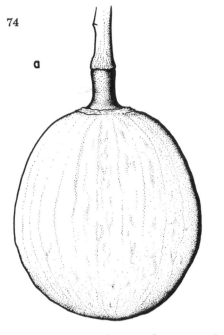

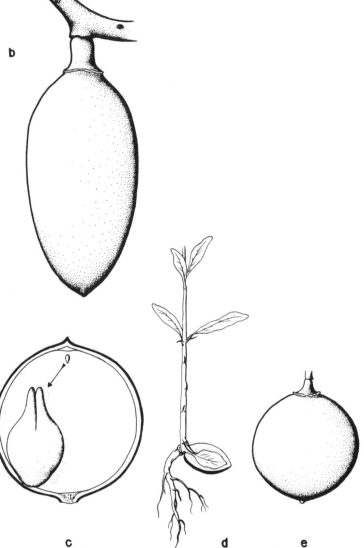

FIG. 3-24. *a, Scorodocarpus borneensis,* fruit (Boden Kloss No. 18745, Borneo, UC; × 1.5); *b, Champereia griffithii,* fruit (UC 346248, Malay Peninsula; × 2); *c, Heisteria longipes,* longisection of fruit (× 5) with enlarged and inverted embryo (Monte Verde, Costa Rica); *d, Ximenia americana,* seedling (after Heckel, 1899; magnification unknown); *e, Agonandra excelsa,* fruit (Hassler No. 7246a, Paraguay, UC; × 1.5).

MYZODENDRACEAE

The most bizarre of Santalalean families almost certainly is Myzodendraceae. Small in number and stature, and isolated geographically, they have scarcely received serious attention since the magnificent and detailed account of the family in Hooker's *Flora Antarctica* (1846) and the work of Skottsberg (1913b). Virtually unknown to most botanists, Myzodendraceae fascinate each naturalist who rediscovers their curious structure and life cycle.

The family is monotypic. The single genus *Myzodendron* has been subdivided into two rather dubious subgenera, together containing no more than ten species. These species are all found in the antarctic beech (*Nothofagus*) forests of South America, about from Argentina's Nahuel Huapi National Park south to Cape Horn (Hauman and Irigoyen, 1923-1925; Thomasson, 1959).

Habit

A *Myzodendron* plant is implanted on its host (almost always *Nothofagus*) in such a way that the host and parasite are always clearly separable (see fig. 3-29). Frequently the stem of the parasite is two or three times as thick as the host branch supporting it, and it is not surprising that the host normally dies beyond the parasitic attachment. The excessive development of the stem of *M. brachystachyum* and others might easily limit their life expectancy, as the larger plants may be broken off during storms or heavy snowfall. There is no sign of epicortical roots, and lateral expansion of the endophyte seems to be minimal. The haustorium expands in a uniformly radial fashion, within the cuplike host placenta which keeps pace with it. All this is reminiscent of the *Psittacanthus* woodrose, but even more so of the main haustorium of *Antidaphne* (Kuijt, 1965b).

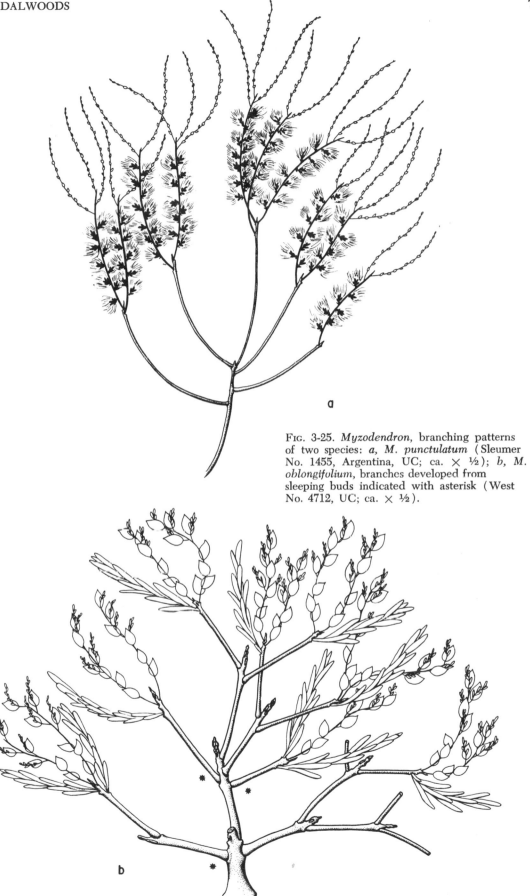

FIG. 3-25. *Myzodendron*, branching patterns of two species: *a, M. punctulatum* (Sleumer No. 1455, Argentina, UC; ca. × ½); *b, M. oblongifolium*, branches developed from sleeping buds indicated with asterisk (West No. 4712, UC; ca. × ½).

Even at the end of the plant's first growing season, the shoot apex aborts. One or more lateral shoots continue vegetative growth in a sympodial fashion in the next growing season. This branching pattern is repeated in each subsequent growing season (fig. 3-25). Its regularity is disturbed only by an occasional "sleeping" lateral bud which expands after one of more years of dormancy.

Vegetative shoots are often thick even when young, giving the leafy species a rather elephantine appearance. This great increase in diameter has curious consequences. The buds in axillary positions become literally buried under secondary tissue, and cannot be discerned in mature shoots. In their stead a dark brown, elongated patch of bark is visible. Presumably, this unusual cortex serves a protective function. The emerging bud has the appearance of an endogenous one, rupturing the brown membrane in an irregular fashion (fig. 3-26,*b*).

The stem of *Myzodendron* shows a number of aberrant anatomical features. The peculiar warty appearance of the stem of subgenus *Gymnophyton* is due to raised stomata above large substomatal cavities. The stomatal apparatus of the other subgenus has not been studied. Vascular structure of *M. brachystachyum* and others is extraordinary in that no closed wood cylinder is ever formed. In two-year-old stems a medullary system of bundles is formed; it duplicates the outer system which, during the same growing season, adds its second xylem increment. After the second year no new bundles are formed, but both vascular cylinders add secondary tissues. Thus the outer cylinder always has one more ring of xylem than the inner cylinder. Xylem differentiation is almost entirely radial, and the bundles remain separate units. In other species, however, the vascular organization is more common.

There are two entirely different types of branches in *Myzodendron*. Innovations are purely vegetative branches; inflorescences are reproductive only. In some species (e.g., *M. oblongifolium,* fig. 3-25,*b*) innovations emerge just below the inflorescences which are nearly terminal on the axis. The old inflorescences abscise, leaving the innovations to carry on vegetative growth. In *M. punctulatum* and some others the positions of innovations and inflorescences are reversed, with the former being nearly terminal.

In the leafy species of *Myzodendron* each type of shoot has its own distinctive type of leaf. The unusual vasculature of the leaves of this genus is further discussed in chapter 9.

Inflorescence and Flowers

The inflorescence is basically spikelike or raceme-like, subtended by a bract. As in some Loranthaceae, the petiole of the bract is sometimes fused with the inflorescence axis and, especially because of its basal constriction, the bract may appear to have a terminal position. In *M. quadriflorum* the flowers seem to be attached to a petiole-like structure, separated from the terminal bract by a constriction. Further condensation, as seen in *M. commersonii* and others, places the flowers in a near-axillary position with respect to the bract. Although most inflorescences are relatively short, those of *M. linearifolium* may reach 15 cm or more, at least in fruit.

Myzodendron is a dioecious genus, although a few very interesting bisexual flowers have been observed by Skottsberg (1913*b*). Staminate flowers are without perianth and consist of nothing but two or three stamens with a central cushion of lobed outline (fig. 3-27,*d*). The anthers are simple structures, at least initially biloculate, dehiscing by means of a terminal slit, and attached to the end of a filament. Pollination is probably by small insects.

The pistillate flower is much more difficult to interpret morphologically (fig. 3-27,*c*). The three-lobed stigma, implanted on the top of the ovary, produces no complications. The lower portion of the gynoecium, however, is enclosed in a peculiar three-parted jacket. Each portion of this jacket is fused with the ovary along its mid-region, but its seams extend sideways in a winglike fashion, leaving three deep grooves which run the length of the ovary. Each groove harbors a young bristle-like organ (seta) which even during flowering may be a millimeter long. During maturation of the fruit, the setae attain a length of many centimeters.

In this unusual flower the setae are vasculated and can scarcely be dismissed as trichomes. The discovery of some bisexual flowers (Skottsberg, 1913*b*) has provided an acceptable morphological explanation. In such bisexual flowers the stamen (sometimes there are two) substitutes for a seta. The seta, therefore, turns out to be a staminode, a highly modified stamen, protruding from grooves formed by three perianth members. The latter are fused with the ovary but not with each other, except at the very base. The position of the stamens also leads to the recognition of a superior ovary, not an inferior ovary as has been previously thought (Smith and Smith, 1943).

Embryology, Fruits, and Germination

Within the ovary the organization is not greatly different from that of Olacaceae and Santalaceae. There are three reduced ovules at the top of a short placental column, each hanging down into a pocket of the ovarian cavity. The upper part of each pocket becomes confluent with the others and, indeed, to the outside, as the stylar canal appears to open up.

Embryologically, *Myzodendron* shows clear affinities with Olacaceae. The ovules, which lack recognizable integuments, have only a single parenchyma layer covering the embryo sac. The antipodal end of the embryo sac forms a long tubular haustorium into the placental column. The first cross wall, isolating the haustorium, is probably formed after the initial division of the primary endosperm nucleus. The tip of the haustorium nearly reaches the vascular bundles at the base of the flower, and ramifies in that region. This entire development, as far as we know it, is scarcely different from that of the Olacacean species which have received embryological attention.

Only one of the ovules matures into a seed, and is lifted up by the elongating placental column. The

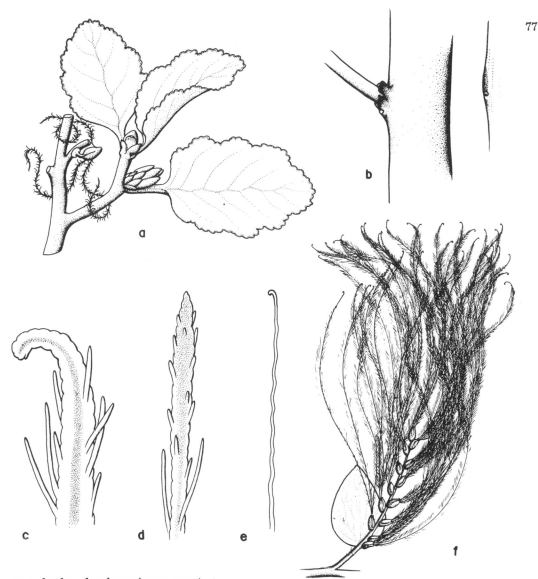

FIG. 3-26. *a, Myzodendron brachystachyum,* germinating seed on *Nothofagus antarctica* (after Hooker, 1846; × 1.5); *b,* same species, showing appearance of branch before (right) and after (left) emergence of lateral shoot (Donat, UC; × 5); *c, M. oblongifolium,* apex of seta (West No. 4712, UC; × 70); *d, M. linearifolium,* apex of seta (Eyerdam No. 10607, UC; × 70) *e, M. punctulatum,* tip of seta hair (Mexia No. 7937, UC; × 70); *f, M. oblongifolium,* bract and fruits (West No. 4712, UC; × 2).

fruit does not undergo any striking change in form except for the extension of the setae. These bristles range in length from about 10 mm (*M. punctulatum*) to 85 mm (*M. oblongifolium*). The setae are always covered by hundreds of long unicellular trichomes. The tips of these unicellular hairs are slightly swollen and curved (fig. 3-26,*e*) in some species. The tip of the seta may be nearly naked and straight (*M. linearifolium,* fig. 3-26,*d*) or with a stiff, recurved hook (*M. oblongifolium,* fig. 3-26,*c*).

Because of the massive and simultaneous development of the setae *Myzodendron* has been called "Barba de Angel" (Chile). The name "Feathery

Mistletoes" has also been suggested ("Federmistelgewächse"; Warburg, 1913). The pistillate plants in the fruiting season are crowned by flowing, beardlike masses of setae. The minute achene is easily carried away by its three setae, which quickly become attached to other host branches. It is believed that the setae serve as anchors while the seeds germinate (fig. 3-28,*c*). Whether hygroscopic activity of the setae plays a role in bringing the seedling to a host surface is not known.

The seedling has a curious structure (fig. 3-28,*b*) in the shape of a small, thick-walled cup with a short stem. The cup corresponds to the radicular pole. The stem is really the hypocotyl with a tubular, slightly notched tip. The tubular portion, completely enclosing the first two foliage leaves, represents a fusion of two cotyledons. The cotyledonary tube remains in the endosperm for some time. Interestingly, the endosperm is green, as in Viscaceae, as shown by material collected by Mrs. R. N. Goodall at Ushuaia, Tierra del Fuego. The cotyledonary tube is eventually ruptured by the emerging foliage

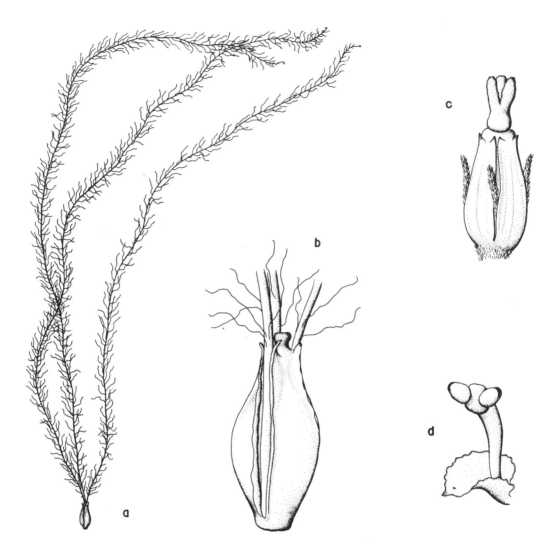

FIG. 3-27. *a, Myzodendron oblongifolium,* fruit (West No. 4712, Chile, UC; × 3.5); *b,* same enlarged (× 25); *c, M. brachystachyum,* female flower, with emerging setae (after Hooker, 1846; × 9); *d, M. punctulatum,* male flower and bract (After Hooker, 1846; magnification unknown).

leaves. The radicle, in emerging from the achene, pushes aside the style and the disk, which fall away. It quickly expands into a trumpet-like organ which is applied to the host surface. The margin of the trumpet extends a thin adhesive membrane radially over the host surface, and penetration follows from the axial part of the haustorial disk. This flaring haustorial mantle again reminds us of *Antidaphne* (Kuijt, 1965*b*). Nothing is known, however, about the internal organization of the haustorium of Myzodendraceae.

MUTUAL AFFINITIES OF SANTALALEAN FAMILIES

The points raised in a recent article (Kuijt, 1968) are briefly summarized here. The new familial ar-rangement is based on the recognition of an ament-like or ament-derived inflorescence in *Antidaphne, Eremolepis, Eubrachion, Myzodendron,* and many Olacaceae. The latter family—or some portion of it— is regarded as the starting point of four separate evolutionary lines. Santalaceae have always been re-garded as one of these, and correctly so. *Antidaphne, Eremolepis,* and *Eubrachion* form a trio distinct from Viscaceae. Whether or not a separate family (Eremolepidaceae) is erected for these three small genera, they may have *Opilia*-like ancestors. *Myzo-dendron* can only have been derived directly from Olacaceae, a fact shown by the position of the setae (modified anthers), which are inserted below the ovary, and are alternate with the perianth segments. Neither of these conditions occurs in Santalaceae,

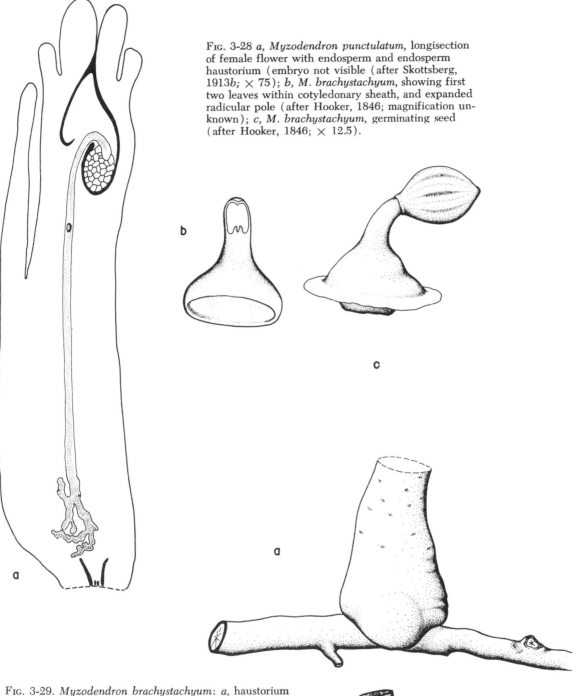

FIG. 3-28 a, *Myzodendron punctulatum*, longisection of female flower with endosperm and endosperm haustorium (embryo not visible (after Skottsberg, 1913b; × 75); b, *M. brachystachyum*, showing first two leaves within cotyledonary sheath, and expanded radicular pole (after Hooker, 1846; magnification unknown); c, *M. brachystachyum*, germinating seed (after Hooker, 1846; × 12.5).

FIG. 3-29. *Myzodendron brachystachyum*: a, haustorium of plant nearly eight years old (Werdermann No. 312, Chile, UC; × 2.5); b, old haustorium showing cuplike woodrose of host (after Hooker, 1846; × ½).

from which Skottsberg (1913*b*) and all other workers have derived *Myzodendron*. Viscaceae may be traceable to Santalacean plants similar to present-day epiphytic parasites such as *Phacellaria*. Loranthaceae, contrary to tradition, cannot have evolved from Santalaceae. The flower of Loranthaceae is dichlamydeous, and can scarcely have a monochla-mydeous ancestor in Santalaceae. Loranthaceae, therefore, forms the fourth evolutionary line taking its origin in Olacaceae, possibly in a *Chaunochiton*-like species. The system of Santalalean families is summarized in the accompanying diagram. For supporting evidence the reader is referred to the original article (Kuijt, 1968).

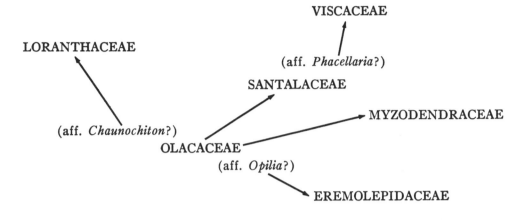

4

Broomrapes and Parasitic Figworts

The allocation of broomrapes and parasitic figworts to separate families (Orobanchaceae and Scrophulariaceae) is a time-hallowed taxonomic liberty taken by most students of these plants, although, in reality, it is very difficult to draw a clear line between the two families. *Lathraea* through the years has been tossed back and forth from Orobanchaceae to Scrophulariaceae-Rhinantheae, leaving one with a certain feeling of uneasiness about this familial distinction. *Hyobanche, Harveya,* and *Buchnera* have caused similar difficulties.

For the purposes of the present account I am accepting the familial status of Orobanchaceae. To assist readers with special interests in these groups I have made a survey of the affinities between Scrophulariaceae and Orobanchaceae at the end of the chapter. A comprehensive review of supraspecific classification in Scrophulariaceae is given by Thieret (1967).

Both Orobanchaceae and parasitic Scrophulariaceae are predominantly temperate in distribution. Some grow in subarctic regions, as *Pedicularis* and *Boschniakia* (northern Canada and Siberia) and *Euphrasia* (Tierra del Fuego), but most species are at home nearer the equator. The Mediterranean climates, especially, seem to have favored the evolution of large genera: *Castilleja* and *Orthocarpus* in western North America, *Harveya* in South Africa, and *Melampyrum* in the southeastern parts of Europe and adjacent Asia. These parasites are generally absent from tropical regions, with a few important exceptions such as *Striga* and *Aeginetia*. Tropical rain forests lack them completely. That this is true of parasitic figworts is understandable, since most of them are photosynthetic terrestrial plants which would be at a great disadvantage in places with reduced light intensity. But it is strange that the broomrapes, not being green and therefore not restricted in this way, have not been able to invade the tropical rain forests. The Balanophoraceae, also nongreen, and of similar habit, have been extremely successful under these conditions. Perhaps it is the dissemination of broomrape seeds, requiring a dry atmosphere and possibly wind, which makes colonization of rain forests difficult.

HABIT

Both groups of parasites—in Scrophulariaceae and in Orobanchaceae — include annuals and perennials. Some genera are annual (with an occasional exception), such as *Striga, Rhinanthus, Euphrasia, Melampyrum,* and *Orobanche.* Others are predominantly or entirely perennial: *Castilleja, Pedicularis, Sopubia*

(Williams, 1960*a*), *Lathraea, Boschniakia, Conopholis.* Neither category accomodates *Tozzia,* which leads an underground existence for some years, then emerges, flowers, sets seed, and dies (Heinricher, 1901*a*). Several genera closely related to parasitic figworts are shrubby: *Raphispermum* and *Radamaea* from Madagascar and some *Odontites* (Wettstein, 1895). These parasites are of medium height. The smallest may well be among the curious long-flowered New Zealand species of *Euphrasia* (see fig. 4-6,*b*; Du Rietz, 1931*a*).

The appearance of most parasitic figworts betrays nothing of their predatory life. All are root parasites. It must have been a great surprise to the botanical world when these plants and another family of similar habit, Santalaceae, were revealed as parasites in the same year (Decaisne, 1847; Mitten, 1847). The known extent of parasitism in Scrophulariaceae is indicated in table 3. The vegetative apparel of parasitic figworts seems in no way different from that of their autotrophic fellows among Scrophulariaceae. When not in flower they are hardly noticed in the field.

TABLE 3

DOCUMENTED PARASITISM IN SCROPHULARIACEAE

GENUS	DOCUMENTATION
Alectra	Botha, 1948; Bedi, 1967
Bartsia	Heinricher, 1901*a*
Buchnera	Lundell No. 14078, Texas; Mexia No. 989, Mexico (both UC)
Buttonia	Boodle, 1913
Castilleja	Heckard, 1962
Centranthera	Barnes, 1941
Cordylanthus	Piehl, 1966
Dasistoma	Pennell, 1928; Piehl, 1962*b*
Euphrasia	Crosby-Browne, 1950; Philipson, 1959
Gerardia	Pennell, 1928, 1929, 1935
Harveya	Marloth, 1932
Hyobanche	Marloth, 1932
Lathraea	Heinricher, 1931
Melampyrum	Piehl, 1962*a*
Melasma	Anonymous, 1924; Marloth, 1932; Saunders, 1934
Orthantha	Heinricher, 1898*a*
Orthocarpus	Cannon, 1909
Pedicularis	Sprague, 1962*a*; Piehl, 1963, 1965*a*
Rhamphicarpa	Fuggles-Couchman, 1935
Rhinanthus	Sablon, 1887; Heinricher, 1898*b*
Schwalbea	Fox No. 4620, North Carolina (UC)
Seymeria	Pennell, 1928, 1929
Siphonostegia	Kusano, 1908*a*
Sopubia	Williams, 1960*a*
Striga	Williams, 1958
Tozzia	Heinricher, 1901*a*

The occurrence of rhizomes is very uncommon, even in perennial species. *Pedicularis canadensis*, and some others of the same genus (Prain, 1891), reproduces vegetatively by means of rhizomes, and the same seems to be true of *Agalinis linifolia* (Pennell, 1928), *Bartsia, Tozzia, Lathraea* (Hegi, 1907-1931), and possibly *Aeginetia* (Tiagi, 1952b). *Alectra parasitica* in India has a greatly swollen and branched squamate rhizome (Rau, 1961), but whether this organ functions reproductively is not known. *Rhamphicarpa* sp. is stoloniferous according to the notes on a herbarium specimen at the University of California (Maas Geesteranus No. 6165, from Kenya).

The aerial stem is characterized by slenderness, though some *Pedicularis* species are somewhat fleshy. The fleshy condition generally seems to correspond to a more advanced, more complete parasitism, and a decrease in chlorophyll content. This is typified in the fleshy holoparasitic broomrapes, but also in the several genera of Scrophulariaceae approaching this habit. The broomrapes *Aeginetia* (fig. 4-1,*e*) and *Orobanche uniflora* lack stems almost entirely; the floral pedicels lift the flowers far above the ground. The closely related *O. fasciculata* has similarly elevated flowers on rather short stems.

The parasitic figworts are the only group of parasitic angiosperms in which variously divided leaves occur. (I have shown some of the variations in figs. 4-2, 4-3, and 4-4.) The basal leaves of *Seymeria, Pedicularis*, and *Gerardia pedicularia* are pinnatifid and have a fernlike appearance, especially the latter two with round, smooth sinuses and scalloped margins. The leaf of *Gerardia densiflora* is made up of threadlike divisions. A great deal of lobing or division is often present near the base of the leaf. Other figworts have leaves which are entire, or coarsely dentate, and even scalelike. The variety of leaf shape is particularly great in *Gerardia, sensu lato*, which has everything from fernlike to three-lobed or linear to scalelike leaves.

The leaves which are part of the inflorescence are often quite different from the lower vegetative leaves. It seems that two opposing trends exist, in this regard, within Scrophulariaceae. In *Seymeria macrophylla* (fig. 4-2) leaves become progressively simplified closer to the inflorescence tip; so most floral bracts are simple and narrowly lanceolate. In contrast, it is the floral bracts in *Melampyrum* which are deeply incised (fig. 4-2), whereas the basal foliage leaves are nearly linear in outline.

The complexity of venation seems to parallel that of leaf shape. A *Pedicularis* leaf has the usual intricate dicotyledonous venation; even in Orobanchaceae leaves of considerable size, the vasculature is much stylized (see chap. 9).

The leaves are frequently covered with small, few-celled glands or other trichomes. An alpine *Castilleja* of Montana bears numerous long white hairs, justifying its name, *C. nivea* (Pennell and Ownbey, 1950). In some species (*Pedicularis*, fig. 8-2) conspicuous hydathodes are present in leaf margins, possibly bearing a relation to the plant's parasitic mode of life (see chap. 8).

In *Tozzia alpina* of Europe the subterranean leaves are crowded and without chlorophyll. They are rather fleshy and their margins are bent forward, leaving a slitlike opening into the cavity thus formed (fig. 4-4,*e*). This cavity is lined with small glandular trichomes of unknown function.

The most extraordinary leaves in Scrophulariaceae are present in *Lathraea* (fig. 4-4,*d*). The youngest leaves are shaped much like those of *Tozzia*. In more mature leaves, however, the upper part recurves and the marginal areas of the upper and lower parts fuse. This "fusion" may, of course, be merely a descriptive formula; and the young, developing leaf may actually undergo a process similar to invagination. If this is so—and the literature is vague in this regard—no actual fusion of separate tissues occurs. The mature pitcher-like leaves are riddled with small chambers and passages, together forming a continuous system. These internal spaces are lined with small glandular trichomes of two types. One is a shieldlike sessile cushion of a few cells with a minute aperture; the other bears two-celled heads on one-celled stalks (Hegi, 1907-1931).

What are we to make of this extraordinary foliar transformation? No wonder these leaves have been described as devices to trap and digest small invertebrate animals. Such animals may often be seen on the glandular surfaces. This interesting idea unfortunately has little evidence to support it (Scherffel, 1888) and has now been generally abandoned. In fact, it is not impossible that at least the stalked glands discourage ants from inhabiting the leaf cavity and thus interfering with its normal activities; similar glands on the flower of *Orobanche* may have the same function, as suggested below.

Another functional interpretation has been made by Haberlandt (1897), who cites the fact that the main haustorial contact is with the host xylem, even to the penetration of the host vessel member by parasitic parenchyma (Heinricher, 1931). Haberlandt suggests that the glands function as hydathodes in the most active growth period (spring) to increase the "pull" on the host xylem. In so doing, a greater amount of photosynthates may be drawn from the xylem of the host into that of the parasite. (These aspects of parasitism are more fully explored in chap. 8). Haberlandt's suggestion seems reasonable as a working hypothesis.

In most parasitic figworts, phyllotaxis is opposite. The regularity of this arrangement often disappears when reaching the inflorescence (*Lathraea*, fig. 4-1,*d*). With more advanced parasitic status there is a shift to alternate phyllotaxis (cf. *Harveya squamosa, Hyobanche, Alectra*). Again, the alternate arrangement characteristic of Orobanchaceae is a logical extension of a trend visible in parasitic Scrophulariaceae.

FLOWERS AND FLOWER BIOLOGY

The floral variability of parasitic figworts is difficult to convey in brief, and I hope that the illustrations

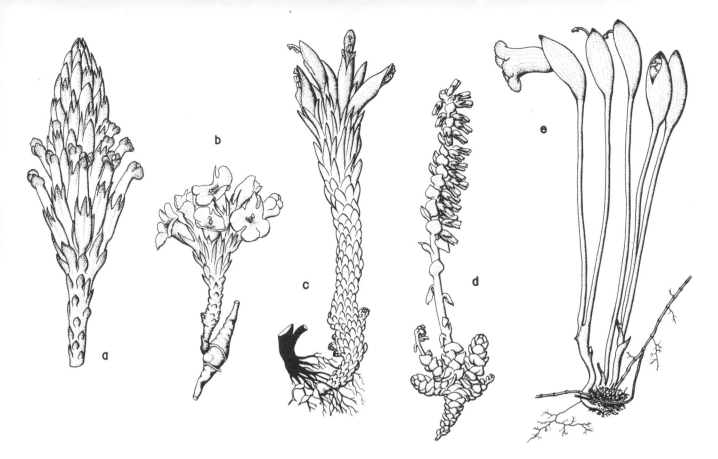

FIG. 4-1. Figworts and broomrapes without chlorophyll or nearly so: *Harveya squamosa*, slightly less than natural size (after Flowering Plants of South Africa, 2: 67. 1922); *b, H. purpurea* on root of *Roella ciliata* (after Marloth, 1932); *c, Hyobanche glabrata* (after Marloth, 1932); *d, Lathraea squamaria* (after Beck v. Mannagetta, 1930); *e, Aeginetia japonica* Sieb. and Zucc., (Suzuki No. 489155, UC) on *Miscanthus sinensis* (× ½) (*b,* and *c,* by permission of heirs of R. Marloth).

FIG. 4-2. Leaves of base (left) and top (right) of inflorescences of *Melampyrum arvense* (upper three: "Bohemia," UC 415094) and *Seymeria macrophylla* (lower four: Cobbe No. 138, Ohio, UC.) (all × ½).

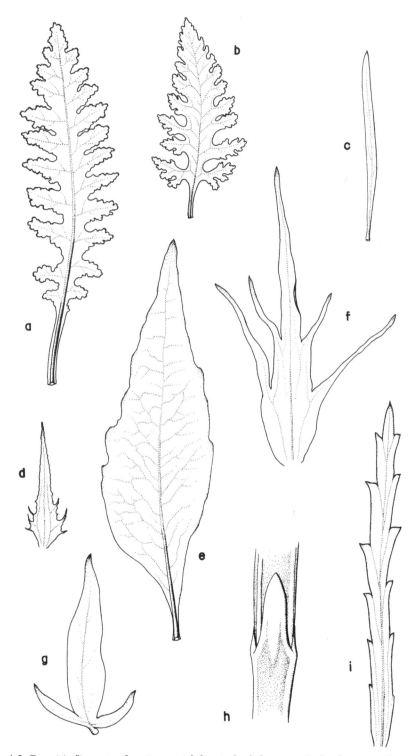

Fig. 4-3. Parasitic figworts, showing variability in leaf shape: *a, Pedicularis canadensis* (Lundell No. 11127, Texas, UC); *b, Gerardia pedicularia* (Umbach No. 6478, Indiana, UC; *c, G. besseyana* (House, New York, UC); *d, Alectra sessiliflora* (Rodin No. 3919, Transvaal, UC); *e, Dasistoma laevigata* (Heller and Halbach No. 1191, W. Virginia, UC); *f, Orthocarpus lithospermoides* (Heller No. 11346, California, UC); *g, Gerardia auriculata* (Graves No. 2111, Iowa, UC); *h, G. aphylla* (Bell No. 17230, North Carolina, UC); *i, Bellardia trixago* (Walker No. 3483, California, UC) (*h,* × 10; all others, × 1.5).

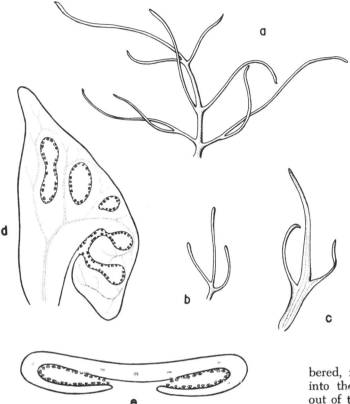

Fig. 4-4. Leaves: *a, Gerardia densiflora* (White-house No. 16700, Texas, UC; × 1.5); *b, Cordylanthus filifolius* (Hall No. 1423, California, UC; × 1.5); *c, Castilleja pallescens* (W. W. Jones, Montana, UC); *d, Lathraea squamaria*, longisection (after Hegi, 1907-1931); *e, Tozzia alpina* transection showing middle region of subterranean leaf (after Goebel, 1897).

will compensate for this difficulty. There is scarcely another group of comparable size among the angiosperms in which evolution has brought about so many strikingly different floral types.

The flowers are generally bisexual and zygomorphic, although many *Pedicularis* flowers are almost grotesquely twisted out of any symmetry. The calyx is usually greenish, with the visual attraction of pollinators falling to the corolla or bracts. In *Aeginetia* (fig. 4-1,*e*) the calyx forms a complete, spathelike floral envelope pierced by the emerging corolla tube, resulting in a slit on one side. The corolla rarely approaches the actinomorphic condition (*Seymeria*: Pennell, 1935), and falls short of it even in *Gerardia, Cycnium* (fig. 4-6,*a*), and *Harveya* (fig. 4-1,*a* and *b*). In *Euphrasia* (fig. 4-5*b* and 4-6*b*) the flower is already more zygomorphic, and is even more so in *Castilleja* (fig. 4-5,*a*) and related genera, in *Melampyrum, Pedicularis* (fig. 4-5,*d*), *Rhinanthus* (fig. 4-6,*c*), and all Orobanchaceae. Both calyx and corolla are either four- or five-lobed and gamopetalous. Each may be rather shallow (as in the calyx of *Orobanche* and the corolla of some *Euphrasia*), but long tubular corollas have developed in some. The extremes are seen in some New Zealand *Euphrasia* (fig. 4-6,*b*) and Chinese *Pedicularis* (Li, 1948, 1949); the tubes of the latter attain a length of 11 cm.

The four stamens are implanted on the petals, often very near the latter's base, and are of two different lengths. The anthers, basically two-cham-

bered, frequently have trigger-like spurs extending into the floral tube. These spurs must be pushed out of the way by pollinators (fig. 4-7,*a*) and cover the intruding animal with a shower of pollen. The same effect is achieved in other ways in *Harveya* and *Hyobanche* from South Africa (Marloth, 1932). In the former, as in *Christisonia* and *Aeginetia* (Beck v. Mannagetta, 1930), one anther cell has been transformed into a sterile spur blocking the floral passage (fig. 4-8,*a*). In the genus *Hyobanche*, as in many other Gerardieae, the sole remaining fertile cell is so large that it, also, protrudes and serves as a trigger (fig. 4-8,*c*). In *Buttonia* both types of stamens are represented in each flower (Hooker, 1871). *Boschniakia* lacks any sort of spur (fig. 4-7,*b*) but, as in many other genera (*Gerardia lanceolata* fig. 4-8,*b*), has many hairs growing from the anthers. It has been suggested that these hairs, by mutual interlocking, make the anthers dehisce as a single unit (Pennell, 1935). A similar function may explain the thick mucilage uniting the anthers of *Aeginetia* at maturity (Juliano, 1935). The dehiscence is by means of a longitudinal slit except in some *Seymeria* where the anthers have a poricidal dehiscence (Pennell, 1925). The attachment of filaments to anthers is normally basifixed, but dorsifixed conditions exist, and even versatile anthers are known, as in *G. lanceolata*.

The pollen appears to be remarkably uniform, although a comprehensive survey is not available. The species studied show nearly spherical, tricolpate pollen without any striking exine modifications (Tsoong and Chang, 1965).

It is unfortunate that this great assemblage of floral types has not been subjected to much careful

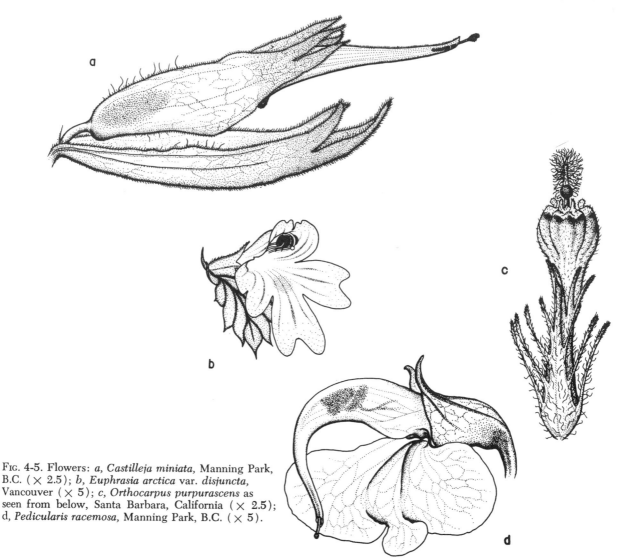

Fig. 4-5. Flowers: *a, Castilleja miniata,* Manning Park, B.C. (× 2.5); *b, Euphrasia arctica* var. *disjuncta,* Vancouver (× 5); *c, Orthocarpus purpurascens* as seen from below, Santa Barbara, California (× 2.5); d, *Pedicularis racemosa,* Manning Park, B.C. (× 5).

work on pollination and pollinators. The divergence in sympatric species of *Pedicularis,* for example, is almost surely due to a high degree of specificity of visiting insects (Li, 1951), a situation reminiscent of orchids. It is easy to see that parasitic Scrophulariaceae and Orobanchaceae are adapted to cross-pollination. Even here there are exceptions in the cleistogamy of a few. *Epifagus* usually bears sterile flowers above, and fertile but eleistogamous flowers on the lower part of the inflorescence; so cross-pollination may be exceptional here (Schrenk, 1894). Some of the smallest plants never reach the soil surface and complete their entire life cycle underground (Leavitt, 1902). In *Lathraea,* also, subterranean cleistogamous flowers have been recorded (Hegi, 1907-1931). Parthenogenesis is known from some populations of *Orobanche uniflora* (Jensen, 1951). An apparently cleistogamous species of *Castilleja, C. cryptantha,* has been described from Mount Rainier (Pennell and Jones, 1937).

Wind pollination does not seem to occur, except possibly in an incidental fashion in *Bartsia* (Jaeger, 1961).

Insects play by far the most important role in pollination. Many genera, such as *Rhinanthus, Pedicularis, Melampyrum,* and Orobanchaceae, have typical bee- or bumblebee-flowers, complete with landing platform, copious nectar, and bright colors. The usual two-lipped condition of these flowers, with anthers and stigma associated with the upper lip; the frequent honey guides (*Euphrasia; Gerardia* [Pennell, 1935]) and contrasting colors around the floral entry (*Melampyrum,* some *Rhinanthus, Orobanche*), all are aspects of the same mechanism. Even in the tubular corolla of *Castilleja,* the knob-like lower lip would seem to point to ancestral bee-flowers (Pennell, 1948). It is an interesting fact that *Pedicularis* and *Bombus* are coextensive geographically (Sprague, 1962b). The extent of involvement of Lepidoptera in pollination is unknown, but it is

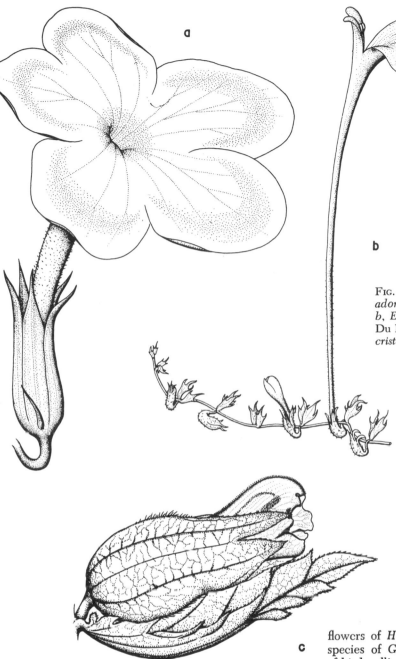

FIG. 4-6. Flowers: *a, Cycnidium* (? *Cycnium*) *adonense* (after Wettstein, 1895; × 1.5); *b, Euphrasia wettsteiniana* f. *longissima* (after Du Reitz, 1931 *a;* × 1.8); *c, Rhinanthus crista-galli*, Manning Park, B.C. (× 5).

likely to be considerable in the long-tubed *Pedicularis* (Pennell, 1948) and perhaps in similar *Euphrasia* in New Zealand. *Orobanche lutea* and *O. purpurea*, notwithstanding their bright colors, are said to lack nectar (Hegi, 1907-1931).

Birds only exceptionally function as pollinators. No broomrape, to my knowledge, is visited by birds, although the reddish *Boschniakia* of the Pacific Northwest is a likely suspect. The hummingbird pollination of a single species of *Pedicularis, P. densiflorus,* has been well documented (Sprague, 1962*b*), although even here bumblebees are active.

In South Africa, *Harveya* is occasionally visited by birds (Marloth, 1932), and it would not be surprising if the same were true of the brilliantly scarlet

flowers of *Hyobanche.* In the northern Andes some species of *Gerardia* (*Virgularia?*) have the aspects of bird-pollinated flowers: *G. lanceolata*, for example, has long tubular or funnel-form corollas which are dark red or purplish, and the flowers are crowded at the end of the raceme, well above the foliage. Pennell (1928) casually speaks of bird-pollination in the South American genera *Virgularia* and *Esterhazya*. A monotypic segregate of *Gerardia, Macranthera* of eastern North America, is said to be hummingbird-pollinated also (Pennell, 1935). Pollination by hummingbirds of *Castilleja* has been observed by scores of botanists, but has been documented only recently (Grant and Grant, 1966).

In some instances there are mechanisms that ensure seed-set after the pollinator has failed to appear. The style may curve down in such a way as to effect self-pollination. In contrast, the entire floral tube may undergo lengthening, bringing the epi-

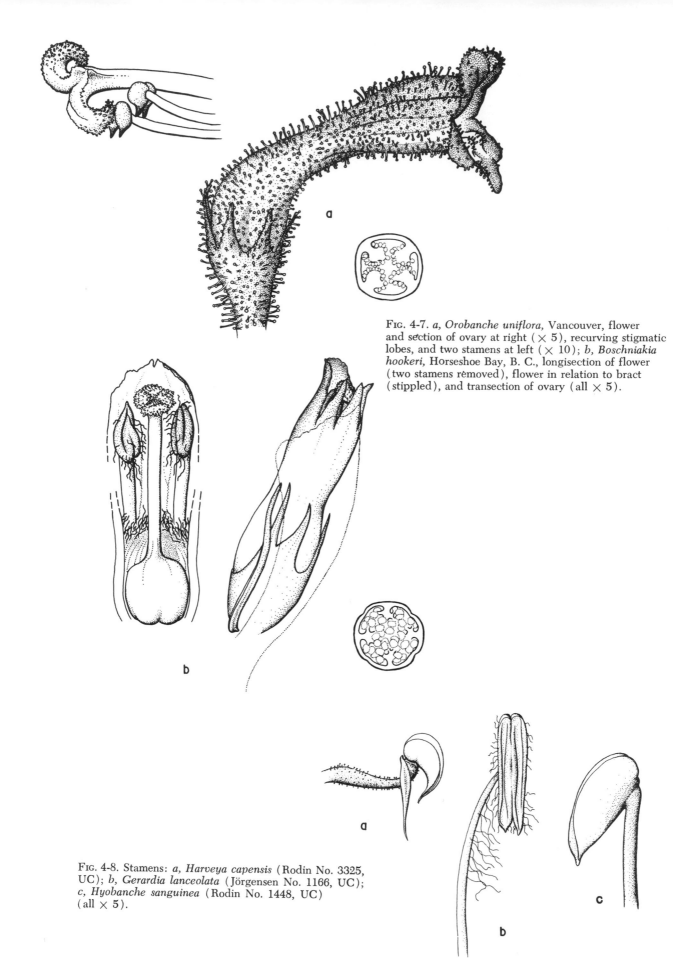

FIG. 4-7. *a, Orobanche uniflora*, Vancouver, flower and section of ovary at right (× 5), recurving stigmatic lobes, and two stamens at left (× 10); *b, Boschniakia hookeri*, Horseshoe Bay, B. C., longisection of flower (two stamens removed), flower in relation to bract (stippled), and transection of ovary (all × 5).

FIG. 4-8. Stamens: *a, Harveya capensis* (Rodin No. 3325, UC); *b, Gerardia lanceolata* (Jörgensen No. 1166, UC); *c, Hyobanche sanguinea* (Rodin No. 1448, UC) (all × 5).

petalous stamens into contact with the style. Each of these two mechanisms of facultative autogamy has developed in *Euphrasia, Rhinanthus,* and probably others (Kerner v. Marilaun, 1888; Wettstein, 1895). In *Orobanche uniflora* I have observed what appears to be a third method of autogamy, one reminiscent of that in Compositae (fig. 4-7,*a*). The two lips of the stigmatic surface curve back and, in so doing, the lower lip reaches the foremost pair of anthers. Whether pollen is actually transferred in this fashion has not been demonstrated, but this would seem likely, as the papillar stigmatic surface brushes against the anthers. *Melampyrum lineare* is said to be at least potentially autogamous or apogamous (Cantlon *et al.,* 1963).

The floral bracts are often brought into the service of pollination. They may be as brightly colored as the axillant flowers, heightening the over-all attractive effect of the inflorescence, as in *Hyobanche,* or they may form a contrasting pattern. In *Melampyrum* the bracts may be a deep purple, the flowers yellow or white (Hegi, 1907-1931). In *Castilleja* the flower is usually greenish, the bract a brilliant orange or red, but in a few species the opposite is true. In the geographical distribution of the latter genus the solitary South American species may be the only one showing self-compatibility (Heckard, pers. comm.).

Extrafloral nectaries, glands secreting a sugary liquid, are said to be present at the base of the lowest leaves of *Melampyrum* (Kirchmayr, 1908). It is thought that this lures ants away from flowers, reserving the floral nectar for the use of effective pollinators. It is possible that the dense covering of stalked glands on the outside of the calyx and corolla of *Orobanche* (fig. 4-7,*a*) provides another method of keeping ants away from the floral entry.

The style and stigma are generally well differentiated, the latter with a papillar surface and often bilobed. In *Dasistoma* we can speak of two distinct styles, although they are basally fused. The ovary is superior and either one- or two-chambered. This distinction has long served to separate Orobranchaceae and Scrophulariaceae, respectively. As Boeshore (1920) pointed out long ago, the distinction is difficult to maintain. In both *Lathraea* (usually placed in Scrophulariaceae, but not always: cf. Hutchinson, 1959) and *Christisonia* (Orobanchaceae) there is an imperfect partition of the ovary. In the latter genus the partition reaches only halfway up the ovary. Arekal (1963) has noted a tendency to unilocularity and parietal placentation in *Orthocarpus luteus.* The former trend, however, has been noted also in the fruit of several nonparasitic Scrophulariaceae (Pennell, 1935). In the parasites, at any rate, the unilocular condition would seem to be an advanced one.

In bilocular species the septum functions as a placenta. In the typical broomrapes placentation is parietal in two, three, or four series of ovules (Beck v. Mannagetta, 1930). The placentae of broomrapes may show much lobing and branching in *Christisonia, Aeginetia,* and *Xylanche.* Such placental ramifications greatly enlarge the surface on which ovules may be formed, and doubtless are related to the great number of seeds characteristic of these plants.

EMBRYOLOGY

The plants dealt with in this chapter have been subjected to much embryological work (Lundquist, 1915; Michell, 1915; Cassera, 1935; Krishna Iyengar, 1942; Tiagi, 1951*b,* 1952*a,* 1952*b,* 1956, 1963, 1965, 1966*b;* Kadry, 1952, 1955; Berg, 1954; Kadry and Tewfic, 1956; Steffen, 1956; Banerji, 1961; Arekal, 1963; Tiagi and Sankhla, 1963).

The ovules are anatropous or campylotropous, and very simply constructed. There is a single integument of several cells in thickness, and a scant one-layered nucellus which has often been resorbed by the expanding embryo sac at the time of maturity. The inner epidermis of the integument forms a nutritive jacket, or endothelium, which is thought to provide food to the embryo and growing endosperm. It is eventually resorbed completely by the endosperm. This endothelium, at least in *Pedicularis palustris* (Steffen, 1956), is 32-ploid by endomitosis. Such a condition is very reminiscent of polyploidy in another nutritive tissue, the tapetal layer of angiosperm anthers (Witkus, 1945).

The embryo sac is of the Polygonum type, although *Melampyrum lineare,* instead of being monosporic, is tetrasporic, and has seven nuclei in the mature embryo sac (Arekal, 1963). In this same species, six to eight delicate tubular processes develop from the micropylar pole and reach into the integumentary tissues. One of these processes frequently enlarges and enters the funiculus, receiving four nuclei. In *Boschniakia* the young embryo sac, in consuming the nucellus, may produce a bulbous, pendent lobe into the ovarian cavity (Tiagi, 1963). Several other genera of parasitic figworts have large micropylar haustoria (Schnarf, 1931), but the *Boschniakia* condition is not known from other members.

The endosperm is cellular from the beginning. Like some of the Santalalean plants described earlier, the endosperm of parasitic figworts and broomrapes sends out haustorial processes into surrounding tissues. Actually, there are haustoria at both chalazal and micropylar endosperm poles. The chalazal haustorium is binucleate, consumes some of the surrounding tissue, but eventually collapses. The micropylar haustorium is usually two-celled, but polyploidy is a prominent feature. In *Pedicularis sylvatica* each of the two cells is binucleate (Berg, 1954); however, work on *P. palustris* has shown the micropylar haustoria to be up to 384-ploid (Steffen, 1956). The extreme in polyploidy is found in the chalazal endosperm haustoria of *Melampyrum pratense* (3072-ploid: Erbrich, 1965). The interesting fact is that a nutritive organ again turns out to be polyploid. Both of the endosperm

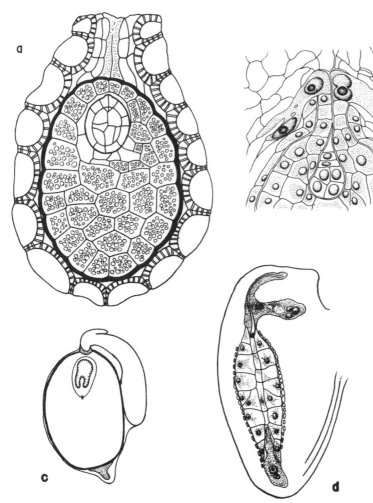

FIG. 4-9. *a, Orobanche hederae,* longisection of seed (from Beck v. Mannagetta, after Koch; × 300); *b, Cistanche tubulosa,* longisection through young seed at micropylar end; micropylar haustoria with hypertrophied nuclei, and proembryo surrounded by endosperm (after Tiagi, 1952a; × 600); *c, Pedicularis sylvatica,* longisection of mature seed with elaiosome (after Berg, 1954; × 22); *d, P. sylvatica,* early stage in embryogeny showing bilobed, 4-nucleate micropylar haustorium invading surrounding tissues; undivided, tubular zygote; binucleate chalazal haustorium; early endosperm cells; and remaining endothelial cells surrounding latter (after Berg, 1954; × 40).

haustoria disappear later, when the embryo begins to mature.

The micropylar haustorium of *Pedicularis sylvatica* is not evanescent; because of its great biological interest it deserves special mention (Berg, 1954). Instead of disappearing, it enlarges and manufactures an internal lattice of cellulosic strands. This peculiar organ bulges from the mature seed, forming a whitish, spongy crest more than halfway down the side. The unequally bilobed condition reflects the asymmetry of the original haustorium (fig. 4-9,c and *d*). A great quantity of oils and starch is stored in this organ. In fact, it is an elaiosome: ants derive nutrients from it and are active in the seed's dispersal. As an elaiosome it is unique in being unicellular, in having a cellulosic, netlike internal support, and perhaps in being endospermic in origin. It is also an interesting example of an organ serving two completely different successive functions.

The sequence of divisions in the young embryo is of the onagrad, crucifer, or caryophyllad type in *Pedicularis sylvatica* (Berg, 1954, *Boschniakia* (Tiagi, 1963), and *Aeginetia* (Tiagi, 1952*b*), respectively. This suggests how little taxonomic reliance can be placed on such embryo typologies. The suspensor may be relatively long, but consists of very few cells or even a single cell, and is resorbed by the growing embryo.

Nutritionally there is a complex exchange of materials, an exploitation of one tissue by another, in the embryology of figworts and broomrapes. The growing embryo sac obtains food from the nucellus. After fertilization both the growing embryo and the endosperm exploit the endothelium. The endosperm goes even farther, penetrating chalazal and micropylar portions of the integument by means of haustorial organs. The final phase, of course, is initiated at the time of germination when the embryo will digest the endosperm before it establishes its parasitic contact with the host. If we realize how common these parasitic relationships are in the reproductive phases of flowering plants, the varieties of parasitism within this one life cycle will not surprise us.

FRUITS AND SEEDS

The fruits of figworts and broomrapes are capsules, opening in various ways (Weberbauer, 1901) and dehiscing as a unit or scarcely opening at all in *Tozzia* (Heinricher, 1901*a*). Sutures may develop at the locules or at the septa of the fruit, and reach down to varying extent. In *Rhinanthus* (fig. 4-10,*a*) only the apex opens; so the large seeds stay in the capsule for a long time. The dry capsule with seeds thus becomes the "rattle" of many European vernacular names for *Rhinanthus*. The capsule of *Pedicularis* splits open mainly along one suture (fig. 4-10,*e*). That of *Gerardia greggii* has fruit walls which recurve when ripe. Other examples are illustrated for *Orthocarpus, Gerardia,* and *Rhamphicarpa* (fig. 4-10,*d, b,* and *c*). Most of the capsules show degrees of "imbibition movements." *Seymeria* capsules, for example, are closed when wet, and open again when dry. In *Striga,* on the contrary, the capsule opens upon wetting, and remains open thereafter. In *Melampyrum sylvaticum,* curiously, the fruit is actually pushed open by the growing seeds, which have a still green and living seed coat at that time. In *Lathraea* the seeds are violently expelled (Beck v. Mannagetta, 1930). The fruits of broomrapes are also capsules, opening longitudinally. The stiffness of the pedicel after death of a broomrape often turns the fruit into an organ like a pepper-shaker vibrating in the wind. In *Aeginetia,* however, the fruit wall merely flakes off irregularly to liberate the seeds (Tiagi, 1952*b*).

In size the seeds range from about 0.3 mm (*Striga, Parentucellia, Orobanche*) to 5 mm (*Rhinanthus* and *Melampyrum arvense*). Some are flat, with a broad marginal wing, as in *Rhinanthus*. All have endosperm, enclosing a small embryo; the compound body is often visible through the testa. The shape of the seed varies from compressed and discoid (*Rhinanthus*) to much attenuated (*Melasma*) and irregular ovate or L-shaped types. *Cordylanthus palmatus* seeds are spirally twisted and lack symmetry.

The number of seeds per flower is enormously variable. Some New Zealand species of *Euphrasia* have no more than two seeds, i.e., one per locule (Du Rietz, 1931*a*). *Tozzia alpina* has only one or two seeds per flower (Heinricher, 1901*a*). In *Melampyrum, Pedicularis,* and *Rhinanthus* (Tiagi, 1966*b*) the number of seeds is only slightly larger. Beyond this point there may be a relation between the number (and, consequently, the minuteness) of seeds per flower and the degree of dependence on the host. At any rate, *Striga,* one of the most advanced figworts, produces minute seeds in great numbers. An average plant of *Orobanche minor* is said to produce more than 100,000 seeds (Hegi, 1907-1931), but *Boschniakia rossica* probably holds the record with more than a third of a million seeds per plant (Gavriliuk, 1965). Yet, on the basis of seeds per capsule, *Aeginetia* is the indisputable winner: one capsule is believed to produce from 40,000 to 70,000 seeds (Ling, 1955).

The seed coats of parasitic figworts and broomrapes show a structural variety which has scarcely been hinted at in the scientific literature (Arekal, 1963). My descriptions and illustrations cannot pretend to do them justice (figs. 4-11, 4-12, 4-13). This meager selection should not be considered representative of these genera. *Cordylanthus palmatus,* for example, may well be atypical for that genus (Heckard, pers. comm.). Seed anatomy here is an untapped reservoir of systematic information.

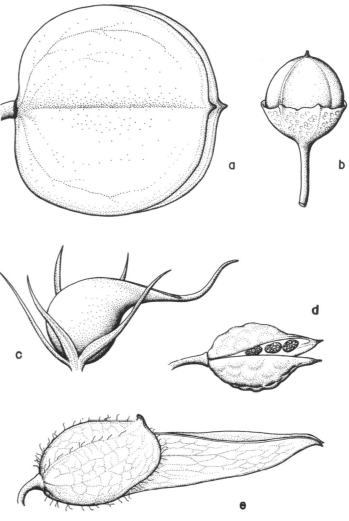

FIG. 4-10. Capsules in figworts: *a, Rhinanthus major* (Kjellmert, Sweden, UC; × 5); *b, Gerardia aphylla* (Bell, No. 17230, North Carolina, UC; × 5); *c, Rhamphicarpa longiflora* (after Wettstein, 1895); *d, Orthocarpus lithospermoides* (Purpus, California, UC; × 5); *e; Pedicularis canadensis* (Dubois, Quebec, UC; × 5).

The seed coat is made up of a sclerified layer with exceedingly elaborate sculpturing. This sclerification seems to take place only on the inner wall and the radial walls of the superficial layer (fig. 4-9,*a*). The outer wall, not receiving such thickenings, appears to collapse and disappear with the death of the cell, leaving a layer of sclerified polygonal cups tightly fitting together to form a protective layer. I am extrapolating somewhat at this point: the seed coat in *Orobanche* develops in this fashion, and I am assuming a similar differentiation in at least such Scrophulariaceae as *Castilleja, Seymeria,* and *Melasma*. However, several other genera (e.g., *Striga, Sopubia,* and *Buchnera*) have seed coats with very shallow depressions which might have a different origin, and the surface of *Melampyrum*

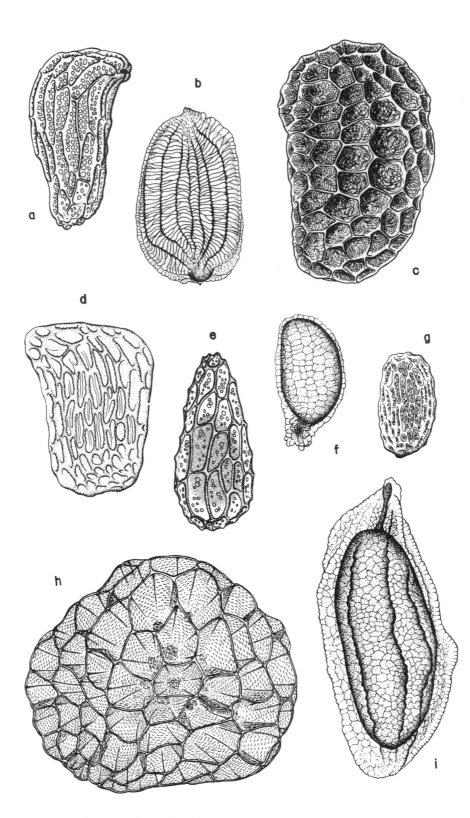

FIG. 4-11. Seeds: *a, Buchnera floridana* (Deam No. 60652, Florida, UC; × 80); *b, Bellardia trixago* (Kellogg, California, UC; × 80); *c, Seymeria texana* (Warnock W. No. 1089, Texas, UC; × 40); *d, Sopubia ramosa* (Bogdan No. VB 571, Kenya, UC; × 80); *e, Orobanche fasciculata* (Eastham, Destiny Bay, B.C., UBC; × 150); *f, Parentucellia viscosa* (Tracy No. 13230, California, UC; × 80); *g. Striga multiflora* (Clemens No. 18180, Philippines, UC; × 80); *h, Boschniakia hookeri* (Horseshoe Bay, B.C.; × 45); *i, Euphrasia curta* (Paulsen, Denmark, UC; ×40).

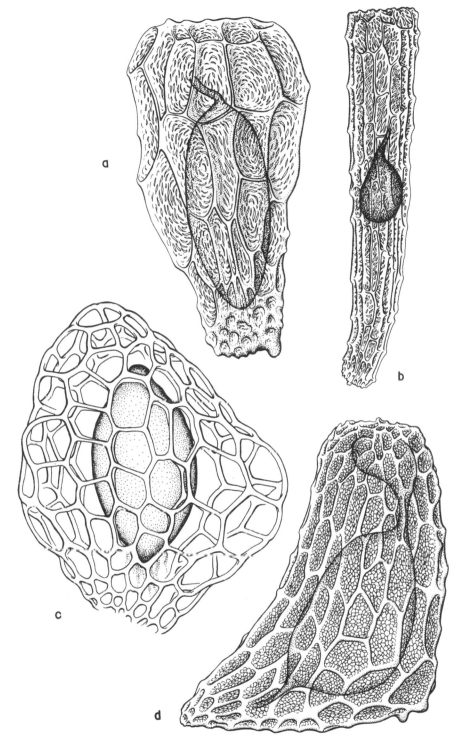

Fig. 4-12. Seeds: *a, Castilleja foliolosa* (Brandegee, California, UC; × 40); *b, Melasma melampyroides* (Box No. 1250, Antigua, UC; × 80); *c, Orthocarpus laciniatus* (Chile, UC 103135; × 80); *d, Gerardia aphylla* (Bell No. 17230, North Carolina, UC; × 80).

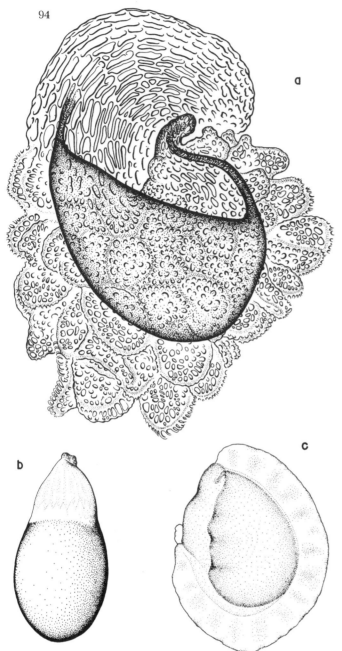

FIG. 4-13. Seeds: *a, Cordylanthus carnulosus* (= *C. palmatus* ?) (Heckard No. 1449, California, JEPS; × 60); *b, Melampyrum arvense* (Suza, Czechoslovakia, UC; × 10); *c, Rhinanthus minor* (Vautier *et al.* No. 506, France, UC; × 10).

The sculpturing on the walls of the seed coat is often so delicate and intricate as to defy accurate description. The pits may be circular or elongated, arranged in serial or parallel fashion, or in no apparent order. In *Melasma* the bars separating pits are studded with granular, refractive tubercles. The bars of *Sopubia* are wavy in outline, but extend into thinner bars which end blindly, as the ultimate veinlets in a dicot leaf. Pits in *Boschniakia* are very uniform in size and in their spiral arrangement. The most elaborate of all pitted types is seen in *Cordylanthus palmatus*: one end of the seed bears many arching crests with perforated sides. These crests are frequently crowned with comblike thickenings which may represent places where the outer testa wall has fallen away.

In other seeds an intricate network of fine ridges covers the depressions of the testa, as in *Gerardia aphylla*. In *Seymeria* these ridges are spiral thickenings running in opposite directions around each cell, with their intersections cutting off pits. The seed coat of *Euphrasia, Parentucellia,* and *Bellardia* may be quite different in nature: I cannot guess what their sculptured surfaces represent.

Among the most interesting is the seed of *Orthocarpus*. The testa represents a cagelike network within which the embryo and endosperm appear to be freely suspended. This network seems to be double, the meshes of the inner one corresponding to those of the outer network. How can such a system develop from a simple ovule? One possibility is that, on all tangential walls of the cells making up the testa, thickenings develop only where they meet radial walls. When the primary walls collapse after the death of these cells, two nearly identical networks, one inside the other, would be left. This is but one indication of the vast amount of interesting anatomical and systematic information awaiting discovery in the seeds of these plants.

In all these details of embryology and seed structure, it is difficult to find a clear distinction between Scrophulariaceae and Orobanchaceae. Tiagi (1956) was unable to find a distinction of this sort separating *Striga* from the broomrapes. The only possibility emerging is the degree of differentiation of the embryo. Broomrape embryos lack visible differentiation, and the many parasitic figworts studied have embryos with root, root cap, and cotyledons. Even here, several important genera (e.g., *Hyobanche* and *Harveya*) are missing from the list. *Lathraea* seems to be intermediate, for the cotyledons are very small in the mature seed, although they enlarge during germination (Heinricher, 1894). At any rate, such a distinction would be closely related to the parasitic habit and could scarcely form the basis of familial separation.

What is the functional significance of these structural details? Such variations could scarcely have come into being unless specific advantages were attached to them. Let us therefore consider the dissemination of seeds in parasitic figworts and broomrapes.

seeds is quite smooth. In *Epifagus,* wall thickenings seem to be lacking in the testa, and the outer wall remains intact (Koch, 1878). The other extreme is seen in *Boschniakia hookeri,* where most of the seed consists of a honeycomb-like testa. The seeds of *Xylanche* (*Boschniakia*) *himalaica* are the same as those here illustrated, but those of *B. rossica* resemble *Orobanche* seeds (Smith, 1933).

DISSEMINATION AND EARLY DEVELOPMENT

It can probably be assumed that seeds in most of these plants are disseminated by a combination of wind and water. It must certainly be true of those with extremely small seeds, such as *Orobanche*, *Striga*, and *Parentucellia*, that seeds can be moved by winds for considerable distances through the air or over the ground. It has been suggested that the many cavities on the testa make it easier for the wind to lift such seeds. The fact that precisely those seeds which are smallest, and thus are most likely to be wind-dispersed, have the least elaborate topography, would seem to deny this. *Boschniakia*, having probably the largest of all broomrape seeds, also has the deepest cell cavities in the testa. I would suggest that such cavities retain air, allowing the seed to be carried along on top of the surface of water; they are therefore adaptations to dispersal by water during heavy rains. Even the smallest seeds are almost surely swept into cracks in the soil, in the same fashion, where they may eventually come into the rhizosphere of a potential host (Garman, 1903; Durbin, 1953). Viable seeds of *Striga*, for example, have been recovered from a depth of 60 inches in sandy soil (Robinson and Kust, 1062).

Several instances of animal-dispersed seeds are recorded. The seeds of *Tozzia* are carried by ants (Hegi, 1907-1931), but appear to lack a specialized elaiosome. A few cell layers of the fruit wall, however, contain abundant starch and may be attractive to ants (Weberbauer, 1901). In *Melampyrum* a conspicuous elaiosomic body is attached to one pole of the seed (fig. 4-13,*b*). Although it has been said (Lundström, 1887) that ants carry these seeds about because of a similarity to their own pupae, this explanation seems too fanciful. It is more likely that the elaiosome is a supply of food. Ant dispersal has been well documented for *Pedicularis sylvatica* (Berg, 1954). The elaiosome in this species is an enormously inflated micropylar endosperm haustorium filled with oils and starch. Pennell (1935) also speaks of ground-feeding birds as possible dispersal agents of *Melampyrum;* the hard seed coat perhaps facilitates a safe passage through their intestinal tract.

An unusual method of dispersal, which is coordinated with the dispersal of the achene of one host species, has been described for *Orthocarpus* (Atsatt, 1965). The most common host, *Hypochoeris glabra*, has spreading bristles on top of the achene which intercept and entangle the netlike testa of *Orthocarpus* seeds. Thus *Orthocarpus* seeds may be carried with the seed of the host, the two germinating together and resulting in a parasitic relationship. The evidence of a significant degree of pairing of the two partners is not available, however; even if such association could be demonstrated, other explanations might be equally valid. The host species in the area studied has been there only since the

turn of the century. It is difficult to see how such coördinated dispersal is anything more than coincidental at this time, although there is no way of telling in which direction it might evolve. The dissemination and life cycles of the two plants might fit together rather nicely, but the lack of host-specificity in nearly all parasitic Scrophulariaceae should not be forgotten.

The study of germination of parasitic figworts is currently a rather active field of research. It has been shown, for example, that in the more advanced groups of plants a stimulant secreted from the host root is required to set germination in motion. The seeds of less advanced plants can germinate without such a stimulant but there may, of course, be other barriers to germination. Further interesting physiological aspects of germination are discussed in chapter 8, but here we shall consider only the directly observable facets of the process.

Germination in most genera is epigeous. The radicle grows into the soil after forming a ring of hairlike structures, and usually branches profusely at an early time, while the two cotyledons withdraw from the exhausted endosperm and spread. In most situations it is not long before the first haustorial contacts are made. The haustoria are exceedingly small, and structurally very different from those of Santalacean haustoria (chap. 7), although they also are lateral in position. Especially in annual Scrophulariaceae it is important to establish a large number of haustoria early in the growing season. In perennial Scrophulariaceae the search for host roots must be renewed each year. *Tozzia*, and apparently also *Alectra* and *Striga*, fail to raise their diminutive cotyledons above the ground (Heinricher, 1901*a*, Okonkwo, 1967). The entire existence of *Tozzia* for several years is holoparasitic and subterranean. When the plant finally reaches maturity a green shoot suddenly reaches up and flowers (fig. 4-16,*g*), and then the entire plant dies. Thus less than 4 percent of its life cycle is spent as a hemiparasite, and all the rest as a holoparasite. *Tozzia* represents an interesting link to *Lathraea*, a holoparasite. In *Lathraea* an even longer period is spent underground, about ten years (Heinricher, 1894). The flowering shoot which finally emerges contains no chlorophyll, and plants do continue to flower for a number of seasons. The cotyledons of *Lathraea* are even smaller than those of *Tozzia*, but enlarge upon germination. According to Botha (1951*b*), *Alectra* spends its first few weeks underground as a holoparasite, after which it emerges. A careful descriptive study of germination in this genus, and indeed in *Hyobanche*, would probably add many interesting details to this account. As in their green relatives, haustoria quickly establish a channel of nutrients from the host. *Lathraea clandestina* forms the largest known haustoria in Scrophulariaceae: they reach the size of a small pea (Heinricher, 1917*b*).

Many features of Orobanchaceae appear to be already foreshadowed in their figwort relatives. The tendency toward decreased differentiation in the

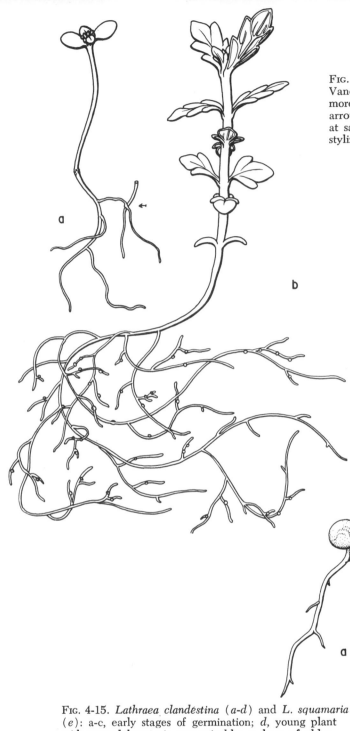

Tozzia and *Lathraea* embryo culminates in a total lack of differentiation in the embryo of the broomrapes. In another very interesting way the behavior of the *Striga* seedling seems to anticipate that of broomrapes (Williams, 1961*b*). Besides chemically stimulating the germination of *Striga* seeds, the host root influences the parasite in two other ways. First, the *Striga* radicle is made to grow toward the host root. This is precisely what happens in the broomrapes also. Unfortunately we do not know what happens when the radicle reaches the host root. Second, the chemical exudates from the host roots bring about striking morphological changes in the seedling parasite: an increased radial diameter, a proliferation of cortical cells, a production of root hairs, and an

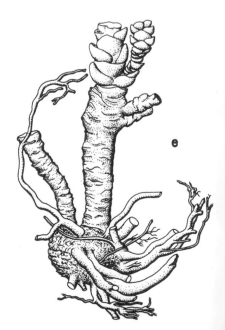

Fig. 4-15. *Lathraea clandestina* (*a-d*) and *L. squamaria* (*e*): a-c, early stages of germination; *d*, young plant with several haustoria recognizable; *e*, base of older plant from which inflorescences have been removed (*a-d*, after Heinricher, 1894, × 1.5; *e*, after Heinricher, 1931; × ½).

apparent breakdown of polarity. Whether the radicle of *Striga* always forms a truly terminal primary haustorium or behaves more like other Rhinantheae, the series of events which take place could lead to a terminal haustorium as in the broomrapes. Herbarium material of *S. gesnerioides* (see fig. 4-19,*a,c*) shows a tuberous, irregular growth attached to the host root, giving rise to flowering shoots. Since even young plants of this species have no secondary roots, the transition to a primary haustorium seems to be complete. *S. senegalensis,* however, does produce secondary roots after its primary haustorium has become established (Okonkwo, 1967). A similar shift from secondary to primary haustoria has been made by the ancestors of nearly all mistletoes, and Myzodendraceae, and possibly some epiphytic Santalaceae. In the latter families this change seems to be related to the achievement of the epiphytic habitat, but in the broomrapes and figworts a completely different explanation must be sought.

The embryo of Orobanchaceae, lacking cotyledons and root cap, is no more than an undifferentiated globe of cells (fig. 4-16,*b*). In *Orobanche minor* no more than about forty-five cells are involved (Vallance, 1951). Upon germination the cells of the radicular pole expand and burst out of the testa. This growth in length continues (it may be merely cell expansion without mitotic activity) in the direction of the host root. Penetration of the latter appears to be almost imperceptible, without extensive damage to the root. Some of the contact cells penetrate the epidermis and cortex and are soon in contact with the stele, thus establishing a functional conduit. Meanwhile the external part of the seedling begins to grow into the small tubercle characteristic of seedling broomrapes often with the ruptured testa still attached.

The early life history of the broomrape genus *Aeginetia* (Kusano, 1908*b*) includes some curious details. In this plant two or three hyaline, globular cells emerge from the micropylar pole of the seed. The number of cells later increases to about fifteen, forming a hemispherical cushion. Several of these cells begin to protrude, becoming papillate and eventually assuming the shape of hairs up to a millimeter in length. They are often septate, sometimes branched. Their apices are said to attach themselves to the host, recoiling or contracting to bring the seedling close to the host. Some, indeed, may twine around the root hairs of the latter. A greater number of these hairlike tendrils are formed when the seed is farther away from the host. These hairs are so reminiscent of the epidermal hairs which often precede the formation of secondary haustoria (see fig. 7-25) in parasitic figworts,[1] and also of the hairs on

primary haustoria of *Striga* and those around the upper radicle of several other related parasites, that one wonders whether all these trichomes may not have a similar function of attachment. No comparable structures are known from other broomrapes, but very few have been studied in this regard.

The events following the formation of the small tubercle have been repeatedly described in the literature (e.g., Kadry and Tewfic, 1956). A crownlike girdle of secondary roots surrounds the apical meristem, which is being differentiated, it is said, endogenously (Kadry and Tewfic, 1956). The development of the secondary roots is depicted for *Orobanche uniflora* in figure 4-17. The roots of this species rarely branch, but become slender white wormlike outgrowths reaching out in all directions. They are apparently devoid of a root cap (chap. 9) and certainly lack root hairs. Whenever a host root touches upon a root of the broomrape, an inconspicuous haustorium soon connects the two physiologically. The appearance of these haustoria is much like those of *Euphrasia* and most other figworts. (Their anatomy is further discussed in chap. 7.)

Unfortunately, we find little in botanical writings about subterranean organs of broomrapes. It is very easy to get the impression that the growth and function of secondary roots, as sketched above for *O. uniflora,* is standard for all broomrapes. It certainly is common, especially in *Orobanche* itself. In a collection of a broomrape of *O. ramosa* affinity (Kuijt and Dempster No. 2119, UC) I have seen this type of root as long as 2 or 3 cm, and those of *O. australiana* appear to be nearly as long (Dempster No. 3617, UBC). The following variations, however, deviate from the pattern above.

In the monotypic genus *Epifagus* of eastern North America a large number of short secondary roots issue from the base of the plant. Some emerge even from between the lowest leaf scales, but most of them form a nestlike mass near the point of attachment. It is surprising to find that these roots, which are somewhat branched and quite stiff, have no haustorial contact with any host roots. As Boeshore (1920) mentioned, they seem to have lost all haustorial power. The function of these roots is a mystery. It has been suggested that they may aid in holding the inflorescence aloft, as they pierce the surrounding dead leaves and other raw humus in all directions; in other words, they may be organs of stability (Schrenk, 1894). It is perhaps significant that root development is profuse in soft, rich soil, but very poor in stony or sandy soil (Cooke and Schively, 1904).

Several other broomrapes have nonhaustorial secondary roots. I have illustrated three stages of the growth of such a root system in *Orobanche grayana,* a broomrape preferring a sandy coastal situation in the Pacific Northwest (fig. 4-18). A young plant shows the typical crown of secondary roots around the host connection, but they already have a stubby appearance and do not elongate. From this crown develops, eventually, a very complex mass of coral-

---

[1] The first recognition of the secondary host contacts of broomrapes may be found in Turner's "The Grete Herball," 1526: ". . . I know that the freshe and young Orobanche hath commyng out of the great roote, many lytle strynges . . . wherewith it taketh holde of the rootes of the herbes that grow next unto it."

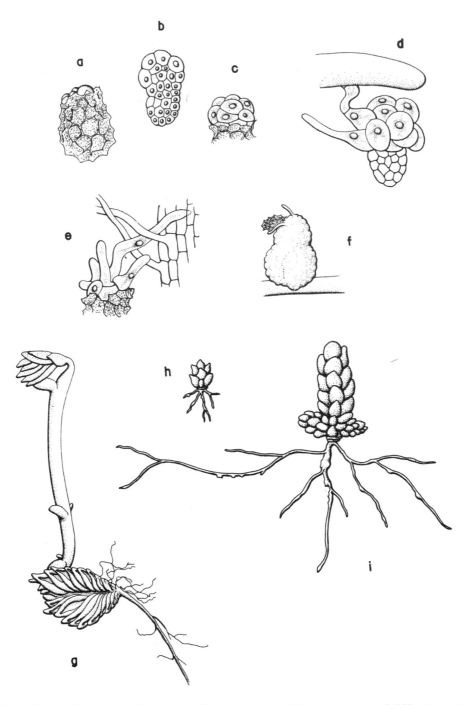

Fɪɢ. 4-16. *a-f, Aeginetia indica,* stages of germination and formation of tendril-like hairs from enlarged cells of one pole of embryo (after Kusano, 1908*b; a-e,* × 65, *f,* × 20); *f,* young seedling established on host root; *g, Tozzia alpina,* transition from subterranean to aerial shoot (after Heinricher, 1901*a;* × 1.7); *h* and *i,* seedlings of two stages of *Lathraea squamaria* (after Irmisch, 1855; × 1.7).

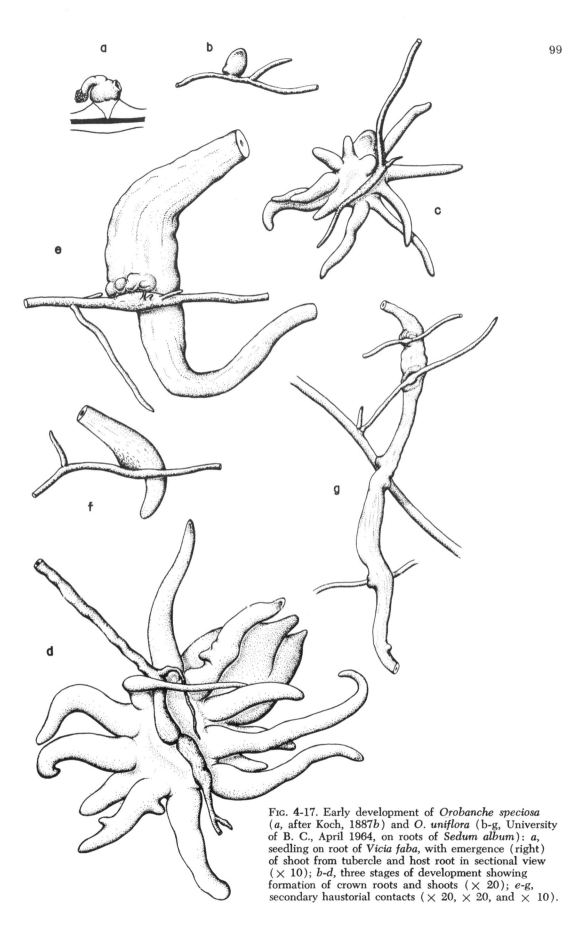

FIG. 4-17. Early development of *Orobanche speciosa* (*a*, after Koch, 1887*b*) and *O. uniflora* (b-g, University of B. C., April 1964, on roots of *Sedum album*): *a*, seedling on root of *Vicia faba*, with emergence (right) of shoot from tubercle and host root in sectional view (× 10); *b-d*, three stages of development showing formation of crown roots and shoots (× 20); *e-g*, secondary haustorial contacts (× 20, × 20, and × 10).

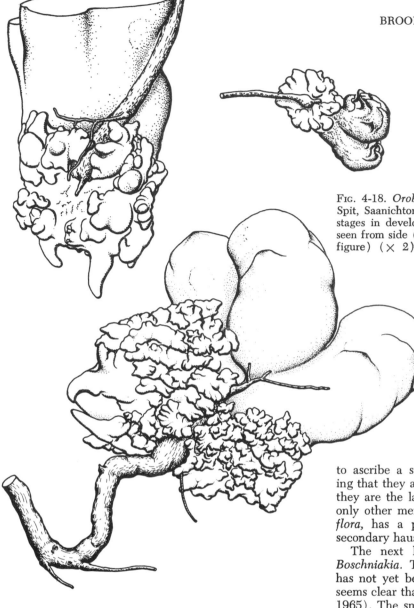

FIG. 4-18. *Orobanche grayana* var. *violacea*, Cordova Spit, Saanichton Bay, B. C. (det. L. R. Heckard), three stages in development of coralloid root system, as seen from side (above two figures) and below (lowest figure) ($\times$ 2).

like, short, fleshy roots. I have not been able to find any sign of secondary haustorial attachments. Stability can scarcely be the function of these organs. Considering the environment, a storage function may be suggested. The nutritive nature of the root mass is attested by the foraging of beetle larvae, which frequently gnaw away large portions. I have observed the same coralloid roots in a collection of *O. pinorum* Geyer ex Hook., kindly identified for me by Dr. L. R. Heckard (Rossbach and Hodgdon s.n., from New Mexico, JEPS). It is possible that the entire section *Myzorrhiza* of *Orobanche* is characterized by this type of root.

When next we look at the base of the mature plant of *Orobanche fasciculata* (fig. 4-19,*e*), scarcely any roots are visible. The slightly tuberous base of each inflorescence may give rise to a small bulbous outgrowth, but some are without even this organ. In all likelihood these are comparable to the coralloid roots of *O. grayana*, but it would be unreasonable

to ascribe a storage function. It goes without saying that they are not haustorial. I would suggest that they are the last remnants of secondary roots. (The only other member of section *Gymnocaulis*, *O. uniflora*, has a profuse root development and many secondary haustoria.)

The next logical step takes us to the genus *Boschniakia*. The early development of this genus has not yet been studied in sufficient detail, but it seems clear that no roots are ever formed (Gavriliuk, 1965). The small tubercle after establishment probably never ceases to enlarge, and may attain a diameter of several inches. In rocky territory the tuber may become flattened or otherwise misshapen, according to the configuration of rocks around it. The surface of the tuber shows a system of fine cracks, and the inflorescences emerge from the top, often several from one tuber. *Boschniakia* is perennial, and it is not surprising that the drain on the host root makes the distal part of the latter die off at an early age. The tuber thus seems to assume a terminal position.

The situation in *Conopholis* is so much like that in *Boschniakia* that a close affinity of the two genera would seem to be indicated. *Conopholis*, perhaps like *Boschniakia*, takes four or five years of subterranean development to reach the flowering condition, after which it may continue to flower for many years (Wilson, 1898). Instead of a single tuber, *Conopholis* produces a great tubercular mass supporting dozens of scaly buds and many inflores-

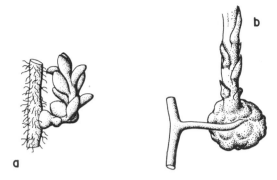

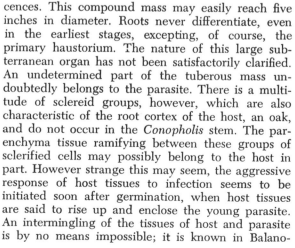

Fig. 4-19. Primary haustorial attachments:
a, Striga "asiatica," young seedling (after
Rogers and Nelson, 1962; × 15); *b, S. ges-*
nerioides, plant in bud (Rodin No. 2827,
S. W. Africa, UC; × 5); *c, S. gesnerioides,*
flowering plant (Halse, Transvaal, UC; × 2.5);
d, Boschniakia hookeri, young flowering plant
on *Arctostaphylos uva-ursi,* Horseshoe Bay,
B. C. (× 2); *e, Orobanche fasciculata* on
Artemisia, Armstrong, B. C., plants in full
flower (× 2).

cences. This compound mass may easily reach five
inches in diameter. Roots never differentiate, even
in the earliest stages, excepting, of course, the
primary haustorium. The nature of this large subterranean organ has not been satisfactorily clarified.
An undetermined part of the tuberous mass undoubtedly belongs to the parasite. There is a multitude of sclereid groups, however, which are also
characteristic of the root cortex of the host, an oak,
and do not occur in the *Conopholis* stem. The parenchyma tissue ramifying between these groups of
sclerified cells may possibly belong to the host in
part. However strange this may seem, the aggressive
response of host tissues to infection seems to be
initiated soon after germination, when host tissues
are said to rise up and enclose the young parasite.
An intermingling of the tissues of host and parasite
is by no means impossible; it is known in Balano-

phoraceae, where tubers of apparently dual com-
position are reported. A careful study of the situation in *Conopholis* would repay the investigator.

Aeginetia may be unique in its subterranean behavior also. It is the only broomrape, as far as I
know, which regularly reproduces vegetatively. It
does so by means of rootlike organs which grow in
various directions to produce new inflorescences here
and there (Juliano, 1935; Tiagi, 1952*b*). Although
these rootlike structures form haustorial attachments,
their anatomy is reminiscent of stems. The viability
of these "roots" when severed from the mother

plant (but possibly still attached by means of secondary haustoria) is such that new inflorescences may develop. It is no wonder that *Aeginetia* is an agricultural pest in many places in the Far East, especially on sugar cane.

Our discussion has gone full cycle through the life history of parasitic figworts and broomrapes. Starting with the parts which are normally visible to the casual observer in the field we have followed these plants through their reproductive phases up to the point where they lift themselves from the ground. In some of the more advanced forms this point marks a transition from a curious and sometimes extended subterranean existence to an aerial existence. An extremely important aspect of parasitic life—the structure and behavior of the haustorial connection itself—has been omitted, but the subject will be given a single treatment for all parasitic flowering plants in chapter 7.

AFFINITIES BETWEEN SCROPHULARIACEAE AND OROBANCHACEAE

The justification of Orobanchaceae as a separate family has been seriously disputed. Further study, in greater detail, has not dispelled these doubts but has magnified them.

It is an interesting commentary on systematic botany that, when a family such as Balanophoraceae stands in isolation, we clamor for links with other families so that this difficult group can be related to them; yet when such links are obvious and abundant beyond reasonable expectation we sometimes prefer to ignore the facts. The latter is true of the figworts and broomrapes. The tradition of regarding Orobanchaceae as a distinct family would seem to be an example of published information outrunning taxonomists.

Have botanists been unduly impressed by the holoparasitic mode of life of broomrapes, as contrasted with the green figworts? One is tempted to dismiss this familial separation, since a nearly complete continuity of morphological features exists (Boeshore, 1920). But perhaps it is better to be tolerant of this failure of the taxonomic system. In other families in this alliance (e.g., Gesneriaceae, Bignoniaceae) there is a great deal of overlap of characters also. A reduction in the number of families will not necessarily express phylogenetic relationships more accurately. The present chapter neither requires nor justifies changes in familial status.

Whether the parasites within Scrophulariaceae form a discrete natural unit is by no means clear. At this time we can only say that all parasites belong to one of two tribes, Rhinantheae (sometimes called Euphrasiae) or Gerardieae (sometimes called Buchnereae). Does this mean that all the members of these tribes are parasitic? We do not know, but the possibility exists. Many of the genera of Gerardieae, meanwhile, are still without the "stigma" of parasitism. The systematic significance of parasitism

within Scrophulariaceae remains an open question. I have summarized the known extent of parasitism in table 3. All Orobanchaceae are parasitic. The structural continuity of the parasites treated in this chapter would seem to indicate that parasitism has evolved only once in this alliance, and that all descendants of the original parasitic stock may eventually be accommodated within one taxonomic framework.

The various facets of continuity between the two groups may be summarized as follows:

1. *Loss of photosynthetic capacity and increased host-dependence.*—This is perhaps the most obvious result of parasitic evolution, although by no means inevitable. It has run its full course in the broomrapes, where no chlorophyll is known. This is true also of *Lathraea* and *Hyobanche*, however, and *Striga gesnerioides* and *Tozzia* are well advanced in this direction. By losing chlorophyll, parasites become more completely dependent upon their hosts.

2. *Fleshiness of stem.*—The green figworts are characteristically slender-stemmed. In some *Harveya* and *Striga* species the base of the stem has become stout and succulent, and this is even more evident in the non-green figworts and the broomrapes. In *Aeginetia* the slenderness of the stem is obviously a derived condition.

3. *Phyllotaxis opposite to alternate.*—This trend may exist even within a single individual. All broomrapes have alternate phyllotaxis. This trend has exceptions in the opposite arrangement of leaves in *Lathraea* and *Tozzia*.

4. *Longevity.* — The perennial condition is undoubtedly an advanced feature. The genus *Orobanche*, however, is annual. *Lathraea, Boschniakia, Conopholis,* and *Sopubia* are perennial. *Aeginetia* may become so, and *Tozzia* is intermediate. The advanced nature of perennial species is thus more clearly visible in parasitic figworts. The short life cycle of *Orobanche, Epifagus,* and many other broomrapes may, in fact, be a case of reversion.

5. *The shift from secondary lateral haustoria to primary terminal ones.*—In the broomrapes the radicular apex of the seedling becomes directly transformed into the haustorial organ; this process is anticipated in the germination of *Striga*. In the broomrapes there is a trend toward the elimination of haustoria on the secondary roots, and even these roots themselves, leading to the condition found in *Boschniakia*.

The two major types of haustoria, primary and secondary, have a completely parallel evolution in two unrelated groups, Loranthaceae and Orobanchaceae. Primary haustoria are derived directly from the apical meristem of the radicle; secondary ones, from subterminal parts of lateral roots. The most primitive Loranthaceae have no primary haustoria; these organs have evolved only in more advanced species. A genus like *Struthanthus* has both primary and secondary haustoria. Some of the most advanced mistletoes of this family lack secondary roots and rely on the primary haustorium. Broomrapes, also, have developed a primary haustorium which, even in species developing secondary haustoria, is always the main absorptive organ. In some broomrapes secondary roots are differentiated but are not haustorial in function; in others, probably more advanced, such roots have disappeared. In the mistletoes this shift to the primary haustorium may be regarded as an adaptation to the epiphytic habitat, but no such explanation will hold for broomrapes. Perhaps in both families the primary haustorium represents a more direct, a more efficient development of haustorial contact.

6. *Anther evolution.*—Anthers are rather uniform throughout these parasites, but in *Buchnera* one of the two anther cells has decreased somewhat in size (Pennell, 1935). In *Harveya* of South Africa one cell has become a sterile spur, and the same is true of *Aeginetia* and *Christisonia*. In *Hyobanche* no trace remains of the second cell, the first having attained a considerable size. I am not suggesting that the sequence above is necessarily an evolutionary one; yet, in the context of the other evolutionary trends, there appear to be no serious objections to the possibility that *Christisonia*, and perhaps even *Aeginetia*, are derivatives of *Harveya*. Should this conception be valid, the familial status of Orobanchaceae would be further undermined.

7. *Septation of ovary; placentation.* — Nearly all broomrapes have unilocular ovaries; figworts have bilocular ovaries. The ovary of *Christisonia* (Orobanchaceae) is two-celled below and one-celled above. In parasitic figworts a tendency toward the one-celled condition has been noted in *Lathraea, Orthocarpus,* and others. Indeed, the same trend appears in other parasites as well (Boeshore, 1920). Placentation has shifted from axile to parietal, from figworts to broomrapes: this trend has been reported from *Orthocarpus*. The parietal placentae of some broomrapes (*Aeginetia, Christisonia*) are elaborately branched; in others, simply bilobed. The shift to parietal placentation is, of course, a corollary of the disappearance of the septum.

8. *Reduction of the embryo.*—The parallel between advancing parasitism and decreasing size of seed seems to hold true generally, but the large seeds of *Rhinanthus, Melampyrum,* and even *Lathraea* and *Tozzia* are exceptions. Nevertheless, the broomrapes have the smallest seeds of this alliance, perhaps followed by *Striga,* also an advanced parasite. Beyond these, however, the trend is obscure, for it is difficult to evaluate relative degrees of para-

sitism among various green figwort genera. Other factors not directly related to the parasitic way of life may also have influenced the size of the seed.

The degree of differentiation of the embryo reinforces the trend in seed size. In all normally photosynthetic parasitic figworts the embryo has fully differentiated cotyledons and a radicle with root cap. In *Tozzia* and *Lathraea* the embryo is far more rudimentary. The broomrape embryo is only a small globe of parenchymatous cells; cotyledons and root cap are never distinguishable.

9. *Number of seeds.*—With the exception of several genera, the number of seeds increases with more advanced parasitism. This increase is related to other trends (discussed below). The broomrapes are at the summit of this evolutionary development, producing astronomical numbers of seeds per flower.

10. *The need for a host stimulant in germination.* —In most parasitic figworts, germination does not seem to require stimulus by the host, although other rather complicated environmental conditions may occasionally be necessary. In four known genera, however, and in the broomrapes, the seed normally germinates only when stimulated chemically by an exudate from the roots of certain vascular plants. These four genera, *Lathraea, Tozzia, Alectra,* and *Striga* (Heinricher, 1894, 1901*a*, 1930; Botha, 1948; Sunderland, 1960*b*), are among the most advanced parasitic figworts. The necessity of such a germination stimulant, therefore, shows a clear relationship to advanced parasitic status.

These evolutionary tendencies, which have spaced the various genera on a multidimensional system of axes of advancement, are not strictly isolated from one another but show several mutual connections. The lack of embryo differentiation is certainly in part the outcome of reduction in the size of seeds. The small size of seeds, in turn, is obviously related to the greatly increased number of seeds produced per flower. In the most advanced genera, vast numbers of seeds must be produced because of the small chance that each seed has of finding itself within the immediate rhizosphere of a suitable host. Germination takes place only within the first several millimeters of certain host roots, and only in the younger portions of these. In an evolutionary sense this germination mechanism represents a higher degree of host-dependence and loss of photosynthetic capacity which has, among other things, eliminated the brief self-sustaining phase of the seedling in search of a host.

5

Rafflesiaceae, Hydnoraceae, and Balanophoraceae

The plants discussed in this chapter are among the most extraordinary products of the plant kingdom. For this reason, and for the sake of convenience, they are combined under a single chapter heading. Although Rafflesiaceae and Hydnoraceae are related—they have sometimes been reduced to a single family—they have no recognizable affinities to Balanophoraceae. All plants of the three families are holoparasitic, and most of them attack only the roots of the host plant.

RAFFLESIACEAE

In 1822 the botanical world was astonished by the description of a most incredible flowering plant. An inhabitant of the jungles of Sumatra, the plant was nothing less than a flower of gigantic dimensions emerging, nearly leaflessly, from the thin root of a liana. *Rafflesia arnoldii,*[1] as the plant was called, has attracted the attention of botanists ever since, not only because it has the largest known flower (nearly a meter in diameter) but also because it represents the epitome of reduction in the higher plants. Evolution had stripped this parasite of irrelevant organs, leaving only an almost mycelial haustorial system and reproductive organs. The nearest parallel to *Rafflesia* might be in the genus *Sacculina* of Crustacea. No wonder some botanists refused to accept such plants as having biological individuality. Thus Trattinick (1828) referred *Rafflesia* and other parasites to a special group which he compared to an asylum. Such plants are merely insane vegetable productions, Trattinick said: to use his own phrase, they are "vegetabilische Verrücktheiten" (see chap. 1).

The family which derives its name from this genus (Rafflesiaceae) is almost entirely tropical. Only a few plants emerge into subtropical or Mediterranean climates. The best known of these is *Cytinus hypo-*

cistis of the Mediterranean, a plant which has been known at least since the time of Dioscorides. A root parasite on *Cistus* species, its flowers are bursts of brilliant yellow and red among the green host foliage. The same genus occurs in South Africa and Madagascar. In southern Japan another root parasite, *Mitrastemon,* has its northernmost station (Jochems, 1928). The smallest plants of the family are found in the genus *Pilostyles,* which parasitizes leguminous shrubs in arid regions of the Near East, parts of East Africa, Western Australia, the American Southwest and adjacent Texas, Mexico, Guatemala (Standley and Steyermark, 1946), and similar areas in South America even as far south as the Straits of Magellan (Schmucker, 1959). All others, however, are of tropical distribution. The more purely tropical genera of the Old World are famous for their large flowers; in the neotropics there is only the small-flowered *Apodanthes,* and *Bdallophyton* and *Mitrastemon* with medium-sized flowers. *Bdallophyton* is the only American endemic; *Pilostyles, Apodanthes,* and *Mitrastemon* (Matuda, 1947) are represented in the Old World also. *Rafflesia, Sapria, Rhizanthes,* and *Cytinus* are restricted to the Old World, the first three genera to the Indo-Malayan region.

The family is notoriously difficult to work with, as normal collecting techniques are usually inadequate. About all one may expect by way of a herbarium specimen of *Rafflesia* is a thick radial slice of the flower, which turns black upon drying. When the number of species in the family is given as about fifty (Harms, 1935), therefore, we can be certain that this is a very rough approximation. Considering the usual course of events when tropical groups are belatedly monographed, it is perhaps not too conservative to reduce the estimate of Rafflesiaceae to about thirty species, distributed over seven or eight genera.

Emergence of Flower

With regard to the way and place of emergence of Rafflesiaceae from their host, very few species have been investigated, but it now seems possible that the flowers or inflorescences of the family are unusual in developing endogenously. By this is meant that these organs originate not just within host tissues but even within the parasitic tissue,

---

[1]Winkler (1927) contends that no true *R. arnoldii* Brown has been collected since the (staminate) specimen was described in Brown's original paper (1822), and that the (pistillate) material forming the basis for Brown's second paper (1834) belongs to another species, *R. titan* Jack, which also has not been collected since. The plants referred to these two species by subsequent authors are said to belong to the related *R. tuan-mudae.*

rather than on its surface. The growing bud thus, at its inception, separates itself schizogenously from the thin superficial layer of parasitic tissue. Later the bud breaks through this protective shell and the overlying host tissues to emerge from the root or stem it inhabits. Both these ruptured tissues are difficult to identify in a plant in flower or fruit, as they form irregular flaps around the base of the parasite. In the large-flowered genera the adjacent host tissues seem to receive a considerable amount of stimulation and may form the cupule which supports the flower. This cupule bears curious reticulations in *Rafflesia* which may be comparable to lenticels. No such hypertrophy takes place on hosts of *Pilostyles* and *Apodanthes*.

Endriss (1902) specifically denies an endogenous origin of the flowers of *Pilostyles ulei,* and was unable to find remnants of the protective membrane. However, I know from personal experience with *P. thurberi* how deceptive this situation can be. In the latter species there certainly is a delicate protective membrane, intimately fused with the adjacent host tissues. In other words, the flower of at least this species, and of *P. haussknechtii,* are truly endogenous. To these may be added, upon the authority of Solms-Laubach (1867-1868, 1876), the genera *Rafflesia, Rhizanthes,* and *Cytinus* (fig. 5-2). No wonder Solms-Laubach believes that this extraordinary developmental feature applies to all members of the family. A reevaluation of the floral development of *Pilostyles ulei* would seem to be required.

Although the host organs attacked are usually roots, there are some exceptions. Even the large-flowered tropical species are occasionally found on stems. In many tropical forests it is very difficult to determine the morphological nature of the parasitized organ. Some of the Rafflesiaceae in question may actually emerge up to 4 meters above the forest floor (fig. 5-1,*b*; v. Steenis, 1934). Heinricher (1917*a*) had earlier shown the ability of *Cytinus hypocistis* to invade the host stem from the root. Indeed, in other species of the same genus in Madagascar a position on the lower part of the host trunk appears to be common (Baker, 1888). In *Apodanthes* and *Pilostyles* only the branches of the hosts are attacked.

In genera like *Rafflesia, Sapria,* and *Rhizanthes* the flowers emerge individually, although buds are often present nearby. We must assume that in most such instances adjacent flowers belong to what genetically is the same individual. What applies to the flowers of these three genera applies, from what I can learn, equally to the inflorescences of *Bdallophyton,* several of which may appear from a single length of host root. In *Cytinus* small groups of inflorescences spring from one or more nearby swollen roots. Here again, it is probable that several roots are invaded by an endophyte derived from a single seed.

In *Pilostyles thurberi* a great mass of flowers appears along a limited portion of the host (fig. 5-1,*a*).

We may assume that this colony represents a single individual. Indeed, a gradual expansion of the parasite away from the center into both downward and upward directions can be traced. Soon the central area, the apparent court of entry, is left barren as no more flowers emerge. Numerous scars remain on this portion as reminders of the flowers produced in the past. Soon the infection ramifies into nearby branches of the host; the floriferous areas eventually become separated, but now progress only acropetally. We can speak of a reproductive zone, therefore, maintaining a certain distance from the shoot apices of the host. In a number of American species of *Pilostyles* and *Apodanthes*—if these genera are, indeed, distinct—a very similar condition prevails, and the same is true of *P. hamiltonii* from Australia (Smith, 1951) and of *P. aethiopica,* where flowers occur only on branches one and two years of age (Soyer-Poskin and Schmitz, 1962). Herbarium material, unfortunately, is rather limited as a source of this kind of information. I would suggest that field observers look for any regularity in the position and pattern of the flowers of these parasites on the branches of their hosts.

The most interesting pattern of shoot emergence is known from *Pilostyles haussknechtii* (Solms-Laubach, 1874*a*). The infected part of the host, *Astragalus,* shows a regular zonation; areas bearing the parasite's flowers succeed those without. Flowers are thus restricted to shoots of a certain age, and actually only to the basal part of that year's growth. Flowers emerge only on the very base of the host's petiole, one on each flank. Thus the number of flowers per host leaf and their position show a remarkable constancy. It is interesting that this emergence pattern is correlated with an endophytic system which reaches right into the apical meristem of the host. Here we may glimpse an uncanny synchronization of growth rhythms of host and parasite. The ultimate parasitic cells are found above the subapical region of elongation and thus are carried upward with the apical meristem which is permanently infected. Through its presence in the host's meristematic region the parasite can select, as it were, the places where flowers will be produced —hence the regularity of flower distribution along the host branch. In my study of the parallel situation in the mistletoe genus *Arceuthobium,* I called this unusual adaptation to the host "isophasic parasitism" (Kuijt, 1960; see chap. 7).

The work of Watanabe (1936-1937) has revealed that another genus of Rafflesiaceae also shows an isophasic pattern. At least some of the Japanese species of *Mitrastemon* are able to induce broomlike malformations on the roots of their hosts, *Quercus* spp. These brooms, although a production of the root system, are again reminiscent of *Arceuthobium* and its brooms. Infection brings about a decrease in apical dominance in the host root. Simultaneously, lateral roots are formed which radiate in all horizontal directions; in fact, the distinction between lateral

FIG. 5-1. Smallest and largest members of Rafflesiaceae: *a*, *Pilostyles thurberi*, on *Dalea emoryi*, Yuma Co., Arizona; *b*, *Rafflesia "arnoldii"* (probably *R. tuan-mudae*) on a liana, Lebong Tandai, South-West Sumatra (courtesy Director, Hugo de Vries Lab., Amsterdam; photo, Dr. H. Schaefer).

and main roots is lost. The resultant horizontally spreading root system, its component members somewhat shorter than normal, form the flattened broom which runs just below and roughly parallel to the surface of the ground. Proximity to the surface is, of course, a necessity for the parasite's reproductive activities.

The *Mitrastemon* flowers which originate from this remarkable malformation do so for only a short distance, beginning about 10 cm from the root tip and extending 30 cm beyond the root tip. Here also we see a reproductive zone, just within the periphery of the broom. Flower scars mark the more central, older portions. At flowering time a fairy ring of *Mitrastemon* flowers springs from the forest floor.

In contrast to the situation in *Pilostyles* and some *Arceuthobium* brooms, the endophyte of *Mitrastemon* brooms probably does not reach into the apical meristem of the host. Watanabe was unable to find the parasite's filaments above a root thickness of a millimeter. (I return to the absorptive system of Rafflesiaceae in chap. 7.)

Flower Structure

In only two genera in this family can we speak of an inflorescence, and even here only the simplest type of spike occurs: in some species of *Cytinus* and in *Bdallophyton* where flowers are axillary along the upper part of the stem, closely appressed to one another (see figs. 5-6,*a* and 5-10,*b,c*). Other species of *Cytinus,* and *Mitrastemon* have a well-differentiated stem with leaf scales, often basally branching in the former, terminating in a single flower (see figs. 5-4c and 5-5,*a,b*). This condition in a greatly contracted fashion predominates in all the remaining Rafflesciaceae. Even in *Cytinus* and *Mitrastemon* the leaf scales are scarcely different from the petals. In the sessile-flowered genera the scale leaves have become so inconspicuous as to escape the notice of the casual observer. They are crowded around the base of the flower. Their number and arrangement are not always accurately known. In *Pilostyles* they occur in two whorls of two to six leaves each. *Rafflesia* and *Rhizanthes* have three whorls, each five-membered. Phyllotaxy is whorled except for *Cytinus* and perhaps some *Bdallophyton* (see fig. 5-10,*c*), although in the latter the whorls may simply be many-membered.

The petals are usually four or five in number, although *Rhizanthes* petals seem to have been subdivided into variable numbers of petal-like organs, sometimes up to sixteen. *Bdallophyton* has up to ten petals, possibly arrived at in the same way. *Rhizanthes* is exceptional also in having its young

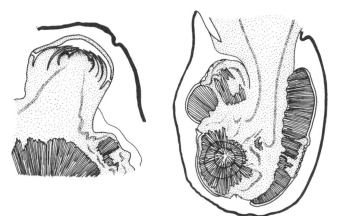

FIG. 5-2. *Cytinus hypocistis*, haustorial system and endogenous formation of inflorescence (after Solms-Laubach, 1867-1868).

petals joining in a valvate manner, while those of others are imbricate. All the larger flowers are fleshy, *Rafflesia* extremely so. The divisions between petals are always rather deep, infrequently reaching the ovary. A peculiarity found only in some species of *Rafflesia* is a circular groove around the base of the central column. The presence of a bulging margin above and/or below this groove is of some systematic importance.

Two of the large-flowered genera (*Rafflesia* and *Sapria*) have an additional modification of the perianth known as the diaphragm. This term is aptly chosen. The diaphragm (see fig. 5-9) is a rimlike extension of the petals, somewhat like the corona of a daffodil but hemispherical in shape and extending across the flower, leaving a central aperture just above the central column of the flower. The inner floral chamber in *Sapria* is strengthened by riblike outgrowths (see fig. 5-4,*d*).

Along the inner surfaces of this floral chamber and the petals above, as well as on top of the diaphragm of *Sapria*, is a variable pattern of trichomes. These appendages are stalked, glandular organs, the tip of which often is divided. The possible significance of these "ramenta," as they are called, is referred to below. *Rhizanthes* also has a dense inner covering of trichomes, but they are more hairlike. *Cytinus* has many stalked, multicellular glands, but other Rafflesiaceae are glabrous.

The flowers of Rafflesiaceae range from a rather dull, dark red-brown color in *Pilostyles* and *Apodanthes* to extremely complicated patterns of contrasting colors in *Rafflesia*. The most brilliant colors are known from *Cytinus* with penetrating hues of yellow, orange, and red but also purple (Marloth, 1913). *Sapria himalayana* has a bright red and white color combination reminding one of *Amanita muscaria* of more temperate regions. The outlines of the color pattern in *Rafflesia* are best seen from the photographs (fig. 5-9). The marbled effect on the

petals is caused by the juxtaposition of light-colored, irregular, and somewhat raised patches on a dark background which is variously described as brick-red, dark brown, or purple-brown. The diaphragm is somewhat lighter, and its raised white patches are smaller and more regular. On the inside of the diaphragm, however, the white patches are large and luminously clear, becoming smaller toward the aperture and eventually coalescing into a white ring.

Most Rafflesiaceae appear to be dioecious. Perfect flowers, however, characterize *Mitrastemon* and probably *Rhizanthes*. Perfect flowers have been reported in isolated instances in *Rafflesia*, and both unisexual and perfect flowers are known from *Bdallophyton* (Harms, 1935). The flowers of one sex often show rudimentary structures reminding one of flowers of the other sex (see below). All evidence, therefore, seems to indicate that the dioecious Rafflesiaceae have achieved this condition rather recently, and via perfect flowers.

The sexual parts of the flower of Rafflesiaceae have undergone so much fusion and elaboration that the resulting compound organ, the central column, has caused morphologists much difficulty in interpretation.

The male flowers of *Pilostyles* may form a starting point for this purpose. From the base of the flower of *P. thurberi* there arises a blunt, cushion-shaped emergence with a central depression (fig. 5-3,*a*). Around the flanks of this short central column are three rows of minute spherical to ovate anthers. There is a fringe of large vesicular unicellular hairs just above the anthers. In *Cytinus* the anthers are longer, extending upward along the column and projecting a short spur beyond. Those of *Bdallophyton* are even longer, with attenuated spurs, giving the appearance of a crown (fig. 5-4,*a*).

In the large-flowered genera a progressive elaboration of the androecial apparatus has occurred. The top of the column has greatly expanded sideward, at times in the shape of an elegant goblet (fig. 5-4,*d*). The expanded part of the column is called the disk. Below its margin, in a single ringlike series, the spherical anthers are arranged. In *Rafflesia* the anthers are sunk in individual chambers separated from adjacent ones by a small partition.

The structure of the anthers is imperfectly known. Those of *Bdallophyton*, *Mitrastemon*, and *Cytinus* are clearly two-celled, and this is apparently true also of the spherical anthers of *Sapria* and *Rhizanthes*. In the first three genera, dehiscence is by means of a longitudinal slit; in the other two, by means of a pore. The anther of *Pilostyles* is one-chambered, opening with a transverse slit, and is perhaps more comparable to a pollen sac. *Rafflesia* has anthers which contain many pollen-bearing chambers, some interconnected but all emptying through the same distal pore. Pollen in this family is devoid of sculpturing, and may have either three or four colps; the apertures are very obscure or absent (Erdtman, 1952). A peculiarity of *Rafflesia* and *Rhizanthes* is the fact that pollen is not powdery but is exuded in a slimy mass which may be derived from the two or three layers of tapetal cells.

From this account two plants have purposely been omitted. The first is *Pilostyles holtzii* Engler, the male flower of which is represented in figure 5-3,*d*. Here we find a staminal cylinder or tube bearing one or two rings of globose anthers on its rim. The staminal tube is completely separate from the central column, which appears to be only a remnant of the gynoecium. In another species, *P. ulei*, a zone of fusion between the central column and the staminal tube is said to be still recognizable (Endriss, 1902). At any rate, it probably is in this fashion that anthers of most Rafflesiaceae have achieved their present position just below the margin of the disk. That this should be so is indicated also by the fringe of vesicular hairs just above *Pilostyles* anthers, representing trichomes of the same type as those on the stigmatic margin of a female flower (fig. 5-3,*a*, *b*). In other words, the disk of Rafflesiaceae is in origin a stigmatic disk, wherever it occurs.

The situation in *Pilostyles holtzii* also provides a logical interpretation for the staminal tube of *Mitrastemon* (fig. 5-5,*a*, *b*). This organ is no more than a greatly lengthened tube of the same sort. Instead of bearing one or two terminal rings of anthers, the staminal tube of *Mitrastemon* has a very large number of pollen chambers just below the summit. The tube is very nearly closed at its apex, leaving only a minute aperture. Pollinating insects do not, as one might expect, approach the stigma through this aperture. The flower is protandrous and, after the pollen has been shed, the expanding young ovary lifts off and discards the entire staminal tube, rendering the stigmatic surfaces accessible to pollinators.

The ovary of Rafflesiaceae is inferior except in *Mitrastemon*. Within the ovary the numerous ovules are arranged in varied patterns (fig. 5-7). *Mitrastemon*, a convenient starting point in this regard, shows a series of radial lamellae which bear the ovules on their surface and nearly reach the center. The *Bdallophyton* ovary may be similar in structure. If we imagine a mutual fusion of lamellae in the axial region we have the condition of *Rhizanthes*. More complex fusion and ramification, leaving a multitude of ovuliferous chambers, many interconnected, seems to have taken place in *Sapria* and, indeed, in *Rafflesia*. A contrasting tendency is seen in the small-flowered genera. *Cytinus hypocistis* still has recognizable placental lamellae which join in the upper part of the ovary but recede greatly near the base. *Apodanthes* has parietal placentation, usually in four longitudinal placentae. In *Pilostyles* the ovules are distributed all along the inner ovarian surface. The new species of *Pilostyles* described by Rose (1909) illustrate several intermediate stages between the last two types. Dr. R. Rutherford tells me that the inner topography of the ovary of *P. thurberi* is quite variable.

I am not implying the evolutionary descent of

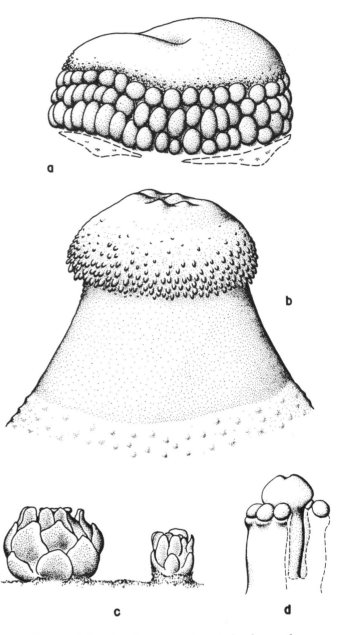

Fig. 5-3. *a* and *b*, *Pilostyles thurberi*, central column of male and female flower (Rutherford No. 278, Yuma, courtesy of collector; × 50); *c*, *P. thurberi*, female (left) and male (right) flowers emerging from same host branch (Rutherford No. 282, courtesy of collector; × 5); *d*, *P. holtzii*, central column of male flower, with part of staminal tube removed (after Harms, 1935; × 15).

Sapria, via *Rhizanthes*, from *Mitrastemon*, but merely wish to demonstrate the possible origin of the two very divergent placentation types of *Sapria* and *Pilostyles* from a common ancestral type. It is interesting that the former type has developed in very large-flowered species; the latter, in very small-flowered ones.

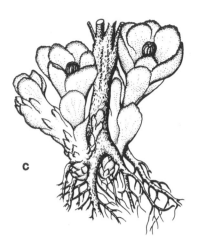

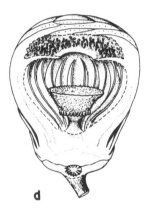

FIG. 5-4. *a*, *Bdallophyton*, female flower and bract, collected in Mexico (courtesy J. Henrickson); below; staminal column of same species (both × 2.5); *b*, *Cytinus hypocistis*, Monchique, Portugal, two views of stigmatic surface (× 10); *c*, *C. dioicus* (after Marloth, 1913, by permission of heirs of R. Marloth; × ½); *d*, *Sapria himalayana*, sectional view of flower (after Hosseus, 1907; × ½).

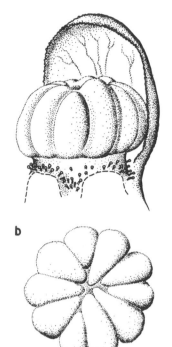

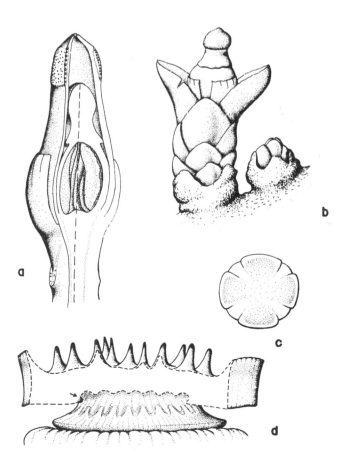

FIG. 5-5. *a* and *b*, *Mitrastemon yamamotoi*: *a*, flower in male phase, partly cut away; *b*, emerging bud, and open flower in female phase, the androecial tube having fallen away (after Makino, 1911); *c*, *Bdallophyton*, pollen grain as seen in polar view (Purpus No. 8546, Vera Cruz, Mexico); *d*, *Rafflesia* "*arnoldii*" (probaby *R. tuan-mudae*), central column of male flower, partly cut away; anthers borne in groove indicated by arrow (after Justesen, 1922).

FIG. 5-6. *a, Cytinus ruber* parasitic on *Cystus incanus,*
Elba; bracts are red, corolla pure white (courtesy
Director, Hugo de Vries Lab., Amsterdam; photo
F. D. Boesewinkel); *b, Rhizanthes* sp. from Mount
Sago, West Sumatra, ca. 1,000 ft. altitude, two buds,
one nearly mature (courtesy Dr. W. Meijer; photo
Arch. Flora Malesiana).

The ovules, so far as they are known, are simple
in structure. A difficulty here is that, at least in
Rafflesia and *Rhizanthes,* the ovules of even an open
flower may not yet have achieved full differentiation
(Solms-Laubach, 1898). In *Cytinus baroni* the
ovules near the apex of the ovary mature first, and
maturation of the remaining ovules progresses downward (Baker, 1888). Development is resumed after
pollination. In *Pilostyles* we find the most complete
ovules (Solms-Laubach, 1874,*b*; Endriss, 1902; Kummerow, 1962) with both integuments represented.
The outer integument even here is variable in prominence. In *Cytinus* it is a small ringlike swelling at
the chalazal end of the ovule. In *Rafflesia* we can
hardly speak of an outer integument, although the
chalazal swelling on the long, curved funiculus is
thought to be homologous to it (fig. 5-8,*a*). The
nucellus consists of a single layer of cells. In development and type of embryo sac it is surprising to
find that these greatly advanced plants behave in
the same fashion as the majority of angiosperms,
resulting in a normal eight-nucleate female gametophyte (Ernst and Schmid, 1913).

In *Bdallophyton, Cytinus,* and *Mitrastemon* the
receptive stigmatic area is radially grooved or par-
titioned. In the large-flowered genera the central
column in the female flower (as in the male) terminates in a large, flaring cup or disk. The upper and
lower surfaces of the disk of *Sapria* are covered with
hairs. In *Rafflesia* the slightly curved, nearly vertical
margin of the disk and the gutter-like depression
just above it are smooth. In *R. arnoldii* and related
species another interesting feature has received
much comment. The top of the disk is occupied by
great, blunt, dark, spinelike processes thought, in
some, to be arranged in several whorls. These spines,
known in the literature as "processus disci," appear
to be somewhat branched and are of various sizes
and irregular arrangement in the original (staminate)
flower of *R. arnoldii.* In others, peculiar multicellular trichomes form small brushes at the apex.

It is very difficult to be certain about the function
of any parts of the female central column of these
Rafflesia species. What constitutes the stigmatic surface? It was originally thought that the spines of
the disk were functional stigmas. This now seems
not to be so (Solms-Laubach, 1876) and we are
at a loss to suggest a function for these decorative
outgrowths. The possibility of their stigmatic origin,
of course, cannot be excluded. With regard to the
functional stigmatic area, the best suggestion seems
to be the smooth outer margin of the disk. It is here
that germinating pollen has actually been observed
in *R. tuan-mudae* (Winkler, 1927).

Flower Biology

It is not surprising that the pollination of such
flowers should have interested even the earliest observers. The discoverer of *Rafflesia arnoldii,* Arnold,
spoke of the carrion-like odor of the flowers, and saw
swarms of flies around them, some laying their eggs
on the floral tissues. But later field observers have
communicated contradictory reports on these points.
Some have emphatically denied both visits by flies
and a carrion-like odor. Justesen (1922), on the
contrary, studying what probably is *R. tuan-mudae,*
observed flies moving from one flower to the next,

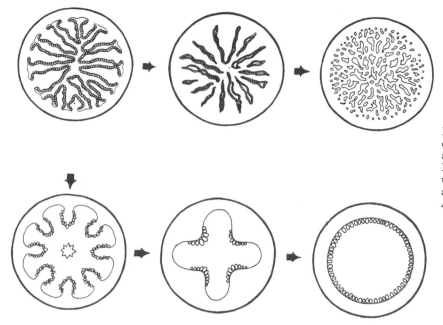

FIG. 5-7. Rafflesiaceae, transections of ovaries showing placentation types and possible derivation (based on Harms, 1935). From left to right: upper series, *Mitrastemon, Rhizanthes,* and *Sapria;* lower series, *Cytinus, Apodanthes,* and *Pilostyles.*

seemingly attracted directly to the staminal grooves, and Bünning (1947) also saw flies in the evening. At least a weak smell of decaying flesh is described by many botanists who have had the good fortune to see fresh flowers of *Rafflesia* under natural conditions. The whole matter is carefully summarized by Winkler (1927), from whom the present data are taken. One possible explanation for the contradictory reports is variation among flowers; another is limitation of the odor to a certain period of the flower's life.

Also related to floral biology is the extremely slow development of the flower. A large specimen of *Rafflesia,* grown from seed in the famous Botanical Gardens at Buitenzorg, now Bogor (Teijsmann, 1856), took one and a half years for several buds to emerge. Yet the functional duration of the flower is probably only two or three days.

The whole reproductive biology of *Rafflesia,* in fact, appears to be unusual. The rarity of the plants is emphasized in nearly every article dealing with these parasites. Their development is very slow and their effective duration is brief. They seem to be limited to Vitaceae (they are known with certainty only from *Cissus* and *Tetrastigma*), further restricting their distribution. There are several references to the effect that sex distribution is often very unbalanced: a local population may consist of flowers of one sex only, or very nearly so (Ernst and Schmid, 1913; Justesen, 1922). Yet fertilization of the extremely large number of ovules in each flower would require a large volume of pollen to be transferred. All these considerations show sexual reproduction in *Rafflesia* to be a narrow bottleneck. Unless apogamy plays a role, only a very precise or efficient mechanism of pollination can restore a precarious balance.

The flies observed by some botanists might be merely incidental to pollination. The fleshy corolla

FIG. 5-8. *a, Rafflesia patma,* longisection of mature ovule (after Ernst and Schmid, 1913; × 70); *b, R. "arnoldii"* (probably *R. tuan-mudae*), seed (after Brown, 1834; × 50); *c, Pilostyles thurberi* longisection of mature seed (Rutherford No. 25 6-B, California, from a preparation by the collector; × 180).

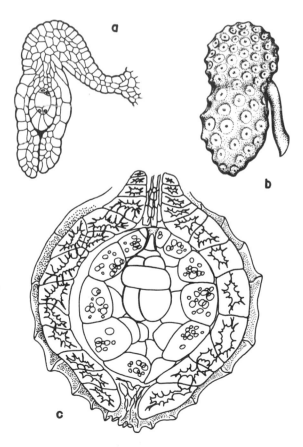

FIG. 5-9. "Queen of Parasites," said to be *Rafflesia arnoldii,* but more probably *R. tuan-mudae;* courtesy Dr. W. Meijer, who took the photographs on Mount Sago, West Sumatra, at 3,000 ft. altitude. Note especially 'processus disci" in lower photograph, and luminous white patches on inside of diaphragm in upper.

It is interesting to note, in passing, how the light-transmitting "windows" in the pitcher of *Darlingtonia* and others are duplicated by the clear white patches on the inside of the *Rafflesia* diaphragm. Justesen (1922) thought that these patches, "shining like diabolic eyes," would intimidate the flies that had already entered the chamber, and drive them to the reproductive area under the rim of the disk, but such a view is rather too anthropomorphic.

Yet we must not neglect any additional or auxiliary aspects of pollination in *Rafflesia.* While studying *Rafflesia* in Borneo, Winkler (1927) noticed that a female flower had filled with rain water during the night. The water level reached the margin of the disk, but within a few hours the water had disappeared. Since there are no leaks in such a massive flower, the water must have been absorbed by the flower. Similar observations were made on a male flower. Taking into account the frequency and regularity of rains, Winkler suggests that the rising and falling water level within the flower may be related to pollination. It is possible that flies brush off pollen on the ramenta and other trichomes of the floral chamber. The globular bodies on the ramenta tips have been thought of as food bodies attractive to insects. (The ramenta have also been thought to channel insects to the most important glabrous portion of the flower, the margin of the disk.) The pollen could be carried up from the ramenta by water to adhere to the margin of the disk, the presumed receptive surface. But the rising water level would probably not reach the anthers of some *Rafflesia* species, as the staminal grooves would remain filled with air even though the water level rose to the top of the disk. It is unfortunate that Winkler did not make the crucial test of finding out whether fresh *Rafflesia* pollen actually floats. Contrary to Winkler's report, Justesen (1922) observed mosquito larvae in standing water within the flower of Sumatran *Rafflesia,* which indicated no more than a very slow water absorption into the flower.

Pollination in other genera of Rafflesiaceae has received little attention. *Rhizanthes* flowers are said to be odorless when fresh (Heinricher, 1905). *Cytinus hypocistis* is said to be pollinated by bumblebees (Harms, 1935). In *Mitrastemon,* fly-pollination has been suspected, and Jochems (1928), indeed, has recorded the visits of fruit flies (*Drosophila*). Watanabe (1936-1937) has shown, however, that pollination by birds of the genus *Zosterops* may predominate. A great deal of nectar apparently accumulates in the cuplike upper bracts from which the birds drink.

Fruits, Seeds, and Dispersal

The fruit of most Rafflesiaceae is little more than an enlarged ovary relying on destruction or decay of the fruit wall rather than on a preformed dehiscent zone to set its seeds free. In *Pilostyles* it is scarcely possible to distinguish a fruit from a mature flower, since the brownish bracts and petals remain unchanged. Adjacent fruits of *Bdallophyton*

becomes a great mass of decaying matter which even in more temperate climates would attract flies. Is it possible that the carrion odor represents no more than an early degeneration of floral tissues, unrelated to pollination? In the related genus *Rhizanthes* the flower is at first odorless, but begins to smell unpleasantly immediately after its brief flowering period (Heinricher, 1905). Two aspects of the flower seem to be in conflict with this notion. First, some flowers emit a putrid odor even when fresh, although this odor never seems strong. Second, the elaborate color scheme of the *Rafflesia* is reminiscent of other fly-attracting plants such as some species of *Aristolochia* and *Stapelia,* as well as *Darlingtonia* and other pitcher plants, and must have the same function.

are basally fused into a compound fruit (fig. 5-10,*b* and *c*) in the manner of a pineapple. In the large-flowered genera, fruits are very inadequately known. *Rafflesia* fruit (fig. 5-10,*a*) has a large fleshy body with a coarsely cracked surface. The interior is a labyrinth of irregular passages and chambers, resembling that of the ovary but somewhat enlarged.

The seeds are extremely small. Even in the large *Rafflesia*, seeds are no more than a millimeter long and less than half a millimeter wide. The shape and contour of the seed reminds one of a two-seeded peanut (fig. 5-8,*b*).

Pilostyles has a slightly pear-shaped seed. Anatomically the seeds are characterized by a single compact layer of sclereids surrounding the embryo and endosperm. A small interruption of this layer occurs at the micropylar end. A similar interruption can be seen in the seed coat of *P. thurberi* on the chalazal pole (fig. 5-8,*c*). This is true also of *P. ulei* (Endriss, 1902), but has been denied for *P. berterii* (Kummerow, 1962). The sclereids have very heavy walls through which a complex system of pits extends.

In *Rafflesia* this sclerenchymatous testa is the outermost layer. The chalazal swelling, often regarded as a rudimentary second integument, also transforms itself to sclerenchyma. A similar body in *Cytinus* seems to be fleshy and may serve as an elaiosome. In *Pilostyles*, however, the second integument is said to grow into the single layer of thin parenchymatous cells covering the sclerenchyma layer (Endriss, 1902). Such a process would seem unlikely (see fig. 5-8,*c*); at any rate, it bears further study.

The endosperm and embryo are among the simplest in the flowering plants. The endosperm, always cellular at maturity, consists of a limited number of cells containing fatty materials. In the center of the seed an embryo is embedded which, in *Pilostyles* (fig. 5-8,*c*), consists of only eight cells. There are two cells which may be comparable to a suspensor, the lower one with a conspicuously thickened wall. Above this two-celled stalk are three tiers of two cells each, making up the embryo proper (Endriss, 1902).

Kummerow (1962) describes a different situation in *P. berterii* (see his fig. 16), in which the embryo is said to consist of many more cells in an irregular arrangement, embedded in a layer of compressed parenchyma cells supposedly representing endosperm and nucellus. I suggest that this author may have made an oblique section through a somewhat immature seed, thus missing both the embryo and the chalazal interruption of the sclerenchyma layer. The so-called embryo might then be endosperm in the process of absorbing the surrounding nucellar tissue.

The nature of seed dispersal in Rafflesiaceae is a mystery. It has been suggested that rats carry *Pilostyles* fruits or seeds (Kummerow, 1962), or that ants and termites break into the fruit and bear away seeds, or even that the seeds of *Rafflesia* are trans-

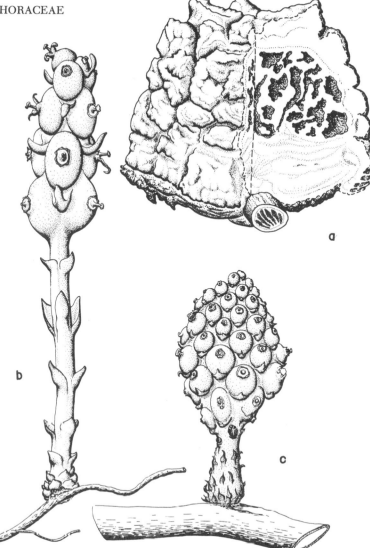

Fig. 5-10. Fruits: *a*, *Rafflesia* "*arnoldii*" (probably *R. tuan-mudae*), partly cut away to show interior (after Brown, 1834; × ½); *d*, *Bdallophyton americanum* (Morrison and Beetle No. 8762, Salvador, UC; × 1); *c*, *B. oxylepis* (Pringle No. 4373, Mexico, UC; × 1).

ported on the feet of wild pigs and elephants. *Rafflesia* must be the only plant said to be disseminated by both elephants and ants.

Similarly, nothing is known with regard to germination or the manner in which entry into the host is effected. Ule (1915) believes that birds eat the fruits of some *Pilostyles* species. The seeds voided by these birds would germinate on the soil to form a mycelial mat which, upon reaching a suitable host, would cause infection. Soyer-Poskin and Schmitz (1962) have reported that *P. aethiopica* seeds germinate even on wet paper. Kummerow (1962) suggests that the seeds of *P. berterii*, carried underground by rodents, would germinate through the root exudations of the host (as in *Oro-*

banche); the endophyte would then travel into the branches. Ule's notion can be discounted as too fantastic. Kummerow's suggestion is not impossible, but would seem erroneous in most cases. The distribution of *Pilostyles* flowers on the host usually indicates a court of infection on the branches, not on the root. In all honesty we should admit our ignorance about the events which fall between seed maturation and the development of the fully established endophyte.

If, in the preceding account, *Rafflesia* has received a disproportionate amount of attention, this is entirely excusable. *Rafflesia*, especially *R. arnoldii*, is the prima donna, not just of its family but of all parasitic plants. The performance and dimensions of this most exotic of flowers are so spectacular that the onlooker cannot fail to be impressed. To the botanist, the mysteries of the plant's existence become even deeper upon closer acquaintance. It is safe to predict that botanists will continue to make pilgrimages into the jungles of Sumatra to gaze upon this extraordinary creation of the plant kingdom.

HYDNORACEAE

The only two genera of Hydnoraceae, *Hydnora* and *Prosopanche*, are as remarkable as Rafflesiaceae. While even the most reduced members of the latter family retain some leaflike organs below the flower, these appendages are lacking in Hydnoraceae. In contrast, Hydnoraceae have a rhizome-like organ which is unknown in Rafflesiaceae. Hydnoraceae have never caught the popular botanical fancy as Rafflesiaceae have done. This is due to their restriction to remote and sparsely inhabited areas, but also to the size of the family. In all probability, there are no more than ten species in *Hydnora* and *Prosopanche* combined.

Hydnora, discovered by Thunberg in South Africa, reaches from there through various East African countries as far as the Ethiopian region. Its occurrence on Madagascar seems well established (Jumelle and Perrier de la Bâthie, 1912), but the *Hydnora* reported from the nearby island of Réunion may have been introduced. *Prosopanche* is restricted to the southern part of South America. One species, *P. bonacinae*, may be found anywhere from Patagonia into Paraguay and possibly southern Brazil. The only other species, *P. americana*, is confined to central and east Argentina (Cocucci, 1965). A single collection from eastern Peru requires confirmation. Notwithstanding their great spatial separation, both genera appear to prefer arid regions.

Subterranean Organs

The rhizome-like organs of Hydnoraceae which traverse the soil are known by a variety of names, thus attesting to the ambiguity of their morphological nature. Here I shall refer to them as roots because they are not articulated, and appear to have a caplike tissue protecting the apical meristem, even

though their structure is anomalous in some ways. They are coarse organs, 1 to 3 cm thick in *Prosopanche*, usually angular, but cylindrical in shape in *Hydnora solmsiana*. On their surfaces, masses of tubercular or vermiform outgrowths are concentrated along the ribs in the angular types, but cover the whole root in other types. Internally we find a cylinder-like arrangement of bundles in *H. africana*, but in some other species of the same genus several concentric cylinders of bundles have been described. *Prosopanche*, at first sight, seems to have four or five normal central bundles, with six bundles lined up in pairs, in a tangential fashion, in each rib (fig. 5-11). It may be, however, that all bundles together form a cylinder with great bays reaching into the ribs, toward the rootlike outgrowths occurring there. In the same genus are a number of secretory ducts of unknown function.

The small lateral organs emerging from these coarse roots are roots themselves. They are endogenous in origin and, even before emerging from the mother organ, show a root cap. Curiously, they are of limited growth and never branch, but form a coarse fringe on the ribs or are scattered over the surface of the large type of root. Schimper (1880) has emphasized that the two types of organ are essentially the same in origin and structure. Thus the root dimorphism is completely comparable to that of *Pholisma* of Lennoaceae, especially since the small roots of *Prosopanche* and *Hydnora* are apparently haustorial in function (Schimper, 1880; Vaccaneo, 1934). I venture therefore, to employ the same terms as I do for *Pholisma* roots (see chap. 6). The large roots which traverse the soil are called pilot roots, the small ones haustorial roots, reflecting the function to "search" for root systems of potential hosts for the former, and that of haustorial union for the latter. Whatever terminology is used, we find here a remarkably parallel evolution of the root systems of two unrelated parasitic groups.

It is from the pilot roots that flowers emerge, either near haustorial contacts (fig. 5-11,*a*) or distant from them (fig. 5-13,*a* and 5-12,*b*). The flowers are not preceded by foliar organs of any sort. A somewhat clavate bud appears among the haustorial roots. This bud breaks forth from the interior of the pilot root in an endogenous fashion, and proceeds to elongate upwardly. A bulbous lower portion often indicates the position of the ovary at an early stage. The young flower eventually reaches the soil surface, allowing scarcely more than its perianth lobes to extend above it.

Flower and Flower Biology

The flower which thus becomes visible to the casual observer is an extraordinary one. It is normally three- or four-parted; the petals are fleshy, with a coarse, cracked, brown exterior. The petals in at least some species part first at their lower end, producing a lateral floral entry for pollinators somewhat in the way of the small flowers of *Phyteuma* of Campanulaceae. The interior of the flower is pure white and

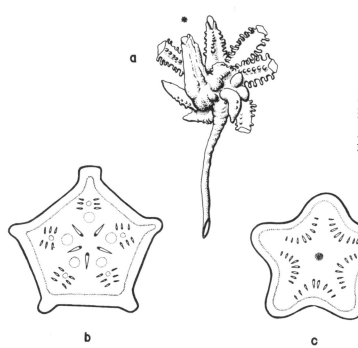

Fig. 5-11. *a, Prosopanche bonacinae,* young plant attached to host root, from which parasite's pilot roots reach out in all directions; basal part of flower indicated by asterisk (after Schimper, 1880; × ½); *b,* and *c, P. bonacinae* and *Hydnora africana,* transections of pilot roots (after Schimper, 1880).

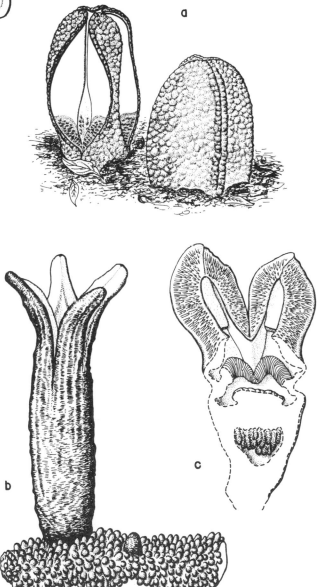

various hues of pink and bright red. Particularly striking is *Hydnora cornii* (Vaccaneo, 1934). The upper two-thirds of its petals are smooth and pure white; the sharply circumscribed lower third is bright red, occupied with bristles pointing into the flower (fig. 5-12,*a*). The entire length of the free petal in *H. africana* (fig. 5-12,*c*) is studded with similar bristles, together forming a cherry-red area in contrast to the rest of the flower. In *H. solmsiana,* which appears to open from the tip (fig. 5-12,*b*), the upper parts of the petals are smooth, but the interior of the flower is a pale salmon-red (Marloth, 1913). When bristles occur they are implanted upon the flat surfaces where adjacent petals were originally joined, and they always point into the flower.

These details of structure and color are obviously related to the flower's pollination. Within the upper part of the petal of *H. africana* (see fig. 5-12,*c*) is a pure white, fleshy body. In *H. solmsiana* and probably others this body is absent, but other floral tissues may have a similar function (Marloth, 1913). It is rich in proteinaceous matter and soon decays. Even before it begins to deteriorate and emit its putrid odor, however, it attracts carrion beetles and probably other beetles. These insects, once having entered the flower's interior, are discouraged from leaving by the bristles, and are thus guided down into the reproductive area. When the flower ages and its petals relax, the remaining insects are set free, and doubtless enter the next flower.

Fig. 5-12. *a, Hydnora cornii,* open and closed flower emerging from soil (after Vaccaneo, 1934; ca. × ½); *b, H. solmsiana,* root with open flower and bud (after Marloth, 1913, by permission of heirs of R. Marloth; × ½); *c, H. africana,* longisection of flower with 1/3 of flower removed (after Brown, 1834; × ½).

116

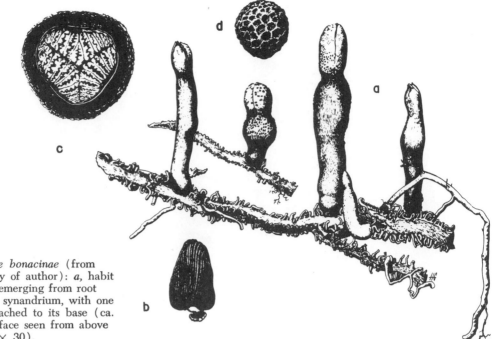

Fig. 5-13. *Prosopanche bonacinae* (from Cocucci, 1965, courtesy of author): *a,* habit of plant, with flowers emerging from root system (ca. × ½); *b,* synandrium, with one interstaminal body attached to its base (ca. × 1); *c,* stigmatic surface seen from above (ca. × 3); *d,* seed (× 30).

The flowers of the two *Prosopanche* species, although they do not have these bright colors, are equally effective in attracting beetles (Bruch, 1923). It is possible that the androecium itself emits the carrion-like smell, but the staminodial bodies directly below the androecium may also be involved. A species of beetle lays its eggs in the floral tube, and the larvae live on the fleshy parts of the plant, apparently without doing appreciable damage to the developing fruit. The adult, while depositing eggs, becomes laden with pollen.

Within the flower, near the entrance, is a very peculiar androecium. In *Hydnora* it is a ringlike compound organ adhering to the perianth tube, showing four bays for as many petals (fig. 5-12,*c*). In the median region of the petals this synandrium is widest; it is narrowest just below the sinuses. The ring thus formed is covered by elongated pollen sacs, in pairs and in a radial pattern. In some *Hydnora* species, however, the ring is nearly interrupted, and the component anthers are upright structures in the shape of a peaked cap. In *Prosopanche* these individual anthers approach each other and form a compound ovoid body (fig. 5-13,*b*) which now nearly closes the passage leading down to the stigmatic surfaces. Below the anthers of *Prosopanche* the floral tube bears bilobed fleshy appendages, of unknown function, one for each petal. They may be attractive to beetles who must force themselves through the center of the synandrium to reach the appendages.

Pollen of Hydnoraceae is very simple. As in Rafflesiaceae, it bears the mark of great reduction: the grains are without sculpture or other distinctive features, and are spherical or oblong (Erdtman, 1952).

The bottom of the floral cavity is occupied by the stigmatic surface. In *Hydnora* the stigma is in the shape of a flat cushion with as many lobes as there are petals, and of radial contours similar to those of the androecium. The stigmatic area of *Prosopanche* is scarcely differentiated, and is little more than the termination of the placental lamellae described below.

The ovary is inferior and unilocular, and very different in the two genera. The ovules of *Hydnora* are borne by individual much-branched placentae suspended from the roof of the cavity. *Prosopanche,* however, has radially arranged placental lamellae which, in contrast to those of *Hydnora,* fill the entire cavity. The continuation of these lamellae forms the receptive area.

The ovule in Hydnoraceae has aroused controversy which may not yet be resolved. Van Tieghem (1897) contended that the ovules, while first scattered over the surface of the placentae, secondarily became embedded in and fused with the latter because of growth of the placental tissue. De Bary (1868) and Chodat (1915), however, hold that true integuments are never present; they speak of the "false integument" said to be no more than the surrounding layer of placental cells. This notion would mean the development of embryo sacs directly in placental tissues, a schizogenous "micropyle" and a "false seed." The situation in *Hydnora,* however, points to ovules with single, massive integuments (Dastur, 1921-1922) and therefore a more normal development. Considering the close affinities between the two genera, we may assume that Van Tieghem's views will prevail in *Prosopanche* also.

Seeds, Fruits, and Dispersal

The seeds are minute and as hard as grains of sand (*Prosopanche*: Burkart, 1963). It has been estimated (Cocucci, 1965) that a single fruit of *P. americana* contains about 35,000 seeds. Information is lacking on *Hydnora,* but we may be assured of similarly

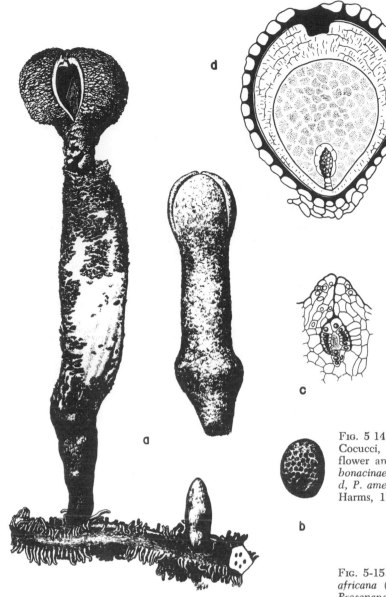

FIG. 5 14. *a* and *b*, *Prosopanche americana* (from Cocucci, 1965, courtesy of author), habit of plant in flower and bud (× ½) and seed (× 15); *c*, *P. bonacinae*, longisection of ovule (after Chodat, 1915); *d*, *P. americana*, longisection of mature seed (after Harms, 1935).

FIG. 5-15. Fruits of Hydnoraceae: *a*, *Hydnora africana* (after Brown, 1834; × ½); *b*, *Prosopanche americana*, (after photograph by Dr. A. E. Cocucci; × 1/3).

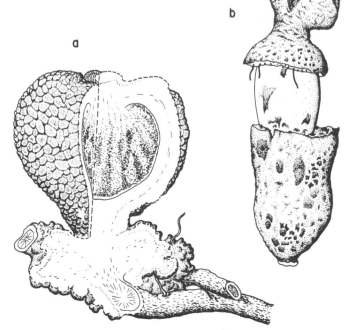

high numbers. It is the seed coat that makes the seed hard; the appearance of this layer is well shown in figures 5-13,*d* and 5-14,*b*. Below the seed coat is a layer of perisperm, presumably also derived from the massive integument. The perisperm of *Hydnora* is only one cell in thickness, but in *Prosopanche*, especially at the chalazal pole, it is more massive (fig. 5-14,*d*). The endosperm is a well-developed tissue characterized by thick walls. The undifferentiated embryo lies in a small cavity within the endosperm. The embryo's early development, at least in *H. africana,* is unusual in that a long chain of proembryonic cells is formed; only the middle ones grow into the embryo (Dastur, 1921-1922).

The fruits of Hydnoraceae are rather massive and fleshy. In *Prosopanche* the ovary simply enlarges to become the fruit and the lamellar organization remains unchanged. The ripe fruit bursts open in a circumscissile fashion, revealing the clear white flesh in which the seeds are embedded (fig. 5-15,*b*). The placentae of *Hydnora* enlarge to fill up the entire ovarian cavity. In both genera the placental tissues contain much starch, and it is not surprising that the fruits are eaten by a variety of animals. *Hydnora* fruits are eaten by baboons (a local South African name is "baviaanskost"), jackals, foxes, porcupines, and other animals (Vaccaneo, 1934). Various African natives, too, including Bushmen (Story, 1958) and Somalis (Vaccaneo, 1934), are not averse to eating them. On Madagascar, *Hydnora* fruits are regarded as among the best of native fruits (Jumelle and Perrier de la Bâthie, 1912). Although the quality of *Prosopanche* fruits seems to be lower, they are eaten by Argentine Indians, who consume them raw or fried (Cocucci, 1965). Armadillos also eat the fruit (Schmucker, 1959).

It is clear that mammals, including man, are the main agents of dispersal, attracted by the edible fruit. The seeds must thus be present in great numbers in the soil in a dormant condition. As in some other highly advanced root parasites, an exudate from the host root is probably required to start the seed's germination. We know as little about this part of the life cycle in Hydnoraceae as in Rafflesiaceae. A fascinating story awaits the botanist who is fortunate enough to have access to viable seeds.

BALANOPHORACEAE

There is little superficial resemblance between Balanophoraceae and the previous two families. Rather than having the large single flowers characteristic of the latter, Balanophoraceae are usually fleshy, club-shaped inflorescences of a strikingly fungoid appearance, emerging from the deep humus of moist tropical forests. Their colors range from pale yellows and browns through various hues of pink and red to deep purple, but green is lacking. These inflorescences which, following Hooker's (1856) monumental work on the family, may be called "capitula," bear myriads of flowers which are among the smallest in all the angiosperms. It is not surprising that these plants have often been regarded as fungi. Thus *Cynomorium coccineum* of the Mediterranean was known as *Fungus melitensis*, or Maltese fungus, during medieval times. Even Hooker, in 1856, found it necessary to devote much space in defense of the notion that Balanophoraceae were indeed Phanerogams.

Balanophoraceae show an over-all pattern of geographic distribution reminiscent of the mistletoes. The genera are restricted to either the New World or the Old World, although intergeneric affinities extend between the hemispheres. In neotropical regions we find the genera *Lathrophytum, Juelia, Ombrophytum, Lophophytum, Langsdorffia, Scybalium, Helosis,* and *Corynaea.* On the African continent there are five endemic genera: *Thonningia, Sarcophyte, Chlamydophytum, Mystropetalon,* and *Ditepalanthus,* although one species of the first genus is known from Madagascar, and *Cynomorium* per-

Fig. 5-16. *Cynomorium coccineum,* Portugal (courtesy Prof. Dr. W. Rauh).

haps occurs along the Mediterranean coast in Africa, as it does in Europe. In the rest of the Old World there are *Balanophora, Rhopalocnemis, Cynomorium, Exorhopala, Dactylanthus, Hachettea, Ditepalanthus,* and *Thonningia,* the latter two on Madagascar. Three of the genera are island endemics: *Ditepalanthus* (Madagascar), *Hachettea* (New Caledonia), and *Dactylanthus* (New Zealand). In Balanophoraceae as in many other plant families, Madagascar has strong ties with the Indo-Malayan area (*Rhopalocnemis* and *Balanophora*) and also with Africa (*Thonningia*).

The family is almost entirely tropical, with a distinct preference in many genera for moist, shaded forests at moderate to high elevations. *Juelia,* found at 3,800 meters in the Bolivian Andes, perhaps holds the record for altitude. *Cynomorium,* in contrast, is apparently restricted to dry coastal habitats and saline areas from the Canary Islands through the Mediterranean to the salt steppes of Central Asia and Mongolia (Chaney No. 41a, at UC). In South Africa, *Mystropetalon* and *Sarcophyte* venture into rather dry subtropical habitats, and the same may be said of *Dactylanthus* in New Zealand. Southern Japan is the most northerly station of *Balanophora.*

A great variety of families of flowering plants are

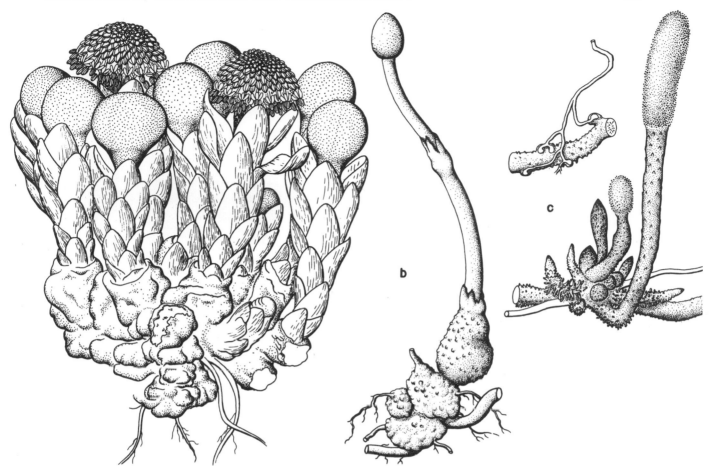

recorded as hosts for various species, but the data do not supply anything germane on host ranges, except that there is as yet no evidence in the family for any degree of specificity (see chap. 8).

Tubers and Rhizomes

The nature of the tuberous organ through which the parasite is in physiological contact with its host has been a matter of controversy. The tuber is least controversial in the genera *Helosis* (Umiker, 1920), *Scybalium, Lophophytum,* and *Dactylanthus* (Moore, 1940), where it appears to be made up of parasitic tissue only. In these genera the tissues of the parasite do not interlace with those of the host, but meet along a surface of rather simple contours. Continuity of xylem from host to parasite has been demonstrated for *Helosis* and may be assumed for the other three genera as well. The tuber is a variable, often irregularly lobed organ ranging from several centimeters to the size of a human head. Its surface is frequently coarsely cracked or tubercular; some species have conspicuous, lenticel-like pustules

FIG. 5-17. *a, Balanophora indica,* plant with male and female inflorescences (after Hooker, 1840; × ½); *b, B. involucrata,* male plant on oak root (after Hooker, 1856; no magnification given); *c, Cynomorium coccineum* on root of *Salsola vermiculata;* part of plant with inflorescences of various ages arising from haustorial contact of rhizome (cut ends) with host root (unshaded); haustorial contacts of rhizome with several branches of root of *Salsola* (after Weddell, 1858-1861; somewhat less than × ½).

and others lack them. Only in *Lophophytum* does it bear scale leaves (see fig. 5-22,*c*) but even here it may represent the base of the inflorescence, as in *Cynomorium,* or a transition between the inflorescence and the tuber.

The tuber of the remaining genera is of singular interest. This unique organ is discussed in detail in chapter 7, but a brief sketch is appropriate here. The *Balanophora* type of tuber is made up of tissues not belonging to merely one of the two partners, but an intricate and harmonious combination of

FIG. 5-18. *Balanophora wilderi* from Rarotonga, branched tubers (after Setchell, 1935).

both. In a mature tuber the vascular strands of the host reach far into this dual organ. This means, of course, that strands of the host have been stimulated to assume an aggressive character—a reversal of roles, as it were, since the host invades the tissues of the parasite.

Another matter that needs exploration is the extent of the endophyte within the host root. Is the parasite limited to the tuber? This is said to be true of *Balanophora globosa* and *B. elongata* (Harms, 1935), but in *B. reflexa* (Beccari, 1869) and *B. indica* (Solms-Laubach, 1867-1868) the endophyte has been demonstrated some distance from the tuber. Thus there emerges the possibility of vegetative reproduction and, consequently, an initiation of tuber formation by the endophyte within, rather than by a germinating seed. Could it be that *Scybalium* (*Sphaerorhizon*) *depressum* (fig. 5-19,*d*), instead of being annual as Hooker (1856) supposed, sends out new crops of inflorescences periodically from a perennial, diffuse endophyte?

Beyond the swollen tuberous organs which hold parasite and host together there are, in several genera, rhizome-like structures which traverse the soil as if searching for new host roots. In Hookers' ac-

count these, as well as the tubers, are referred to as rhizomes, but since to me they appear to be different organs I shall call only the elongate organs "rhizomes," and even here I hesitate. "Rhizomes" are to be sought in *Helosis*, *Langsdorffia*, *Scybalium jamaicense*, and *Thonningia*. Actually, in *Thonningia* the existence of a root cap has been demonstrated (Mangenot, 1947), and the possibility of these organs in the other three genera being modified roots cannot be excluded. They represent coarse, cylindrical, branching organs which, in some of these genera, establish secondary haustorial contacts, where another tuber is formed by the mutual interaction of host and parasite. A rather complicated, anastomosing subterranean network can thus result, with the inflorescences emerging from various places along the rhizomes. The establishment of such secondary unions has been studied only in *Thonningia* (Mangenot, 1947). Hooker (1856) speaks of "corrosion" of the host bark, and implies that the hairy covering of the rhizome may be involved in *Langsdorffia rubiginosa*.

The rhizome of *Cynomorium* (Weddell, 1858-1861) traverses the soil horizontally until it contacts a vigorous host root, where haustorial contact is established. A "nest" of new inflorescences and some new rhizomes emerge at the host connection (fig. 5-17,*c*). The original rhizome, meanwhile, has also curved upward to form a terminal inflorescence.

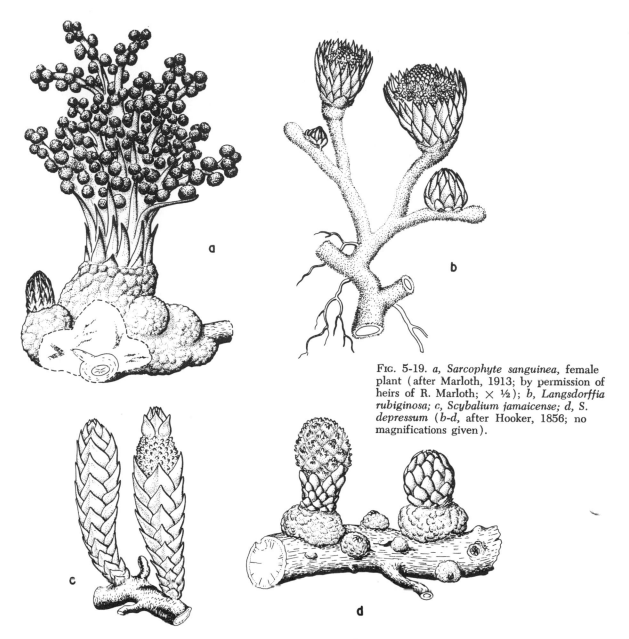

Fig. 5-19. *a, Sarcophyte sanguinea,* female plant (after Marloth, 1913; by permission of heirs of R. Marloth; × ½); *b, Langsdorffia rubiginosa; c, Scybalium jamaicense; d, S. depressum* (*b-d,* after Hooker, 1856; no magnifications given).

Haustorial contacts are made by means of small lateral roots on the rhizome and even the base of each inflorescence. These roots fuse terminally with the host root, forming a simple endophytic wedge or cone to reach the host stele. Frequently the host tissues respond by lifting the invading root on a cushion-like outgrowth (fig. 5-21,c). Similar small lateral roots provide for a degree of vegetative reproduction in *Dactylanthus* (Moore, 1940).

The fleshy root of *Thonningia* has a curious way of attacking host roots (Mangenot, 1947; fig. 5-21,a). When crossing a potential host root it swells up in such a way as to engulf the host root. At maturity the root of the host gives the impression of having penetrated a tubercular portion of the *Thonningia* root, which is far from the truth. The engulfing parasitic root tissues establish an interesting haustorial contact (see chap. 7).

Leaves

The leaves of Balanophoraceae are simple, scale-like organs, ranging from the nearly acicular basal leaves of some species to the very broad, clasping scales of *Scybalium jamaicense* and the large concave scales covering the flowers of some *Balanophora* species. They may be spiral in arrangement or may be whorled. In *Balanophora* we frequently find a few whorls of about three leaves each, forming sheathlike collars around the inflorescence axis (fig. 5-17,b). They are often much reduced in the capitulum itself, as in *Scybalium, Thonningia,* and others. They may, however, be virtually confined to the capitulum, as in *Rhopalocnemis* and *Corynaea.* In a number of genera, including the latter two, the scales of the capitulum are transformed into peltate, often hexagonal organs which drop off before anthesis. *Ombrophytum, Lathrophytum,* and *Chlamy-*

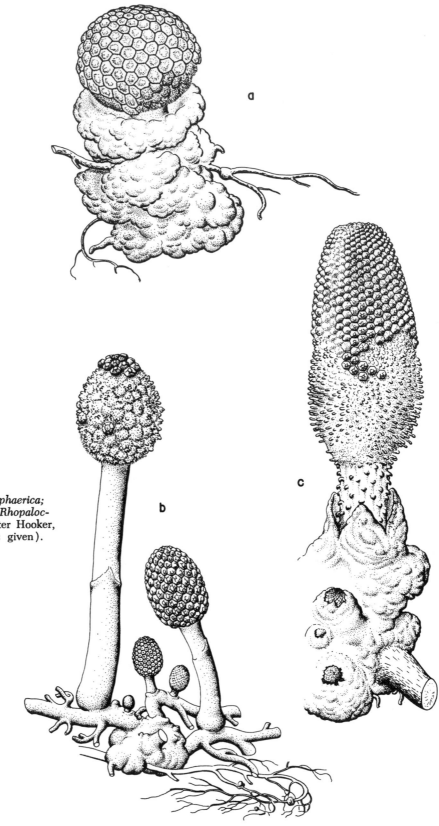

Fig. 5-20. *a, Corynaea sphaerica; b, Helosis mexicana; c, Rhopaloc-nemis phalloides* (all after Hooker, 1856; no magnifications given).

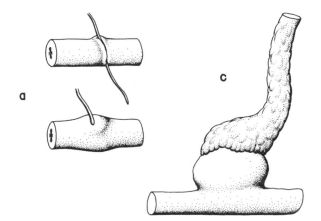

FIG. 5-21. *a*, Two stages in capture of a small host root by root of *Thonningia sanguinea* (after Mangenot, 1947; × 1); *b*, habit of same, male plant (after Mangenot, 1947); *c*, secondary haustorial contact of rootlet of *Cynomorium coccineum* with root of *Salsola* (after Weddell, 1858-1861).

dophytum are leafless, unless the peltate secondary axes of the first are foliar in nature.

Inflorescences

In the origin of inflorescences from the mother organs—whether these be tubers or rhizomes—we find one of the curious morphological features of the family. I am referring to the endogenous origin of inflorescences or flowers, which we have noted in the other families in this chapter. This process is of great ontogenetic interest, for it implies a schizogenous "cutting out," as it were, of some of the organs involved, and is therefore fundamentally different from events in other higher plants. The endogenous development of inflorescences is visible, in varying degrees, in the collar-like volva around the base. The irregularly lobed volva represents the protective cover of the tuber or rhizome after it has been ruptured by the emerging organs. In several genera, however, the volva grows with the primordial inflorescence. The volva of *Ombrophytum* and of the related genera *Lathrophytum* and *Juelia* grows into a large, spathelike organ sheathing the base of the inflorescence. Even more extraordinary is the situation in *Chlamydophytum*, where the entire inflorescence matures endogenously before it breaks through

FIG. 5-22. *a*, *Chlamydophytum aphyllum*, female plant just before rupture of volva, in sectional view (after Harms, 1935; × 1/3); *b*, *Lophophytum mirabile*, bisexual inflorescence, lower portion female, upper ¾ male (after Harms, 1935); *c*, *L. weddellii*, young inflorescence with bracts still attached, and tuber (after Hooker, 1856).

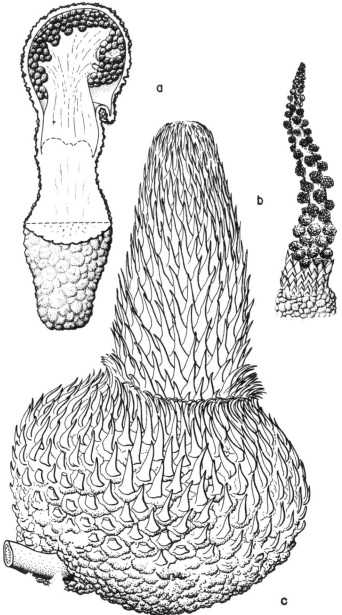

FIG. 5-23. *Thonningia sanguinea* (courtesy Prof. Dr. W. Robyns, from his Flore des Spermatophytes du Parc National Albert, I, fig. 6, p. 109, 1948).

the coarse volva (fig. 5-22,*a*). These are the genera in which foliar organs seem to be lacking, except for those belonging to the flowers themselves. Even a perianth is absent in *Ombrophytum, Lathrophytum,* and *Juelia,* and in the female flowers of *Sarcophyte* and *Chlamydophytum.* The genus *Exorhopala* is said to be characterized by its exogenous inflorescence development (v. Steenis, 1931), but this fact does not seem to have been clearly established.

The inflorescence types of Balanophoraceae are represented in a diagrammatic fashion in figure 5-24. The most elaborate type, as seen in *Sarcophyte,* consists of a fleshy axis occupied with scalelike bracts in spiral arrangement. The upper bracts subtend lateral, leafless branches which, in the male plant, are simple racemes. In the female plant the flowers are organized in globular heads along the lateral branches, each head representing a fusion of many flowers. A similar inflorescence occurs in *Chlamydophytum,* where even the bracts on the main axis are lacking.

In *Hachettea* the entire main axis of the inflorescence is occupied by long orange scale leaves. In the axils of these leaves, in the upper part of the plant, are flower-bearing lateral branches similar to the male ones of *Sarcophyte.* In male plants the flowers are stalked; in female plants sessile. In both sexes the flowers are supported by minute bractlets. If we now imagine all fertile axes to be moved to the summit of the main axis, no longer subtended by bracts, we have the inflorescences of the related *Dactylanthus.* The bracts near the flowers in this genus are bright orange with purple markings, giving the appearance of a capitulum of Compositae.

In *Lophophytum* the trend is quite different. The bracts of the fertile region have here become peltate, imbricate structures obscuring the young flowers. When the fertile axis elongates and the peltate scales are dropped, the flowers can be seen to be crowded on short lateral spikelets, one spikelet per scale. The *Lophophytum* inflorescence finds its logical extension in those of the related genera *Ombrophytum, Lathrophytum,* and *Juelia.* The inflorescence axis of these genera has lost all scale leaves. The tip of the spikelet has now expanded in umbrella-like fashion (fig. 5-25,*b*), apparently taking over the protective function of the scale leaves. The female capitulum of *Balanophora* carries the trend farther. Its flowers are inserted not only on the stalk of the spikelet—which here is a clavate organ—but also around its base, on the main inflorescence axis. The male inflorescence of this genus lacks any vestige of lateral branches, and the

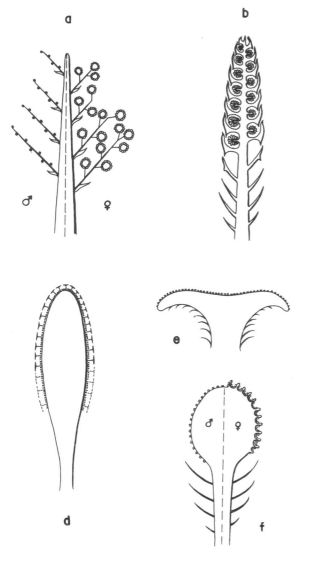

FIG. 5-24. Inflorescence types: *a*, *Sarcophyte*; *b*, *Lophophytum*; *c*, *Dactylanthus*; *d*, *Rhopalocnemis*; *e*, *Scybalium fungiforme*; *f*, *Balanophora*; *g*, *Hachettea*.

flowers are regularly arranged directly on the fleshy main axis. In *Rhopalocnemis* and related genera the spikelet has, on the contrary, become a greatly flattened organ whose hexagonal surface fits in precisely with adjacent ones to form a heavy armament. The flowers which they cover are inserted on the main axis only. This armament, like the scales of *Lophophytum* and *Cynomorium*, which seem comparable, drops to the ground when the flowers mature.

The "spikelet" upon which the female flowers of *Rhopalocnemis*, *Ombrophytum*, and related genera are borne has also been interpreted as a modified peltate leaf. This notion has much to recommend it, as the peltate scale of *Ombrophytum* sp. (Mexia No. 6337a, from Peru, at UC) is supplied by three vascular traces in a leaflike fashion, with the median trace reaching up to the tip of the scale. It is not impossible that the *Balanophora* "spikelet" has been derived from a similar structure.

Flower Morphology

It is difficult to discuss the distribution of the sexes in plants with gregarious inflorescences, for we know so little about the actual limits of individuals. If several inflorescences arise separately but from one host root we may be concerned with as many individuals, but they also might arise from a single endophytic system. It is certain, nevertheless, that true dioecism exists in *Sarcophyte*, *Chlamydophytum*, and some *Balanophora* species. We know, too, of instances in which unisexual inflorescences of both types emerge from a single plant (*Scybalium fungiforme*, *Langsdorffia*). In the bisexual inflorescences all the male flowers may be at the base (*Rhopalocnemis*, some *Balanophora*). The opposite is true of *Mystropetalon*, *Scybalium depressum*, and other *Balanophora* species (Setchell, 1935), as well as the genera *Lophophytum*, *Lathrophytum*, and *Ombrophytum*. In *Helosis*, *Cynomorium*, *Juelia*, and *Corynaea*, the male and female flowers are mixed. In two "mixed" genera, Fagerlind (1938*a*, 1938*b*) has described interesting patterns of sex distribution within single capitula. In *Helosis* the stalk of the armament scale is surrounded first by two concentric circles of female flowers (fig. 5-25,*f*), while male flowers, often aborted, occupy positions under the corners of the hexagonal scale. *Ditepalanthus* has

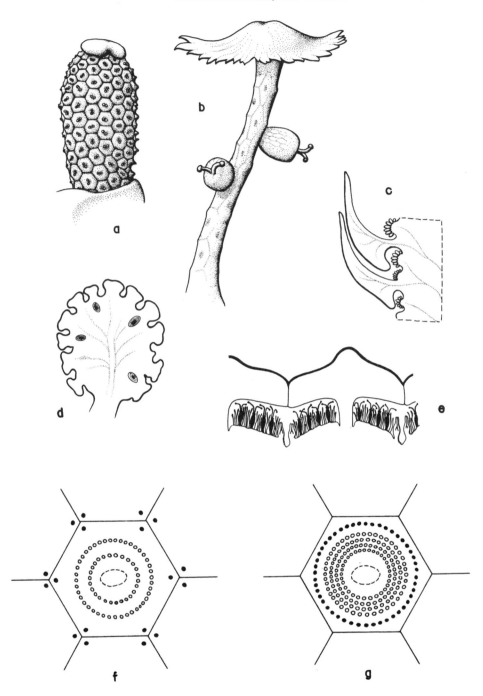

FIG. 5-25. *a, Juelia subterranea,* female spikelet (after Asplund, 1928; × 4); *b,*
Lathrophytum peckolti, female spikelet with two remaining flowers (after Eichler, 1868*b*;
magnification unknown); *c, Cynomorium coccineum,* sectional view of part of inflorescence
with protective bracts and flowers (after Hooker, 1856); *d, Sarcophyte sanguinea,* longisection
through inflorescence unit, showing position of embryo sacs· (after Griffith, 1845); *e,*
Ditepalanthus, sectional view of part of inflorescence with protective bracts, and flowers
(after Fagerlind, 1938*a*; compare with *g*); *f, Helosis cayennensis,* diagrammatic representation
of sex distribution under hexagonal scales (solid lines; base indicated as broken line);
small circles represent female flowers, black dots male ones) (after Fagerlind, 1938*b*);
g, same, for *Ditepalanthus* (after Fagerlind, 1938*a*).

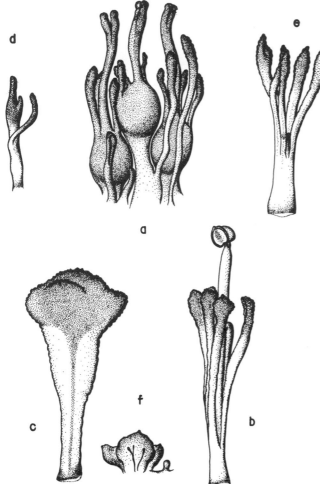

FIG. 5-26. (Left). *Cynomorium coccineum*, Praia da Roca, Portugal (× 10): *a*, group of female flowers; *b*, male flower; *c*, bract; *d*, sterile female flower; *e*, sterile male flower; *f*, small leaflike organs from base of inflorescence.

FIG. 5-27. (*Below*). Male flowers: *a*, *Hachettea austrocaledonica* (after Baillon, Dictionnaire de Botanique, unnumbered plate; magnification unknown); *b*, *Ditepalanthus afzelii* (after Fagerlind, 1938*a*; magnification unknown); *c*, *Mystropetalon thomii*, with hairy bract and one of the two prophyllar organs visible (after Marloth, 1913, by permission of heirs of R. Marloth; × 4.5); *d*, *Juelia subterranea* (after Asplund, 1928; × 8); *e*, *Dactylanthus taylori* (after Hooker, 1859; magnification unknown); *f*, *Lathrophytum peckolti* (after Eichler, 1868*b*; magnification unknown).

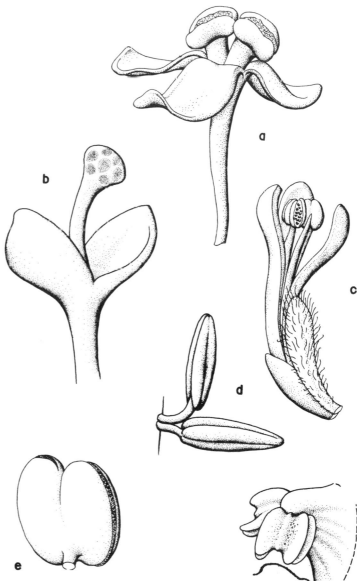

four concentric rings of female flowers followed by a single ring of males (fig. 5-25,*g*). Among bisexual inflorescences, proterogyny seems to predominate. *Cynomorium* is the only member of the family in which perfect flowers have been reported.

The variability of the male flower of Balanophoraceae is so great that it has been a stumbling block in regarding the family as a natural one. The most "normal" of flowers, those of *Cynomorium* and *Mystropetalon*, have several simple strap-shaped petals and one or two stamens with rather long filaments topped by two-celled anthers (figs. 5-26,*b*; 5-27,*c*). The male flower of *Hachettea* (fig. 5-27,*a*) is scarcely different from that of *Mystropetalon*, except for the contraction of the filament and the one-celled condition of the anther. The related *Dactylanthus* has reached the extreme in reduction; the male flower is represented by no more than a single two-celled anther and two minute, filamentous perianth members (fig. 5-27,*e*). In the American genera *Lophophytum, Lathrophytum, Ombrophytum,* and *Juelia* only two small stamens remain (fig. 5-27,*f*).

The interpretation of the androecium of other Balanophoraceae requires more than the notion of

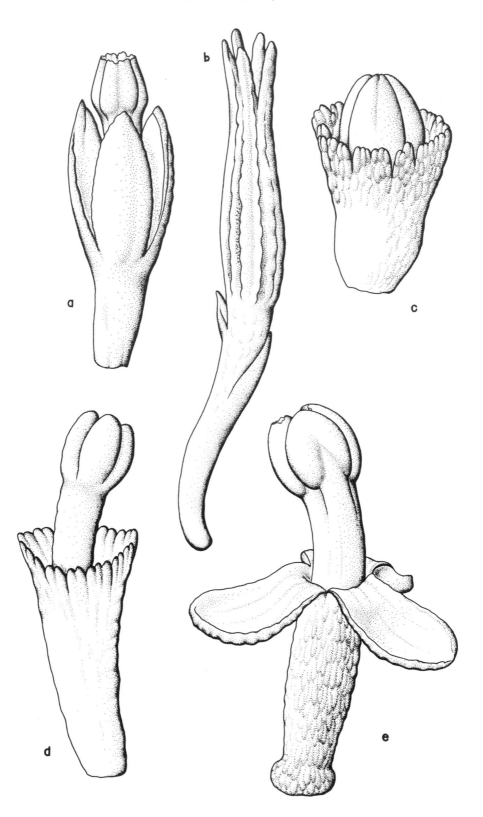

Fig. 5-28. Male flowers: *a, Scybalium jamaicense; b, Thonningia sanguinea; c, Corynaea sphaerica; d, C. crassa; e, Helosis mexicana* (all after Hooker, 1856; magnifications unknown).

reduction. Cohesion of different parts of the stamens has evolved in several genera. The lower parts of the filaments in *Helosis* and *Scybalium* form a tubular organ crowned by two independent two-celled anthers (fig. 5-28,*a* and *e*). The same is true of *Corynaea*, where no sign of separate filaments remains (fig. 5-28,*d*). In *Ditepalanthus, Rhopalocnemis,* and *Exorhopala* the stamens are consolidated into a single androecial column with a number of pollen sacs at the top (fig. 5-27,*b*). The number of pollen sacs in the last two genera—and therefore the original number of stamens—is not clear, as van Steenis' (1931) data do not quite tally with the analysis of Goeppert (1847) and Hooker (1856). The synandrium of *Ditepalanthus* is supplied with two, rarely three, traces, indicating as many stamens (Fagerlind, 1938*a*). In all these genera dehiscence is through apical pores.

In *Balanophora* and *Langsdorffia* similar fusions characterize the androecium, but dehiscence takes place by means of transverse or longitudinal slits (fig. 5-29,*a, b,* and *c*). *Balanophora* is extremely variable in this regard. In some species we can speak of individual sessile anthers with transverse slits. In others the pollen sacs are linear and arranged in parallel fashion at the summit of a staminal column. Many species of *Balanophora* are intermediate between these two extremes, showing variable numbers of pollen sacs of different shapes and arrangements on a short column. *Thonningia*, also, has a number of long, slender pollen sacs fused together on a column; the entire compound organ is in the shape of an attenuated, grooved cone (fig. 5-28,*b*).

Sarcophyte and *Chlamydophytum*, which undoubtedly are closely related, have diverged strikingly in their androecial organization, both from each other and from the rest of the family. The stamen of *Sarcophyte* at first sight reminds one of the entire synandrium of some advanced *Balanophora*. It is a globular body made up of fifteen to twenty pollen sacs which radiate from a solid center (fig. 5-29,*e* and *f*). Each compartment dehisces by means of its own apical pore or rupture, and the anther is set on a separate short filament which, in turn, is in an axillary position to a petal. The androecium in *Chlamydophytum* is a collection of small, sessile, multilocular anthers, one each at the base of the seven petals plus two or three placed directly on the floor of the flower.

The pollen of Balanophoraceae, in keeping with the highly advanced status of the family, is much reduced, leaving few conspicuous features (Erdtman, 1952). The spherical or oblong grains are devoid of exine sculpturing (except in *Cynomorium*) and have three colps or pores which are, in some genera, distinguishable with difficulty. The pollen of *Mystropetalon* is of a unique type: triangular, square, or pentagonal in polar view, but almost always square in equatorial view (Marloth, 1913; Erdtman, 1952).

The female flower is an exceedingly small structure, and has been reduced to the bare essentials.

So little is left of its various organs that the concepts of ovules, placentae, carpels, and even flower are difficult to apply. The whole developmental history of the female flower in many Balanophoraceae may be summarized as follows: a protuberance develops from the inflorescence axis within which a subepidermal cell becomes the embryo sac, and the epidermal covering grows into a pistil-like structure (Lotsy, 1899).

The most elaborate pistillate flowers occur in *Mystropetalon* (fig. 5-30,*b*) and *Hachettea* (fig. 5-30,*c*). Their small globular ovaries are crowned by a three-lobed cup representing the perianth, from which the long curved style extends. Three stigmatic lobes can be distinguished. The flower is elevated on a pedicel-like structure which later becomes a white elaiosome. All but the upper style are enclosed by a pair of prophyllar organs, and subtended by a floral bract. In *Cynomorium*, which would seem to have close affinities with *Mystropetalon*, the young female flower has two or three minute perianth members on top of the ovary, but subsequent growth of the latter often results in a displacement which disposes the perianth members at various levels. Around the flower, various bract-like organs assume the same size as the perianth members, and are often difficult to identify. The ovary terminates in two styles which may adhere for some distance.

In certain other Balanophoraceae, remnants of the perianth are still recognizable in the female flower. In *Hachettea* (fig. 5-30,*c*) there is no more than a lobed rim, but in *Thonningia* (fig. 5-30,*e*) a distinct floral tube occurs, becoming fleshy in fruit. In *Helosis, Corynaea, Rhopalocnemis,* and Lophophyteae the two styles of each flower are set in a depression on top of the ovary. In *Scybalium, Helosis, Corynaea,* and *Rhopalocnemis* a few cells on top of the ovary have become somewhat papillar to form two crests alternating with the styles. These cellular crests, almost by default, have been thought of as perianth members. The perianth of the female flower is lacking in a number of genera. In *Sarcophyte* and *Chlamydophytum* organic fusion has occurred between a number of ovaries, together forming a small capitulum of more than a hundred ovaries each (fig. 5-25,*d*). The female flower in Lophophyteae is represented by no more than an ovary and two styles. *Balanophora* presents the extreme in floral reduction: the entire female flower is reduced to a structure like the archegonium of a moss, without a trace of either vasculation or perianth (fig. 5-30,*a*).

The fact that some Balanophoraceae have one style and others have two styles was the basis for Hooker's (1856) taxonomic dichotomy of the family, a division which has not received much support subsequently. The truly monostylic genera are *Mystropetalon, Langsdorffia, Thonningia,* and *Balanophora*. *Cynomorium* would seem to be a transition form. In *Sarcophyte* and *Chlamydophytum* there is no style.

The number of styles per flower is related to the

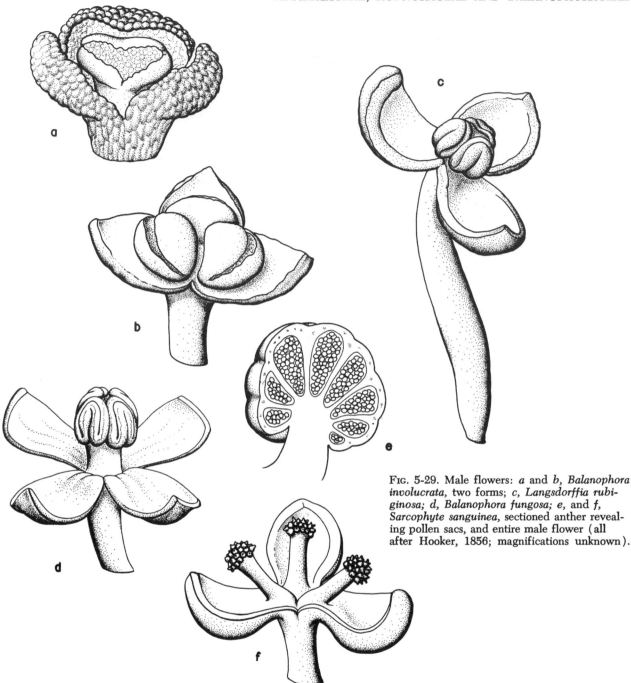

FIG. 5-29. Male flowers: *a* and *b*, *Balanophora involucrata*, two forms; *c*, *Langsdorffia rubiginosa*; *d*, *Balanophora fungosa*; *e*, and *f*, *Sarcophyte sanguinea*, sectioned anther revealing pollen sacs, and entire male flower (all after Hooker, 1856; magnifications unknown).

number of locules in the ovary and the number of carpels. The minuteness of the ovary makes it difficult, however, to ascertain its condition. In *Sarcophyte*, the three ovules occupy as many lobes of the ovarian cavity, indicating a tricarpellate ovary. In *Mystropetalon* the ovary may have one, two, or three locules; the three-lobed stigma would seem to confirm an originally tricarpellate condition. *Lophophytum* has two locules and two styles, and therefore presents no difficulties. *Helosis*, however, has

two styles surmounting a uniloculate ovary, and *Cynomorium* produces similar problems. *Balanophora* has one locule and one style. In all the genera there is only one ovule per locule, and only one develops into seed per flower. It may well be, however, that conventional ways of determining the number of carpels making up an ovary are not adequate in flowers at this level of reduction. There is a pedantic air about the statement that the female flower of *Balanophora* has one carpel; at any rate,

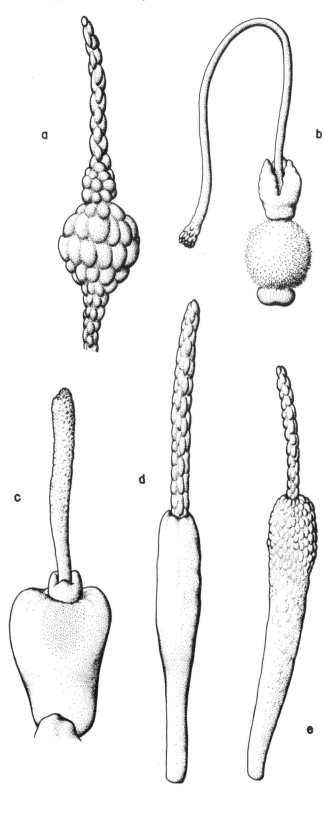

FIG. 5-30. Female flowers: *a, Balanophora involucrata; b, Mystropetalon thomii; c, Hachettea austro-caledonica; d, Langsdorffia rubiginosa; e, Thonningia sanguinea* (*c,* Schlechter No. 218, New Caledonia, UC, × 21; all others after Hooker, 1856; magnification unknown).

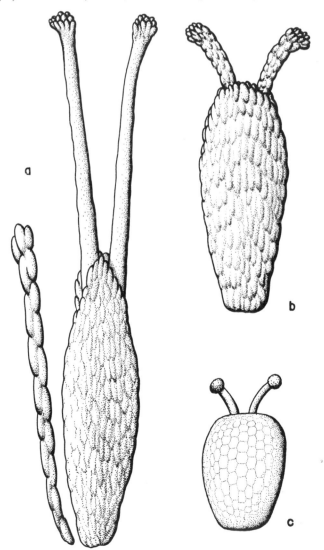

FIG. 5-31. Female flowers: *a, Helosis mexicana* (with paraphysis); *b, Scybalium depressum; c, Lathrophytum peckolti* (*a* and *b,* after Hooker, 1856; no magnifications given; *c,* after Eichler, 1868b, unknown magnification).

there is nothing obvious about it. Perhaps, after all, Lotsy (1899) was right in pointing to the meaninglessness of such notions.

Embryology

The ovule, in keeping with the female flower, is greatly reduced, and is usually little more than an embryo sac. In some genera the ovules are said to be fused with the ovary wall, leaving no ovarian cavity (*Balanophora* and *Langsdorffia;* Fagerlind, 1945*a* and *c*). At any rate, we can rarely speak of integuments or nucellus. The "ovules" are sometimes erect on the floor of the cavity, but more normally are suspended from the roof by a stalklike cell (fig. 5-32,*b*).

The embryology of Balanophoraceae had a controversial beginning in Treub's (1898) and Lotsy's (1899) work on *Balanophora elongata* and *B. globosa.* According to their work, one of the eight nuclei of the mature embryo sac produced the endosperm and the other seven degenerated. The embryo, in turn, was derived from a single cell of this endosperm. Subsequent work by Ernst (1913) on the same two species has given a different view of these events. The embryo sac, according to this author, is formed from diploid nuclei only, of which there are eight. One of the polar nuclei develops into the few-celled endosperm. Its first division is unequal, resulting in a very large haustorial cell, and a more normal cell which is the progenitor of the endosperm. The embryo is believed to be a direct product of the diploid egg cell—a case of somatic parthenogenesis or apogamy. The possibility of occasional fertilization, however, still exists. Ernst goes farther, probably too far, when he believes the same to be true of several other genera of Balanophoraceae.

The information and interpretation by these three authors and several others seemed to conflict with earlier reports of seeing pollen tubes in the style (Hofmeister, 1858) and fertilization itself (Van Tieghem, 1896). It is fortunate, therefore, that Fagerlind (1938*b*, 1945*b*) has been able to clarify the situation somewhat. In Fagerlind's view there is no evidence of bi- or tetrasporic embryo sacs in Balanophoraceae. Monosporic embryo sacs develop from the apical cell of the tetrad, at least in sexual species of *Balanophora.* The original archesporial cell never develops into the embryo sac. Fagerlind's studies have led him to the position that apomixis in the family, so far as is known at present, is limited to *B. globosa* and *B. japonica,* and possibly *Juelia.* In *B. japonica* a high chromosome number has been reported (2n=94—112; Kuwada, 1928), a frequent

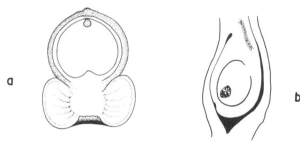

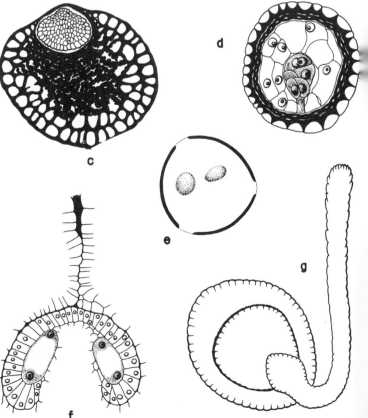

Fig. 5-32. *a, Mystropetalon thomii,* longisection of fruit and basal elaiosome (after Marloth, 1913; × 15); *b, Cynomorium coccineum,* longisection of older flower showing embryo and endosperm surrounded by massive nucellus and integumentary layer (after Juel, 1910); *c, C. coccineum,* section of embryo and endosperm of mature seed (after Juel, 1902; × 55); *d, Balanophora globosa,* mature seed in sectional view (embryo shaded), (modified from Lotsy, 1899, and Ernst, 1913; × 210); *e, Corynaea crassa,* pollen grain (from liquid-preserved material collected by Miss Patricia Kern in Costa Rica; ca. × 700); *f, Rhopalocnemis phalloides,* longisection of ovary showing two embryo sacs (after Lotsy, 1901; × 125); *g, Cynomorium coccineum,* longisection of germinating seed with erect radicle (after Weddell, 1858-1861).

feature in apomicts, in contrast to that of *B. dioica* (2n = ca. 36). Apomixis in *Juelia* has been suggested because of its largely subterranean habits. A type of entomophily would seem to be indicated in others. Moore (1940) reports a rather sweet, heavy perfume and the visits of small flies in *Dactylanthus*, but observations for other species are lacking.

In all frankness we must therefore admit our ignorance of Balanophoraceae in the domain of embryology, pollination, and fertilization. We have reliable information of this sort for very few species; many genera are as yet completely unexplored. Embryological data may be expected to add significantly to the systematics of this family. Fagerlind (1945c), for example, demonstrated rather intimate affinities between Balanophoreae and Langsdorffiae, two tribes which even Harms (1935) considered to be separated from one another by a great distance.

Fruits, Seeds, and Dispersal

A longitudinal section of a mature seed of *Balanophora globosa* is illustrated in figure 5-32,*d*. The embryo is suspended by a short unicellular stalk, and consists of about six cells arranged in three tiers of two cells each, the successive pairs alternating with each other. This minute embryo is surrounded by large endosperm cells filled with oily material. Then follows a double layer of cells with heavy secondary walls: the inner layer with small, sclereid-like cells, the outer one with cells thickened on all walls except the superficial ones. The seed, especially the embryo and endosperm, is almost the same as that of *Pilostyles thurberi* (fig. 5-8,*c*), a similarity which I regard as convergence (see also chap. 9).

In other genera fruits may be more spindle-shaped, ovate, or cylindrical (Harms, 1935). Internally, additional parenchyma and sclerenchyma tissues often surround the seed. *Sarcophyte* and *Chlamydophytum* have multiple fruits made up of many fused and ripened ovaries, together forming a fleshy mass. In *Thonningia* the free perianth tube swells to equal the size of the rest of the fruit, and may perhaps serve as elaiosome. In all genera the mature embryo is very small, quite undifferentiated, and surrounded by cellular endosperm.

As in the previous two families, we know extremely little about the dissemination of seeds or fruits of Balanophoraceae. I have alluded to the possibility of animal dispersal agents in *Thonningia*. In *Mystropetalon* this type of transport is beyond question. The mature fruit, a black spherical body, is supported by a fleshy white ring of fatty tissue (Marloth, 1913). Not only have ants been observed to carry the entire fruit into their nests, but fruits have been recovered, without the elaiosome, from ant nests.

The dispersal of other Balanophoraceae is purely conjectural. The fleshy multiple fruit of *Sarcophyte* and *Chlamydophytum* leads one to suppose that they may be disseminated by fruit-eating animals.

Other genera, as Ridley (1930) suggested for *Balanophora*, may well have the smallest of all known fruits. The same author emphasized the importance of wind-dispersal in *Balanophora*. Although wind may be a significant agency for some species, the general preference of Balanophoraceae for dark, moist tropical forests rules out wind as a dispersal mechanism for most of them.

The recorded observations on germination in the family are limited to Weddell's (1858-1861) early work on *Cynomorium*. Quite unexpectedly, Weddell was able to germinate seeds of this parasite in humid saline soil kept at 30°C. A delicate whitish radicle grew to a length of 4 mm. The mystifying fact is that these radicles grew directly upward, some actually projecting above the soil for 2 mm. The stem apex remained in the endosperm. A single seedling, resting on the soil surface, was seen to grow directly toward the radicle of a nearby *Melilotus* seedling. The radicle of the parasite, from a longitudinal fissure just below its tip, produced two lateral rootlets which grew straight up again. Unfortunately, the seedling died and this puzzling sequence of events has not been further elucidated.

The seeds eventually become incorporated in the soil, through which they may be moved by animals of various sorts and by water. Notwithstanding Weddell's work on *Cynomorium*, it would seem safe to guess that, as in other parasites with greatly reduced embryos and seeds, a chemical stimulus is needed in order to bring most Balanophoraceae seeds to germinate. I would urge botanists who have access to these seeds to try to germinate them in contact with young host roots. This is easily done in Orobanchaceae, and there is no reason to suppose that seeds of Balanophoraceae would not yield in a similar fashion. The unraveling of the story of germination and host entry will, sometime in the future, provide an interesting supplement to the present chapter.

THE AFFINITIES OF RAFFLESIACEAE, HYDNORACEAE, AND BALANOPHORACEAE

The search for affinities in these three families has been long and erratic. The main source of difficulty has been the extreme degree of reduction in many vegetative and reproductive features. In an evolutionary sense, reduction has washed out the trail behind these parasitic families, and their origin—and therefore their systematic positions—becomes a matter of guesswork with little hope of verification. The resultant dwindling of the number of useful characters for phylogenetic speculation is sometimes overshadowed by the elaboration of morphological additions, as in the diaphragm of *Rafflesia* and *Sapria* and in the androecium of *Pilostyles*, *Mitrastemon*, and even *Sarcophyte*. The botanist attempting to establish bridges to other families is frustrated both by the subtraction of useful characters and the addition of irrelevant or misleading details.

I do not pretend to be able to succeed where many botanists more at home in these families, have failed. I shall only indicate the major concepts of the past. For more detailed information the reader is referred to *Die Natürlichen Pflanzenfamilien*, 16b. Some of the earlier notions attempted to exclude these parasites from flowering plants altogether. Thunberg at first thought of *Hydnora* as a fungus related to *Hydnum*, from which it derives its name. The same fate befell many Balanophoraceae.

The association of Rafflesiaceae and Balanophoraceae has a long history, going back to the Rhizantheae of Endlicher (1836-1850) and Lindley (1836) or even to Bartling (1830). Where, in the systematic arrangement of plants, these parasites belonged, was by no means clear. Bartling, perhaps following Brown (1822), placed them in the Dicotyledons. The knowledge of an undifferentiated embryo made Endlicher move them to a position following Protophyta (ferns and cycads) and preceding the monocotyledons. Blume (1830) moved Rafflesiaceae and Hydnoraceae back to cryptogams, near *Marsilea*, even reverting to cryptogamic terminology. The Rhizantheae in Lindley's and Endlicher's sense was discounted as an artificial alliance by Griffith (1845), who pointed out that it was sanctioned neither by natural affinities of the constituent groups nor by uniform circumscription. It was also Griffith who brought Balanophoraceae into the dicotyledons. After this time Rhizantheae as an alliance is scarcely mentioned in botanical literature, although the taxonomic notions which it entailed survive in various forms to the present day. In Harms' (1935) treatment the three families of this chapter are discussed successively in the same order.

The notions about affinities of Rafflesiaceae would seem to have been subtly but strongly influenced by Linnaeus' placement (1753) of *Cytinus hypocistis* in the genus *Asarum* of present-day Aristolochiaceae. When Cytineae and Rafflesiaceae were proposed, it seemed natural to place them near Aristolochiaceae, a practice commonly accepted today. Other families which have been thought to be related to Rafflesiaceae are Nepenthaceae (Brongniart, 1824; Bentham and Hooker, 1862-1883; Hallier, 1901); Passifloraceae (Brown, 1822), Annonaceae and Nymphaeaceae (Hallier, 1901), Cucurbitaceae (Hunziker, 1926), and even Burmanniaceae and Taccaceae of the monocotyledons (Beccari, 1869). I am in no position to weigh these various ideas, although it seems unlikely that the suggested affinities with the monocotyledons will be considered seriously by contemporary botanists. With regard to the supposed link to Aristolochiaceae, however, I would like to make the following comments, unfortunately negative for the most part.

It seems fairly clear that the staminal tube of some Rafflesiaceae — *Pilostyles holtzii* and *Mitrastemon* — is primitive in context of the family. (This is not to say that the details of these organs are necessarily primitive.) If the staminal tube as a relatively primitive feature is acknowledged, it follows that the direct ancestors of Rafflesiaceae had superior, not inferior, ovaries. This, in turn, would make the traditional affinities with Aristolochiaceae less likely, or at least more remote, for there we have only inferior ovaries. The evolutionary sequence proposed for the ovaries of Rafflesiaceae also takes its inception in *Mitrastemon* (fig. 5-7). In a family so greatly advanced that conceptions of primitivity border on the ludicrous, *Mitrastemon*, notwithstanding its unusual floral biology and parasitism, would seem to be structurally nearest the family's point of origin. Future workers might do well to look for a family with perfect flowers and superior, multilocular ovaries.

Aristolochiaceae simply do not fit this prescription. I am tempted to dismiss that family as showing too few similarities to Rafflesiaceae, and too many profound differences from the latter to deserve recognition as a related family. Besides the historical association of one genus, *Cytinus*, with members of Aristolochiaceae, there is a deceptive resemblance in the androecial apparatus of the two families. Should we not expect that the most primitive plants of two supposedly related families show the greatest resemblance? The genus of Aristolochiaceae occupying this position, according to Schmidt (1935) and others, is *Saruma*, characterized by twelve completely free stamens. Now it may be held that Rafflesiaceae having a free staminal tube are closest to having such an androecium. But this would imply the separate evolution of the androecial column in the two families, and admit the irrelevance of this compound organ with respect to their mutual affinities.

I wonder if botanists in the past have not been subtly influenced by the great size of many *Aristolochia* flowers, their veined and marbled color patterns, and their carrion smell. These are precisely the most memorable features of *Rafflesia*, but can be regarded as criteria of affinity no more than similar conditions in *Stapelia* and others.

To this isolated family, Rafflesiaceae, *Hydnora* and *Prosopanche* have been traditionally attached. Early botanists such as Brown and Hooker had little doubt about the matter. Hooker states that the two genera are more closely allied to *Cytinus* than to any other plant.

We cannot take comfort from a truly close affinity, however, and a separate familial status is unquestionably justified. Among fundamental differences between Rafflesiaceae and Hydnoraceae we may reckon the total leaflessness of Hydnoraceae; its root dimorphism in contrast to the total rootlessness of Rafflesiaceae; and differences in seed anatomy and androecium. We must therefore conclude that, if the two families are related, they are only remotely so.

Weddell (1850) added an extraordinary note to the early association with Rafflesiaceae and Hydnoraceae in Rhizantheae by regarding the entire female flower of *Balanophora* as equivalent to the ovule in *Rafflesia*, a notion scarcely deserving serious consideration. Since Eichler (1878) the matter has been compounded by adding Santalaceae and Loranthaceae to the same alliance. In *Die Natürlichen Pflanzenfamilien,* three orders are given: Santalales (including all families discussed in chaps. 2 and 3), Aristolochiales (Aristolochiaceae and Rafflesiaceae), and Balanophorales (Balanophoraceae). The work of Fagerlind (1948*a*) on the gynoecium of the various Santalalean and Balanophoracean groups points to a supposed affinity of Balanophoraceae with Phoradendreae and Arceuthobieae — suggestions anticipated by Van Tieghem (1896), whose Loranthineae contain only mistletoes and Balanophoraceae. In a later paper (1959) on the Santalacean genus *Exocarpos,* Fagerlind goes even farther and implies that the entire flower of *Balanophora* is equivalent to the ovarian placenta of *Exocarpos. Balanophora* would thus be left in a gymnospermous condition. The error of this view is fully demonstrated by the ovarian cavity of some Balanophoraceae, by the two distinct styles of most, and by the undoubted perianth of the female flowers in *Cynomorium* and *Mystropetalon.*

It is certainly going too far to say that Fagerlind's articles have thoroughly demonstrated a relationship of Balanophoraceae to the green members of Santalales, as Cronquist recently stated (Cronquist, 1965). We must not forget that the above-mentioned supposed alliances are all based on the structure of the gynoecium, the organ that seems to have suffered most reduction in the parasites, short of the absorptive system itself. It seems to me, therefore, that we are least likely to find truly significant phylogenetic criteria in the gynoecium, and that we are nowhere else more easily blinded by evolutionary convergence. It is my feeling that the similarities emphasized by Fagerlind are in this category. In various degrees we see reduction of ovules in other parasitic groups. (I return to this common evolutionary process in chap. 9.) There is no recognizable natural affinity of Balanophoraceae to any other parasitic group, including Rafflesiaceae, Hydnoraceae, and the entire Santalalean order (see also Schellenberg, 1932). We have continued to ignore Hooker's (1856) warning that Balanophoraceae and Rafflesiaceae have no characters of systematic value in common. Expanding this to apply also to Santala-

lean families, I would agree with him that the strongest reason for allying them is not that they present characters in common but that none is considered to be allied to any other known order of phanerogams.

In these families, it is altogether too easy to sever their connections with others. The real difficulty is to demonstrate acceptable alternate affinities.

The number of dicotyledonous families theoretically possible as a base from which to derive the floral organization of Balanophoraceae is countless. I can only say that Hooker's (1856) emphasis on Halorrhagaceae as possible relatives of Balanophoraceae, especially *Gunnera,* might have been nearest the mark, after all. It is amazing that this suggestion did not find a place in the most recent treatment of Balanophoraceae (Harms, 1935), especially since Juel (1910)) essentially confirmed the notion. The explanation may be that Hooker himself later abandoned the idea in the *Genera Plantarum.* I would hope that this suggestion be reëntered in botanical circulation, and that it be given an up-to-date evaluation.

The natural unity of the family has been seriously questioned at times. Van Tieghem split Balanophoraceae into a series of separate families, and others followed suit. *Cynomorium* has fared worse than other genera, for even today it is placed in a separate family, Cynomoriaceae. Although my acquaintance with Balanophoraceae is a shallow one, I am unable to support this separation. *Cynomorium,* from what I can gather, bears a close relationship to *Mystropetalon,* a genus no contemporary botanist would want to see removed from the family. It is, indeed, true that within Balanophoraceae *sensu lato* there are some apparently deep systematic fissures. We seem to be faced with an extremely rapid evolutionary divergence, with respect to individual features, between closely related forms. The androecial apparatus of *Chlamydophytum* and *Sarcophyte* have diverged radically from one another, yet the other structures in plants of the two genera show them to be very closely related. Our best choice is to agree with Harms (1935) on the advisability of holding the family together in its present form until a constituent element can be demonstrated to be closely related to another family of angiosperms.

6

Cuscuta, Cassytha, Lennoaceae, and Krameriaceae

The remaining parasitic flowering plants are placed in two genera (*Cuscuta* and *Cassytha*) of otherwise autotrophic families (Convolvulaceae and Lauraceae, respectively), and in two very small families (Lennoaceae and Krameriaceae). They are discussed in a single chapter only for purposes of economy. These parasitic groups undoubtedly represent four independent starting points of the parasitic habit. *Cuscuta* and *Cassytha* are almost without chlorophyll; these twining parasites attack any aerial parts of a great variety of hosts. Lennoaceae and Krameriaceae are root-parasites; the former is holoparasitic but the latter has abundant chlorophyll.

CUSCUTA

The dodders are among the best known of the higher parasites, because of their extraordinary appearance and behavior, and their prominence in areas where botanical science has flourished longest. With the broomrapes they probably share a place in the public awareness of angiospermic parasites second only to the mistletoes. The dodder's rapid development and its stranglehold on and damage to the host have earned it a place in the superstition of many Western countries. The German "Teufelszwirn" and Dutch "Duivelsnaaigaren" are vernacular names of this sort. Another popular German name for dodder on clover, "Kleeseide," is very descriptive of its silky sheen. The smallest species resemble a shiny yellow cobweb cast over the host plant, but other species are much coarser.

Cuscuta is a genus of cosmopolitan occurrence. In Sweden it reaches the 64th parallel, and it has even found its way to Greenland (Schmucker, 1959). The largest number of species is recorded from the Western Hemisphere, where no country from southern Canada to Chile and Argentina is without the genus (Yuncker, 1932). In southern Argentina, *C. pauciflorum* is known from as far south as 47° Latitude (Hunziker, 1949-1950). Dodders are abundant in Europe and Africa also, but less so in Australia and the Indo-Malayan region. The genus has not been found in the Philippine Islands. In fact, one of the generalities we can make about the geography of dodders is that they have their greatest development on large bodies of land, and are usually infrequent on islands. Yuncker (1920) underlines this fact when stating that no dodder so far found in the West Indies is peculiar

to those islands, that all of them are members of predominantly continental species. The species in New Zealand may not be indigenous (Cheeseman, 1925). It would be interesting to subject the geography of dodders to an up-to-date study which might find an explanation for their continental preference.

Many species, because of the similarity of the size of the seeds to those of commercial crops, especially leguminous ones and flax, have been introduced in seed lots to countries where they did not at first occur. The North American *C. gronovii* is a weed in several European countries (Yuncker, 1932); *C. campestris,* from the same homeland, has reached Africa, Europe, South America, China, and Australia. *Cuscuta suaveolens* from South America has invaded all other continents. Two European dodders, *C. epithymum* and *C. epilinum,* are now found nearly throughout the world, and an Asian species, *C. approximata,* has been reported from England. Modern seed hygiene has done much to reduce the spread of dodder seed from one country to another, but complete elimination of this seed is extremely difficult. We must therefore expect to find these unbidden immigrants from time to time. A number of the above-mentioned species are now naturalized in their new homelands, and must be counted with the natural vegetation of such countries.

Yuncker's (1932) revision of *Cuscuta* lists 158 species. Many of these are very difficult to identify, or to separate from neighboring ones, and we can probably expect the total count to decrease appreciably. It is unlikely, however, that all species have been reported. The taxonomic characters of the genus are confined almost entirely to the flower, fruit, and inflorescence, as the vegetative parts of dodders show great uniformity.

There is general recognition of three subgenera within *Cuscuta* (Yuncker, 1932), based on the morphology of the gynoecium (see fig. 6-6). In subgenus *Cuscuta* the two styles are distinct, and the stigmas tend to be much longer than broad. Originally this subgenus was Mediterranean. The styles are distinct, or sometimes lacking, but the stigmas are globose or somewhat peltate or convoluted in subgenus *Grammica.* The latter subgenus is present on all continents, but has most species in the Americas. The smallest of the three subgenera is *Monogyna,* which, as its name suggests, has a single, or united, style. *Monogyna* has a single species in

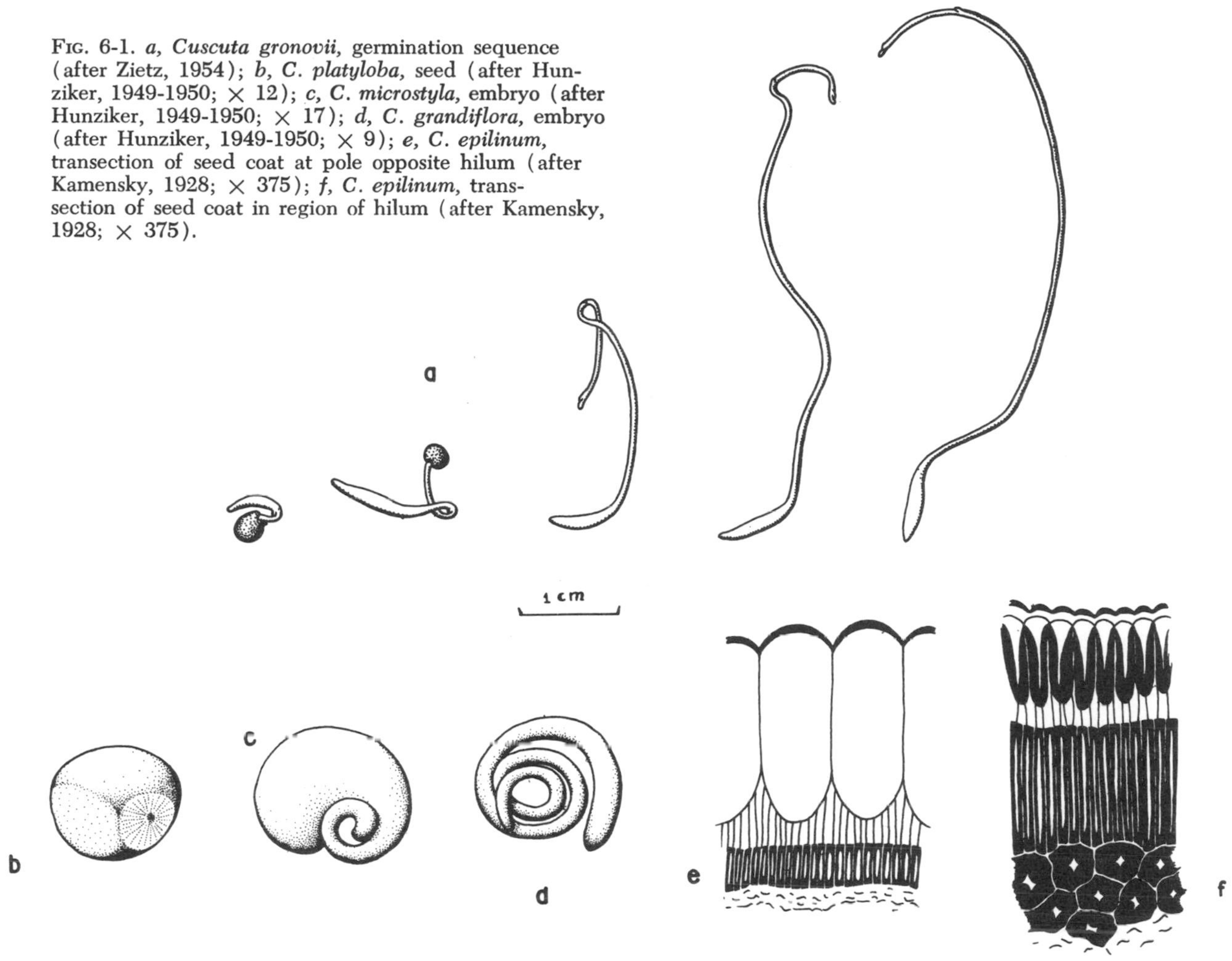

FIG. 6-1. *a, Cuscuta gronovii,* germination sequence (after Zietz, 1954); *b, C. platyloba,* seed (after Hunziker, 1949-1950; × 12); *c, C. microstyla,* embryo (after Hunziker, 1949-1950; × 17); *d, C. grandiflora,* embryo (after Hunziker, 1949-1950; × 9); *e, C. epilinum,* transection of seed coat at pole opposite hilum (after Kamensky, 1928; × 375); *f, C. epilinum,* transection of seed coat in region of hilum (after Kamensky, 1928; × 375).

the southern United States (*C. exaltata*), but the others are distributed throughout Europe, Asia, and Africa. Whether these three generic subdivisions will stand the test of time is impossible to say. It is conceivable, however, that the fusion of styles into a single structure has occurred in different places in the genus. This, if true, would leave *Monogyna* an artificial entity. At any rate, future students of *Cuscuta* would not be wise to take its present taxonomy for granted. It is interesting to note that in *C. japonica* the degree of stylar fusion is variable (fig. 6-6,*a* and *b*).

Seed and Germination

Each flower has four ovules, but one or more of these may abort, resulting in a variable number of seeds per flower (fig. 6-1,*b*). It stands to reason that the shape of a single surviving seed is somewhat dif-

ferent from that of two, three, or four seeds maturing together. The differences are especially evident in the flat surfaces of seeds which have developed alongside one another. Thus a dodder seed may have one or two flat surfaces or none. In size, dodder seeds are also quite variable.

The seed coat is rather complex (Koch, 1874; Kamensky, 1928) (fig. 6-1,*e* and *f*). It consists of the epidermis and two palisade layers; at least one layer may differentiate to become a solid layer of compact columnar sclereids. There are some taxonomically useful characters here in separating individual species (Kamensky, 1928), but they are too inconvenient to be frequently resorted to. The area of the hilum usually has an anatomically different seed coat from that of the opposite pole (fig. 6-1,*f*). Within the seed coat is, first, a variable layer of parenchymatous perisperm, collapsed at maturity;

second, a layer of proteinaceous cells of unspecified origin; and third, the starch-bearing endosperm enclosing the embryo.

The embryo structure of *Cuscuta* would seem to foreshadow the parasitic behavior of the older plant. It is completely without cotyledons. Its shape is that of a coil in one plane: the outer end is club-shaped or rarely ball-like (Yuncker, 1942); the inner end is thin and provided with one or two minute scale leaves. The latter are alternately arranged and probably not comparable to cotyledons.

The seeds appear to be dispersed rather haphazardly. Dodder seeds can pass through the intestinal tract of sheep unharmed, retaining their original viability. This can scarcely surprise readers acquainted with the methods employed to bring about germination in the laboratory (see below). Animals as dispersal agents are probably unimportant in dodders. In general, dodder dispersal is of an unspecialized kind.

In some species of the western United States (Yuncker, 1942) and Hungary (Dorner, 1867) the seeds may germinate directly in the capsule. I have observed this phenomenon in the Arizona desert near Yuma, and suspect it may be characteristic of some dodders. It would seem to be a direct adaptation to parasitism on woody plants, where it is both unnecessary and inefficient for each generation to climb the host anew. Some seeds of the unidentified Yuma dodders had fallen to the ground and germinated there, however, thus allowing the dodder to be washed or otherwise transported to new hosts. It is quite possible that the ball-like radicular pole of the seedling of *C. nevadensis, C. denticulata,* and *C. veitchii,* which supplants the endosperm in part, facilitates this kind of epiphytic germination. We can only speculate on whether the radicle in these three species is retained in the seed and fruit, leaving future observers to clarify the issue in the field. Epiphytic germination of *Cuscuta* should not be confused with the perennial behavior of some species.

Considering the fact that it takes treatment with concentrated sulfuric acid (Gaertner, 1950) or grinding with glass powder (Walzel, 1952b), sometimes followed by a cold treatment (Tingey and Allred, 1961) to bring about germination in the laboratory, it is a miracle that any natural germination takes place in the field. This apparent paradox has been clarified somewhat, however, by Gaertner (1956), who finds that cold storage will facilitate abundant germination in *C. europaea*. Even if such germination requirements are generally valid for dodders, we have no explanation for such species as *C. salina* and *C. sandwichiana* (Sakimura, 1947), which, in nature, are never subjected to a comparable regime. In at least one species, *C. campestris,* a mechanism exists allowing germination of a few seeds at a time for several years (Dawson, 1965). The precocious germination of some desert species poses another example of dodders lacking strict germination requirements. Even for these famous or notorious plants, we still do not have a compre-

hensive picture of reproduction which includes vegetative reproduction and germination as it occurs in nature.

Early Development and the Coiling Habit

Germination in the dodders is a fascinating example of the modification of this process with regard to the parasitic way of life (fig. 6-1,*a*). The radicle protrudes and curves toward the ground, often penetrating it partly. The remainder of the embryo, still enclosed within the endosperm, now digests the latter rapidly, transferring most of its nutrients to the radicular end, which assumes a clavate shape. The growth of the middle region at this time produces a characteristic loop, which is retained even when the seedling finally discards the exhausted seed. Both the opening up of the loop and the later twining of the seedling are under the control of two light reactions (Lane and Kasperbauer, 1965).

It is now necessary for the seedling to find a suitable host, to tap it for nutrients to allow further growth. The seedling, after having grown for a while, becomes erect, thus eliminating the loop, and begins to circumnutate actively. This spiraling movement often brings it into contact with a host.

It has recently been demonstrated, however, that dodder seedlings may find a host by means of quite different, additional growth responses. Consider the young nettle plant in figure 6-2,*b* (Fritsché *et al.,* 1958). Around the edge of the pot which accommodates the plant a number of seeds of *C. europaea* have been placed, following the usual germination treatment. The emerging dodder seedlings all grow directly to the potential host. There must be a stimulus inherent in the nettle plant to bring the dodder seedling to its base.

One might think that the stems of the seedlings were negatively phototropic, as are the radicles of mistletoes. It has been shown, however, that this is not so (Spisar, 1910). The stem of *C. gronovii,* at

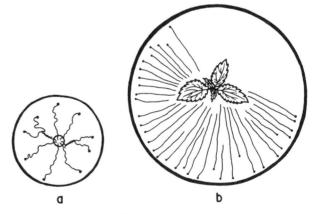

FIG. 6-2. *Cuscuta* seedlings, growth responses: *a,* to water (in vial in center) (after Tronchet, 1958); *b,* to a living plant of *Urtica* (after Fritsché *et al.,* 1958). Species used in *a* is *C. europaea;* in *b, C. gronovii.*

least, is positively phototropic, with a very rapid reaction to unidirectional light. The correct explanation of the seedling's behavior seems to be that it grows toward a source of moisture (fig. 6-2,*a*), but also toward certain chemicals (Bünning and Kautt, 1956). The latter response was suggested first when seedlings were observed to grow toward some host species but not toward others. Even individual leaves, or leaf extracts, will stimulate such growth up to a distance of about 10 cm. The substance Eugenol calls forth a negative tropism of dodder seedlings in higher concentrations, a positive response in lower ones. Nor is the effect of chemicals limited to direction, as some even speed up the seedling's growth. There can, therefore, be no doubt that the dodder's search for hosts involves far more than a mere circumnutation. With the aid of these auxiliary means the seedling is able to grow to a length of 35 cm in seven weeks (*C. gronovii*; Spisar, 1910). If it has not found an acceptable host by that time it succumbs.

The early behavior of the dodder plant has received much attention, but the published results are contradictory. Errors in observation or formulation are sometimes involved, but at other times there may be genuine differences among species, as in the direction of spiral growth. For the sake of clarity, therefore, I am limiting the following account to the recent work of Zietz (1954) with *C. gronovii*, to which I add a few remarks from other sources.

What is known about the direction of the spiral growth of *Cuscuta*? There is no direct connection with the coiling of the embryo and that of the older plant, as the seedling is virtually straight at one time. The later growth movement is an exaggerated circumnutation. Zietz states it to be to the left only (clockwise), although Spisar (1910) had spoken of occasional anticlockwise spirals. In *C. reflexa*, also, a clockwise growth direction has been reported (Singh, 1933). I have not been able to confirm this direction in any species represented in the Herbarium of the University of California at Berkeley, including specimens of *C. gronovii* and *C. reflexa*. This material uniformly shows directions to the right, *anticlockwise*, or *dextral*. Hunziker (1949-1950) agrees with me in this respect, and reports these directions of twining for all Convolvulaceae of similar habit. I have therefore little choice but to consider Singh and Zietz in error. A part of the difficulty, no doubt, lies in the triple terminology current to describe spirality. How confusing the matter can be is shown in the work of Fritsché *et al.* (1958): a sinistral growth of *C. europaea* is discussed but a dextral one is illustrated!

The twining growth of dodder is upward only. I have spoken of a strong, positive phototropism, and to this Zietz adds an equally pronounced negative geotropism. These two responses make it impossible for the plant to entwine horizontally placed objects.

In reality, there are two entirely different types of coiling, the one alternating rhythmically with the other in each plant. The one type of coil is a tight one, is placed almost perpendicular to the length of the host organ, and bears most haustoria. The other coil is loose, has a steep pitch, and rarely has haustoria.

The main value of Zietz' work is that it attempts to provide a harmonious picture of the interrelationships of the two types of coils. I would summarize it as follows. The wide coiling eventually brings the dodder branch into contact with a new host branch. Since thigmotropic responses interfere with growth processes, it is conceivable that such a response in the circumnutating dodder branch will slow down the transport of nutrients to the apex, slowing down apical growth. A reduction in nutrients, in turn, as shown by some of Zietz' experimental evidence, brings about a change to narrow coiling, which appears to be a prelude to haustorial formation. The functional haustoria again bring a rich supply of nutrients to the shoot apex, which is thus allowed to resume its wide coiling in a renewed search for host branches.

Soon after the first haustorial contacts have been made the club-shaped radicle dies, and all contact with the soil is irrevocably lost. Occasionally a seedling actually pulls its radicle out of the ground, from the first coil, although this may represent no more than a shriveling of the hypocotyl and radicle.[1]

General Habit

Dean (1942) has taken the trouble to calculate the total length of branches produced by a single representative dodder plant, and comes up with the astounding figure of nearly half a mile. Some dodders in Mexico, and in Indonesia (*C. reflexa*; Renner, 1934), are able to cover an entire tree with their yellow net of slender twigs. Scarcely a host plant in the vicinity of the dodder escapes attack unless it is resistant. In the western Mediterranean, *C. alba* is known to reach even into standing water, where, completely submerged, it attacks aquatics like species of *Juncus, Oenanthe, Isoetes*, and even *Chara* (Glück, 1911).

The structure of the stem and leaves is very simple. The leaves are minute scalelike organs of the same color as the stem. There is a single vascular bundle in the median region with at most a few phloem strands by way of laterals. Vascular reduction appears to have proceeded farthest in leaves of the subgenus *Grammica* and is least advanced in *Monogyna* (Mirande, 1900). In the stem are three or four vascular bundles made up of normal xylem and phloem, apparently lacking fibrous development. There are only a few stomata on the stem (Pant and Banerji, 1965).

Dodder haustoria sometimes cause rather severe swellings of the host. Dean (1937*b*) has studied and illustrated such hypertrophies. They are usually girdle-like, as they correspond to a haustorial coil.

---

[1]An Indian proverb states that the person finding the root of dodder will have access to all the riches of the earth (Watt, 1889-1890).

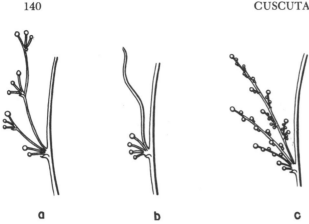

FIG. 6-3. Inflorescence types: *a, Cuscuta chlorocarpa;* *b, C. arabica; c, C. japonica* (based on Mirande, 1900).

FIG. 6-4. *Cuscuta salina* parasitic on *Salicornia* sp. (B.C.)

Frequently the increasing girth of the host branch causes a rupture of the dodder stem between haustoria, which are thus isolated from each other. The isolated haustoria do not necessarily die even when engulfed by host hypertrophy. On *Euphorbia* the dodder may, in this fashion, exist endophytically for some time and suddenly burst forth with a group of flowers from the inner tissues of a host swelling (Denffer, 1947). Anatomical studies have shown that flowers originating in this manner are truly endogenous, breaking through the ·outer tissues of the haustorial organ of the cushion formed by it. Some dodders consistently produce most of their flowers endogenously from their stems (Yuncker,

1920). This seems to be true of such dodders as *C. glomerata* and *C. cephalanthi*. The flowers here are restricted to haustorial coils, and burst forth in rows. Occasionally, through injury or otherwise, a haustorial coil may give rise to a number of small lateral stems. These laterals are also endogenous, emerging from the flanks of the mother stem in two or three series which may be presumed to stand above the vascular bundles (see fig. 6-7,*a*). This is perhaps the natural way of propagation from haustorial coils for perennial plants (see below).

Flowers and Flower Biology

The inflorescence types of *Cuscuta* are diagrammatically illustrated in figure 6-3. The inflorescence may be regarded as an inverted scorpioid cyme showing various degrees of contraction (Mirande, 1900). The primary lateral branch above a leaf scale, whether a vegetative or a floriferous branch, is followed by a somewhat smaller one below it, and so on, in decreasing size and complexity toward the subtending leaf scale. It represents a serial system of axillary buds progressing downward. In more complex parts of the inflorescence a determinate system also prevails: the terminal flower matures first, the lower ones later. In many dodders, however, the simplicity of the inflorescence scarcely allows this sort of analysis.

The typical dodder flower is five-parted (Tiagi, 1966*a*), although flowers with four-, three-, or even two-parted perianths occur (Yuncker, 1920). In size the flowers of different species range from 1 to 6 mm. The petals are membranous, rarely fleshy, and may be white or pale hues of yellow or red. Each is set in a five-toothed cup representing the calyx. Within the perianth members, and alternating with them, are five epipetalous stamens which do not deviate from the standard angiosperm stamen. The center of the flower is occupied by a rather large, bilocular, globular ovary made up of two carpels, with two ovules each. Two, one, or no styles may be present on the ovary. Stigmas may be spherical, peltate, or much elongated (fig. 6-6). When two styles are present they are separated by a cavity on top of the ovary which is not, however, continuous with the ovarian cavity itself.

The flower is so simple that it would have little to offer by way of taxonomic variables were it not for the staminal scales. These are found opposite the stamens and alternate with the petals. In many species these scales are membranous flaps of tissue fringed with a series of finger-like processes of variable length (fig. 6-7,*f*). In other species the scales can more properly be described as two membranous wings attached to the base of the filament.

The morphological nature of the scales has been disputed. Some earlier workers felt that the scales were staminodial (Babington, 1844) or petaloid (Cunningham, 1898). The majority of students, however, have seen the scales as I have described them, as outgrowths of the base of the filament. Indeed, some other Convolvulaceae bear tufts of hairs or scalelike appendages in similar positions. There is no evidence in the literature of any separate vascular supply in even the largest scales. At most the scales may be thought of as enations probably involving only superficial tissues.

The function of staminal scales is a mystery. Their prominence and variability would seem to indicate more than a meaningless remnant. The fact that the scales shrink soon after fertilization (Johri and Tiagi, 1952) may point to some unknown function in attracting pollinators.

Pollination, as seed dispersal, does not seem to have developed a high degree of precision in *Cuscuta*. Yuncker (1920) records the visits of wasps and other hymenopterous insects, but observations of this sort are rare. The pollen is nearly spherical, with three or four colpae or twice as many rugae, but lacking exine sculpturing (Erdtman, 1952).

Embryology and the Fruit

The ovules of *Cuscuta* are anatropous and are provided with a thin nucellus and single integument (Tiagi, 1951a). An interesting gynoecial modification is the so-called placental obdurator (fig. 6-7,d). This is a large group of mucilaginous, papillar cells which are very conspicuous at the time of fertilization. The obdurators stretch from the funiculus around the ovules and upward on to the septum which divides the ovary into two cavities. It is believed that the cells of the obdurators provide nourishment for the growing pollen tubes. They are crushed during embryo formation, and disappear from sight.

Embryologically, the few dodder species studied so far show fundamental variations. Both the Allium type and the Polygonum type of embryo sac occur in the genus, the former in *C. reflexa*, the latter in all other known species (Johri and Tiagi, 1952). After fertilization the embryo sac continues to expand, consuming the adjacent cells of the integument. The endosperm is of the free-nuclear type. When approximately a thousand nuclei are formed

they take up a peripheral position, and cell walls are formed in a centripetal direction. The peripheral endosperm cells in *C. reflexa* break down somewhat later, and their component nuclei fuse to form polyploid nuclei.

The development of the embryo and suspensor also seems to be variable. In some species of subgenus *Cuscuta* there are multicellular suspensors of uninucleate cells. In *Monogyna*, in contrast, the suspensor consists of a few large vesicular cells which are coenocytic (fig. 6-7,c; Fedortschuk, 1931; Johri and Tiagi, 1952). This feature seems to be of limited taxonomic use, since intermediate types occur. Such suspensors suggest haustorial action.

The most important characteristic of the mature embryo is the absence of cotyledons. Upon occasion an additional embryo develops from one of the synergids (Johri and Tiagi, 1952) but perhaps never survives.

While the seed differentiates, the ovary enlarges to accommodate it. In so doing, the ovary often bears the entire perianth aloft. Even at an early

FIG. 6-5. *Cuscuta salina* flowering on *Salicornia* sp. (material from Saanichton Beach, B. C.; × 10).

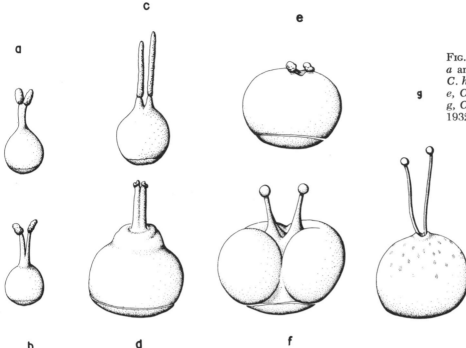

Fig. 6-6. Gynoecial variation: *a* and *b, Cuscuta japonica; c, C. haussknechtii; d, C. exaltata; e, C. friesii; f, C. grandiflora; g, C. cuspidata* (after Yuncker, 1932; magnifications unknown).

stage a line of dehiscence becomes evident at the base of the ovary, surrounding it entirely. Along this zone the top of the fruit wall separates from the base to liberate the seed or seeds.

Perennation

In most of the botanical works mentioning *Cuscuta* it is said that its members are annuals. The extent to which this is true is uncertain. That the greater part of each plant dies off each year in areas of pronounced seasonal change is beyond doubt. There are indications in the literature, however, that individual haustorial coils with their haustoria survive the unfavorable season. The subject has been summarized by Dean (1954). In certain areas of Yugoslavia up to 100 percent of all infected host plants carry perennating fragments of *C. epithymum* through the winter (Stojanovich, 1959). Although the growth habit of dodders makes it difficult to ascertain the amount of perennating, enough is known to convince us that at least in some species the phenomenon is more than accidental or occasional, while in others, even in the same climatic area, plants are strictly annual. Perhaps the explanation of the seasonal resurgence of such plants as *C. salina* and *C. sandwichiana* will be found in their ability to over-winter. These species grow in regions where known germinative mechanisms are not in operation. Perennation may play a role here, although germination may be less difficult than in more inland species.

Systematic Position of Cuscuta

The genus *Cuscuta* within Convolvulaceae occupies an isolated position. No other member of that fam-

ily is parasitic, or even has adventitious roots on its stems. For this reason a separate subfamily, Cuscutoideae, has been erected to accommodate these parasites.

Many students of *Cuscuta*, however, have felt that the dodders were sufficiently different from Convolvulaceae to warrant separation as Cuscutaceae. The first to suggest this was Bartling (1830), but probably most subsequent workers have not agreed with him.

In the last analysis this kind of decision seems to be based on temperament. No matter what new evidence is added, the alternative position is never quite defeated, and the question of family or subfamily remains equivocal. Even plant embryologists can reverse their opinions in this regard. Tiagi (1951a) states that the embryology of *Cuscuta* is essentially the same as that of other members of Convolvulaceae. However, after studying *C. reflexa* with Johri (Johri and Tiagi, 1952), these authors maintain that familial separation is warranted because of embryological and other differences.

Very few botanists question a close affinity of dodders to Convolvulaceae; even those who insist on a separate familial status do not remove the new monotypic family from the vicinity of Convolvulaceae[2]. The important question is one of affinity, where little diagreement exists. My personal feeling is that the problem beyond this becomes one of consensus, and I would therefore favor the inclusion of *Cuscuta* in Convolvulaceae for the present.

---

[2]Hutchinson (1959) is an exception: he places the taxa several orders apart. In the 1926 edition of his book *Cuscuta* is still found in Convolvulaceae.

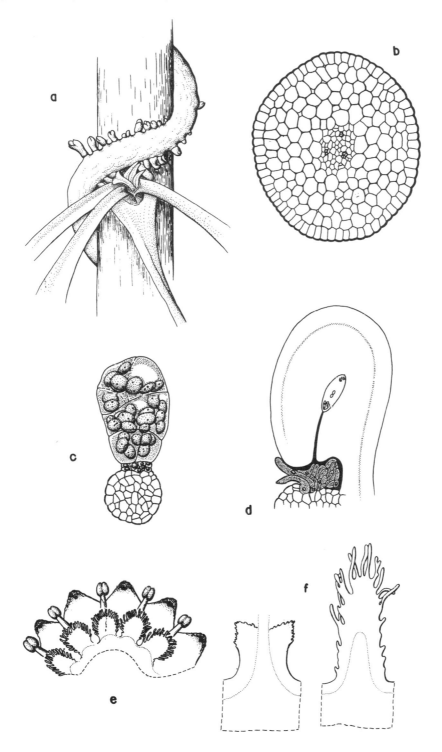

FIG. 6-7. *a, Cuscuta glomerata,* two files of endogenous lateral shoots originating from haustorial coil of injured plant (after Goebel, 1908); *b, Cuscuta* sp., transection of young stem (after Koch, 1874; × 122); *c, C. reflexa,* vesicular coenocytic suspensor cells with proembryo (after Johri and Tiagi, 1952; × 124); *d, C. reflexa,* longisection of ovule and mature embryo sac (placental obdurator shaded) (after Johri and Tiagi, 1952; × 62); *e, C. indecora,* perianth laid open (after Hunziker, 1949-1950; × 5); *f, C. exaltata* (left) and *C. gronovii* (right), staminal scales (after Yuncker, 1932).

CASSYTHA

Surely one of the most extraordinary illustrations of convergent evolution in all the plant kingdom is that involving *Cuscuta* and *Cassytha*. In each case, a single genus has broken away from the completely autotrophic members of its family and has set out on a parasitic course. In both *Cuscuta* and *Cassytha* the flower has lagged behind, as it were, in this divergence, and is still typical of the family as a whole. The remainder of the plants has been so radically modified as a result of its parasitic way of life, however, that we get the impression of typical flowers of Convolvulaceae and Lauraceae having been somehow grafted onto an unrelated parasitic body. Among other things, the two genera illustrate the lagging behind of some characters and the racing ahead of others in the course of evolution.

The similarity in habit has given rise to such Australian names as "scrub dodder," "bushdodder," and "laurel dodder." *Cassytha* material is often misidentified as *Cuscuta,* and future taxonomists will have to glance through folders of the true dodders if they wish to find all *Cassytha* specimens present.

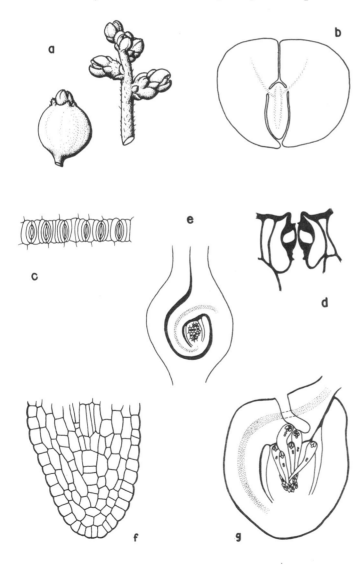

The genus is a small one. Kostermans (1957) speaks of perhaps twenty species, but this is probably too liberal an estimate (Sastri, 1962). No one seems to have taken the trouble to monograph the genus; the nearest approach is Hart's (1925) brief analysis of Victorian species. The greatest diversity of species is found in Australia, a few species occur in Africa, and a single species, *C. filiformis,* is almost ubiquitous in tropical areas in both the New and the Old World.

Cassytha is predominantly a coastal genus, although some Australian species reach the Blue Mountains and have invaded the plains. *Cassytha filiformis* is perhaps most pronounced in its limitation to coastal areas. It is typical of the beach vegetation of many atolls and other tropical regions, but its distribution is not known in detail. The Pacific and much of the Atlantic coasts of Costa Rica are apparently devoid of *Cassytha* (Standley, 1937). So far as I know, the only explanation advanced for the maritime preferences of *C. filiformis* is Wiens' (1962) statement that *Cassytha* fruits may be transported by ocean currents. It would be interesting to know whether the fruits do, in fact, float in sea water and retain their viability. Under ordinary circumstances the seeds remain viable for two years (Mirande, 1905a).

Fruit and Seed

The fruit is almost completely surrounded by a fleshy calyx tube which imparts a red (McLuckie, 1924), yellow, or white color (Taylor, 1950) to the fruit. Birds are said to be attracted by the flesh and may thus disseminate the seeds (Koidzumi, 1917, as cited in Wiens, 1962).

The fruit wall is composed of distinct layers of various types of cells (Boewig, 1904; Mirande, 1905a; Sastri, 1962) among which are columnar layers, sclerenchyma, and cells contributing to mucilage canals. Within this, the single seed has a stony epidermal layer, followed by a central zone of clear columnar cells, and a narrow inner layer with thick mucilaginous walls. The embryo is thus well protected.

The embryo is fully differentiated, having an epicotyl, a radicle, and two very large, fleshy cotyledons. The two cotyledons, which have replaced the endosperm as a storage tissue, are so nearly fused that some earlier workers (McLuckie, 1924) thought the cotyledons to be endosperm.

FIG. 6-8. *a, Cassytha filiformis,* fruit (× 4) and spike (× 2.5) (after Mirande, 1905a); *b, C. pubescens,* sectional view of mature embryo (after Sastri, 1962; × 10); *c,* and *d, C. filiformis,* series of stomata and stoma in section (after Schmidt, 1902; no magnifications given); *e, C. filiformis,* longisection of ovary, showing archesporial cells in ovule (after Sastri, 1962; × 43); *f, C. filiformis,* longisection of apex of lateral root (after Mirande, 1905a; × 144); *g, C. filiformis,* longisection of ovule with several embryo sacs growing toward lysigenous cavity (after Sastri, 1962; × 166).

Germination and Establishment

The germination of *Cassytha* is very similar to that of *Cuscuta* (fig. 6-9,*a*). The radicle bursts through the surrounding tissues and immediately turns toward the ground, penetrating the top surface. The radicle at this point stops growing but, just behind its tip, forms three to five lateral roots which soon overtake it and anchor the seedling. The laterals are said to be exogenous and have the tunicate organization characteristic of the angiosperm shoot (Mirande, 1905*a*), but McLuckie (1924) speaks of a poorly developed root cap. They branch only rarely. While the lateral roots are developing, the nutrients of the cotyledons are being transferred to the basal portion of the primary root, which begins to swell up. Eventually the shoot apex emerges and the exhausted cotyledons are shed along with their protective layer.

The seedling of *Cassytha* lacks the loop typical of *Cuscuta* seedlings, and begins to circumnutate immediately. The direction is the same as that of *Cuscuta*: anticlockwise. The early growth may be as much as four to five inches a week, after which growth slows somewhat. Seedlings in the greenhouse may reach 25 to 32 cm in about two months without a host. Even then, the roots are still alive and functional. If a suitable host has been successfully attacked in the meantime, the roots of the seedling die off in about seven weeks. The plant has now lost all contact with the soil, and continues as a strictly epiphytic parasite. A herbaceous plant often serves as a "starter," but *Cassytha* finds its full development on woody plants.

General Structure

The leaves of *Cassytha* are of about the same size as those of dodder. The mucilage cells typical of *Cassytha* stems also occur in the leaf scale. The vascular supply is reduced to a single median strand.

In considering the stem of *Cassytha* we must first emphasize the perennial nature of the parasite. This means that, in contrast to *Cuscuta*, we may expect some cambial activity. The inner structure of the stem may be seen in figure 6-9,*b* and *c*. The stelar region in its primary state is, on the whole, typical of woody dicots. There is the usual cylinder of vascular bundles, each of the collateral type. Fibers are formed early in the region of the outer phloem. Certain cells between these fibers and the youngest phloem break down to form mucilage canals. The metaxylem has conspicuous, wide vessel members, with large pores. The medullary rays soon become sclerified, and a continuous cylinder of xylem is now present. The cambium, nevertheless, remains limited to the original bundles; when secondary growth takes place, bars or flanges of wood are being formed in those places only. The xylem body of the *Cassytha* stem is thus very different from that of either *Cuscuta* or typical lianas.

The stomata of the stem are all oriented transversely on the stem, and in this regard are like those of mistletoes. They are formed in longitudinal series (fig. 6-8,*c*), and in section show small ledgelike outgrowths into the stomatal aperture (fig. 6-8,*d*).

Flower and Embryology

The inflorescences are in the axils of certain scale leaves of the mature plant. They are simple, spikelike groups of flowers in which each flower is subtended by a median bract and two smaller prophyllar bracts.

The flower, by all accounts, is typical of Lauraceae. The calyx and corolla have three members each, with petals slightly larger than the sepals. The androecium is said to consist of four whorls, each of three stamens. However, the first and third whorls, being opposite the sepals, are formed in that sequence, that is, before the second and fourth whorls opposite the petals (Mirande, 1905*a*). The fourth whorl is made up of sterile, staminodial organs. Another peculiarity is that the anthers in the

Fig. 6-9. *a, Cassytha filiformis*: germination sequence (after Boewig, 1904; ca. × 1); *b*, transection of adult stem (after Mirande, 1905*a*; × 22.5); *c*, transection of older stem (after Mirande, 1905*a*; × 13.5). Flanges of secondary xylem (hatched) radiate from continuous cylinder of primary xylem (stippled) and are capped by phloem and sclerenchyma, the latter two separated by a flattened mucilage duct.

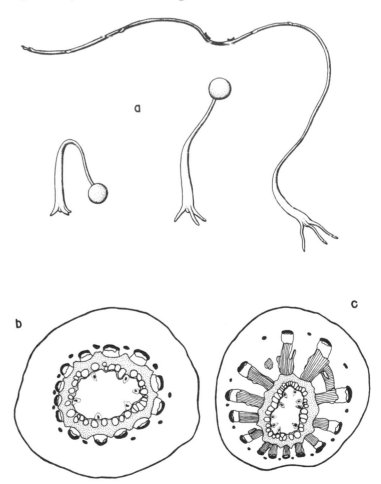

first and second whorls dehisce introrsely, the third extrorsely. The pollen is liberated, as in all Lauraceae, by means of hinged flaps (fig. 6-10,b) revealing the two pollen sacs. The pollen is spheroidal, lacks apertures, and bears numerous wartlike projections scattered over the surface (Erdtman, 1952). Information on pollination is lacking.

The center of the flower is occupied by a single pistil-like gynoecium which is believed to be the remnant of a fusion of three carpels, only one of which continues to form the style and stigma (Mirande, 1905b). The single ovule is suspended from the top of the ovarian cavity. The ovule is anatropous and crassinucellate, and is surrounded by two integuments. At maturity, however, the funiculus has so elongated and arched that it has pushed the micropylar region against the base of the funiculus, so that it faces upward (fig. 6-8,e).

In the center of this ovule, between twenty-five and forty megaspore mother cells differentiate in several tiers (fig. 6-8,e; Sastri, 1956). Most of these complete their meiotic cycle. The functional megaspore (usually the chalazal one) elongates upward while it differentiates into an eight-celled embryo sac of the Polygonum type. Some of the embryo

sacs reach the integuments: one to four of the largest grow out of the nucellus and into the funiculus. The others do not complete so full a development. At least in *C. filiformis* and *C. pubescens,* a rather puzzling cavity develops downward into the funiculus (fig. 6-8,g), which nearly meets the largest embryo sacs. This is not, as might be thought, by way of preparation for the pollen tube: the latter follows the stigmatoid tissue of the style into the funiculus and directly into the embryo sac, without entering the lysigenous cavity. The egg in only one embryo sac is fertilized; the rest are obliterated in subsequent development.

The endosperm is of the cellular type and is eventually supplanted by the massive cotyledons. The food stored in the endosperm is mainly in the form of oil globules, but whether such oil is visible in the cotyledons is not known.

Systematic Position of Cassytha

John Lindley (1853) proposed the Cassythaceae as separate from Lauraceae, but of all subsequent workers only Mirande (1905a) has accepted this separation. The most prominent contemporary student of Lauraceae sees no difficulty in leaving *Cassytha* in Lauraceae (Kostermans, 1957), albeit in the status of a subfamily, Cassythoideae. *Cassytha,* in Kostermans' view, approaches the genus *Cryptocarya* in its floral organization. In the latter genus, also, an enlarged perianth tube completely encloses the fruit. More recently the embryological evidence also has been weighed (Sastri, 1962). It is true that *Cassytha* shows a number of features that deviate from other Lauraceae, such as a secretory type of anther tapetum, the production of numerous embryo sacs and the excessive elongation of some beyond the nucellus, and the cellular endosperm. It is, nevertheless, Sastri's opinion that when these differences are weighed with the similarities the genus is not out of place in Lauraceae.

LENNOACEAE

Lennoaceae is one of three small strictly American families of angiospermic parasites; the other two are Myzodendraceae and Krameriaceae. From the time of the discovery of the first species (La Llave and Lexarza, 1824) the parasitic nature of the plants seems to have been recognized. Since the dividing line between saprophytes and parasites was by no means clear at that time, we do not know whether the "Herba parasitica carnosa" of the original description really meant what it said. After all, even Erasmus Darwin (1825) refers to ivy, *Clematis, Tillandsia,* and *Epidendrum* as parasites!

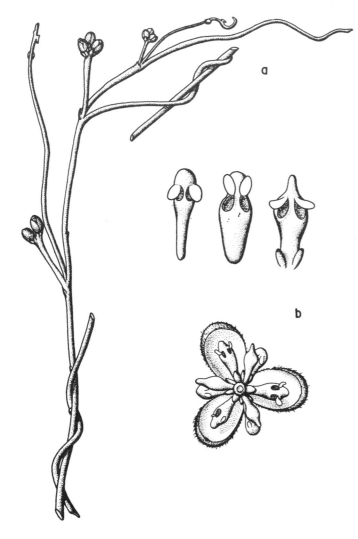

FIG. 6-10. *a, Cassytha glabella,* habit (after Ewart, 1930; ca. × ½); *b, C. pubescens,* open flower and stamens of (left to right) outer, middle, and inner whorl (× 12 and ca. × 26, respectively; after Black, 1948, by permission, The Handbooks Committee).

Three genera are presently recognized: *Ammobroma*, *Lennoa*, and *Pholisma*. Since no systematic treatment exists we are not certain how many species are involved. *Ammobroma* has two species at present (*A. sonorae* and *A. culiacana*), and there may be two species in each of the other genera (*P. depressum*, *P. arenarium*; Templeton, 1962; *L. caerulea*, *L. madreporioides*). A not too conservative treatment might easily leave only a single species of *Lennoa*. From my casual observations I would not be surprised if *Pholisma* and *Lennoa* were eventually united in one genus, for their inflorescence and flower differences seem to be very slight.

Lennoaceae are extremely xerophilous plants. *Pholisma depressum* along the coasts of southern California and Baja California inhabits sand dunes; *P. arenarium* inhabits inland deserts at higher elevations in southern California. *Ammobroma sonorae*, as its name indicates, is lower Sonoran in distribution; *A. culiacana* is similarly xerophilous. In the extremely arid Sonoran desert the fleshy lower stem of *A. sonorae* has been a food of local Indian tribes, who baked it in the manner of a potato (Gray, 1854; Thackeray, 1953). *Lennoa* is also used as a vegetable in the San Luis Potosí region (Dressler and Kuijt, 1968). *Lennoa* demonstrates an amphitropical disjunction, having been reported from deserts of Colombia and northern Mexico (Blake, 1926).

Subterranean Organs

It is unfortunate that nothing is known about the process of germination, or about the way in which the young seedling first manages to reach the root of an acceptable host plant. Even for the mature plant we have very inadequate information. The underground parts (only the top of the inflorescence emerges) of *Lennoa* and *Ammobroma* remain largely unknown. A recent paper (Kuijt, 1966a) provides some details of this sort for *Pholisma depressum* which I shall summarize and supplement with additional observations on *P. arenarium*. The latter material was collected in Joshua Tree National Monument and kindly placed at my disposal by Dr. R. V. Thorne. To what extent these findings apply to *Lennoa* and *Ammobroma* is still an open question.

Pholisma depressum exhibits an interesting type of root dimorphism. I have called the two types of root "pilot roots" and "haustorial roots" in accordance with what I think are their principal functions. The pilot roots are coarse and fleshy, and easily broken. They form a sparsely branched horizontal system at a depth of about two feet. Their function seems to be to reach new host roots. The density of host roots in these coastal dunes is quite low, and the tips of pilot roots often extend eighteen inches or more beyond the nearest haustorial union. The pilot roots themselves, however, are not involved in haustorial contacts, but rather give rise to haustorial roots in strategic places; they function as bridges from host root to host root.

The haustorial roots are short outgrowths from pilot roots. They arise, endogenously, like lateral pilot roots, but are more slender, and often curved. The pilot roots normally show a few longitudinal fissures from within which haustorial roots emerge, often several in a group. The wound thus initiated is widened by the subsequent secondary growth of the pilot root.

The haustorial root, probably guided in its origin and growth direction by a chemical stimulus from the nearby host root, grows directly toward the latter. What happens when contact is made has not been studied in detail, but the apex of the haustorial root apparently has the capacity to digest its way into the host's tissues and become established as a haustorial organ. There is no haustorial disk at any time. Even a small host root may be attacked by a series of haustorial roots (fig. 6-12,*a*), an attack which the host root is not likely to survive unless it responds with marked hypertrophy.

Such hypertrophy, indeed, is typical of the victimized organ (fig. 6-11,*a* and *b*). The swollen part of the host probably corresponds to the extent of the endophyte. Very commonly the host root aborts beyond the court of infection, leaving the parasitic attachment nearly terminal to it. *Pholisma arenarium* has quite different underground parts (fig. 6-13). It is conceivable that the difference is phenotypic, an adaptation to edaphic conditions, for the plants were growing in a bed of gravel. It is also possible, however, that these represent genuine taxonomic differences from the coastal *Pholisma*.

Again there is a striking root dimorphism. The pilot roots, however, form extensive longitudinal series of laterals when branching. The lateral roots are crowded together, indeed, appressed to one another, resulting in flat plates or mats of pilot roots only the tips of which are free. I cannot be certain of the orientation of these plates, but the direction of the shoots issuing from them strongly suggests a horizontal position in the soil. At any rate, *Pholisma* roots appear to be ageotropic.

A partial explanation of this curious formation may lie in the pronounced thigmotropism of pilot roots (fig. 6-12,*c*). It is not uncommon to find a pilot root growing for some distance closely appressed, nearly fused with a stout root of the host. On the ventral side is a large number of haustorial connections, placed in series. Such roots, as do the epicortical roots of mistletoes, seem to sense the direction of the host organ, and grow along it. Thus, when a series of adjacent lateral roots develops contiguously, as a reflection of the meristematic tissue associated with a single bundle in the parent root, this same strong thigmotropic response to each other may force them to form the plates I have described.

Haustorial roots arise anywhere on these root mats. The course of a small host root as it has grown over the surface of the mat is marked with large numbers of haustorial roots (fig. 6-13,*d*) attacking it. A single thin host root trailing over this intercepting

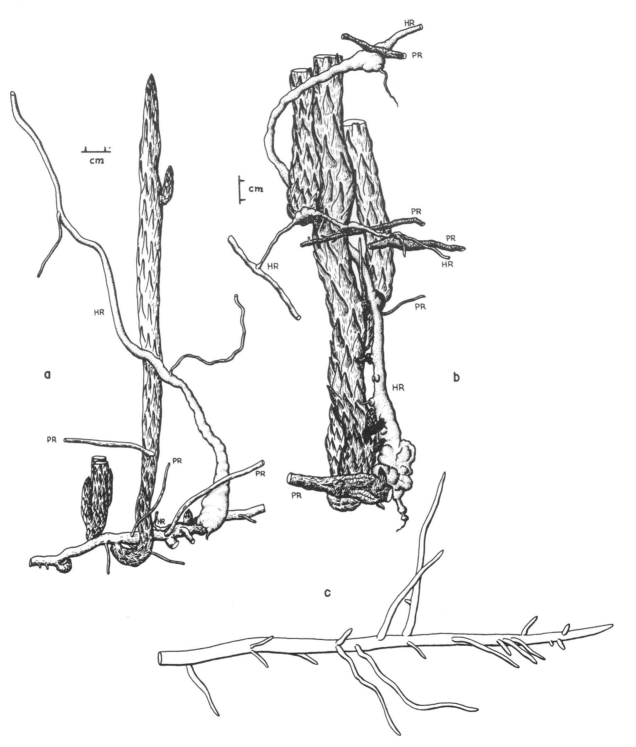

FIG. 6-11. *Pholisma depressum* on roots of *Corethrogyne filaginifolia: a,* large pilot root, giving rise to lateral pilot roots and several inflorescences, attached to swollen host root (light); note formation of horizontal pilot root from base of large inflorescence; *b,* host root attacked in several places by small roots emerging from subterranean parts of inflorescence; one host root and several *Pholisma* pilot roots visible; *c,* branching system of pilot roots (HR = host root, PR = pilot root) (after Kuijt, 1966*a;* × 1).

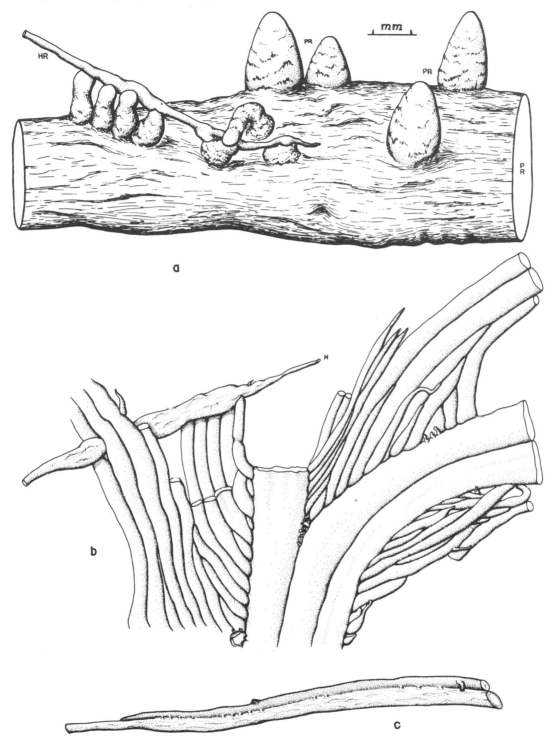

FIG. 6-12. *a, Pholisma depressum,* small host root attacked by series of haustorial roots emerging in series from a pilot root; four young pilot roots have been initiated nearby (after Kuijt, 1966a); *b, P. arenarium,* development of root mat with constricted and parasitized host root (H). Many small haustorial roots on surface of mat have been omitted (material collected in Joshua Tree National Monument by R. V. Thorne; × 2.5); *c, P. arenarium,* thigmotropic root with many haustorial connections, only a few of which may be discerned (same collection; × 2.5).

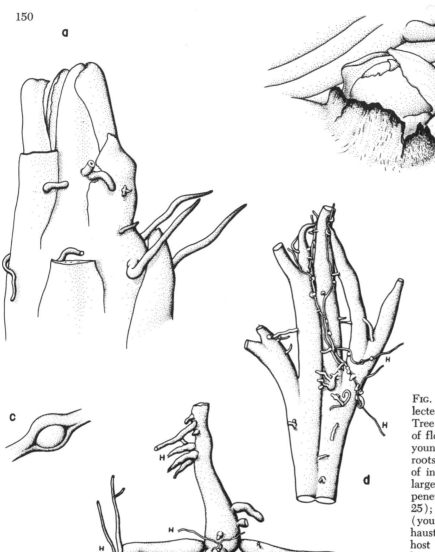

FIG. 6-13. *Pholisma arenarium* (collected by R. V. Thorne in Joshua Tree National Monument at time of flowering): *a*, subterranean tip of young inflorescence with subapical roots (× 5); *b*, endogenous origin of inflorescence bud from flanks of large root of root mat (× 25); *c*, penetration of one root by another (× 25); *d*, dichotomizing pilot roots (young root mat) with numerous haustorial roots attacking slender host roots (× 2.5; H=host); *e*, large host root (below) approached and attacked by *Pholisma* root, two branches of which show thigmotropism (H=host; × 2.5).

surface may provoke the formation of dozens of small, wormlike haustorial roots, many of which penetrate the host tissues. Occasionally a host root is caught between the growing fingers of a root mat where, both in consequence of constriction and invading haustorial roots, it swells up. Self-parasitism is common, much more so than in *P. depressum;* this results in a degree of organic fusion, and doubtlessly reinforces the root mat. There is no sign of root hairs anywhere.

I do not wish to give the impression of a rigid dimorphism in the roots of *Pholisma*. Some roots seem to be intermediate: the young inflorescences sometimes produce roots which seem to be too small for pilot roots but too large for haustorial roots. I have observed such roots almost immediately behind the growing inflorescence apex (fig. 6-13,*a*). That they are haustorial in nature is shown by the fact that they penetrate the leaf scales of the inflores-

cence if these are in the way of their progress. It is possible that these roots are also the result of chemical stimulation from nearby host roots.

Although the pilot roots function, by way of vegetative reproduction, as many rhizomes or stolons, there is no doubt that they are roots (Kuijt, 1967*b*). Their apical meristems are covered with a characteristic root cap. Both types of roots arise endogenously, although this is true of the inflorescence as well. The anatomy of the pilot root also demonstrates it to be a root. It can probably be assumed that the haustorial roots, in their structure, represent a simplified version of the pilot roots.

The functional significance of these curious plates of roots is a mystery. It is true that all host roots are intercepted in this way over a limited area, but would not the same purpose be served if the same length of root were distributed in a more normal pattern? The only other suggestion I can offer is that such roots may serve as storage organs for nutrients and water; evidence for such a function might be adduced also from the rather limited xylary development in the pilot root.

Field observations on *Ammobroma* and *Lennoa* would add much to this account. I am inclined to think that at least the South American *Lennoa* are similar to *Pholisma arenarium* in root arrangement, for an early generic synonym, *Corallophyllum* (Fournier, 1868), was based on a fleshy, pinnately branching root which was mistaken for a succulent leaf.

Inflorescence and Flower

In *Ammobroma culiacana,* and perhaps also in *A. sonorae,* the inflorescences grow directly out of the infected host root (Dressler and Kuijt, 1968). In both species of *Pholisma,* however, the inflorescences arise not from the endophyte but from the pilot roots. Sometimes several stalks emerge near well-developed haustorial contacts, but at other times no such preference is evident and inflorescences are spaced along the pilot root. They arise from the interior tissues of the latter, as is evident from a ruptured collar around the young bud (fig. 6-13,*b*). Occasionally an older inflorescence also gives rise to new ones, in axillary positions, but a given inflorescence probably does not survive beyond one growing season.

The structure of the stem is simple. With the exception above, it remains unbranched until it approaches the surface of the soil. It is covered with rather large, thick, brownish leaf scales which in the aboveground portion may take on a pale hue of pink or purple. The stem is very fleshy and brittle because of its relatively weak vascular development.

The flower-bearing part of the inflorescence rests almost directly on the soil surface (fig. 6-14,*a*). Because of its crowded condition, the inflorescence is difficult to study, and has not been carefully analyzed in either *Ammobroma* or *Lennoa.* In over-all shape the flower mass is ovate in *Pholisma arenarium* but more hemispherical in *P. depressum.* *Lennoa* appears to have a less crowded inflorescence which assumes a more irregular shape not unlike that of *P. depressum.* In *Ammobroma* we find a regular, flat-topped inflorescence approaching a capitulum. From the side it is very reminiscent of the pileus of a mushroom or, to remain within the parasitic flowering plants, of *Scybalium fungiforme* of Balanophoraceae.

The structure of the inflorescence of *P. depressum* has received some attention from Copeland (1935), who found the ultimate portions to be essentially cymose, and probably always monochasial; in *Lennoa,* however, dichasia have been noted. It may be that these are only differences in degree, and perhaps position within the inflorescence.

The flowers of Lennoaceae resemble those of Boraginaceae and Hydrophyllaceae. The floral envelopes are studded with capitate glands bearing one-celled heads or, in *Ammobroma* sepals, with

long hairs giving the inflorescence a woolly appearance. It is possible that these trichomes make it difficult for crawling insects to reach the flowers, and thus reserve the nectar for flying insects. This is only a guess, however, as no precise observations on floral biology of the family have been published. The faintly sweet-scented flowers of *A. culiacana* are visited by flies, beetles, and small butterflies (Dressler and Kuijt, 1968).

The calyx consists of four to eight long slender members which are fused with the receptacle below the insertion of the perianth, but are mutually distinct. In *A. culiacana* the sepals are fused for about two-thirds of their length. In contrast, the corolla is completely gamosepalous. Around the throat of the corolla tube we observe an attractive plicate pattern, with inward and outward folds alternating. The inside of the tube is glabrous; the outside usually bears small trichomes. In color the five or six petals range from pink to pale tints of magenta.

The androecium is made up of five to eight sessile or short-stalked anthers of the normal four-celled angiospermic type, alternating with the corolla lobes. They dehisce longitudinally, and are implanted around the floral entrance. The longitudinal extension of the floral tube takes place at its base, carrying the anthers upward. Fusion with the corolla is so complete that, at least after extension, no xylem supply is recognizable below the anther. In *Lennoa* the anthers are said to be implanted at two different heights, in an alternating fashion (Copeland, 1935).

We are fortunate in having a careful recent analysis of pollen structure of all three genera (Drugg, 1962). *Ammobroma* and *Lennoa* have three-colporate grains, and the latter also has an occasional four-colporate grain. In *Pholisma* four-colporate grains predominate, but three- and five-colporate ones are also found. Exine sculpturing is similar, and the construction of apertures is virtually identical in the three genera. Pollen structure is useful in this family in confirming certain systematic notions as discussed below.

The middle region of the flower is occupied by a simple pistil with a long, straight, solid style bearing the nearly undifferentiated to capitate stigma which occupies the center just below the floral entry. The ovary is superior, depressed-globose, and made up of about eight carpels. These carpels project an equal number of wedgelike outgrowths into the single central cavity. Each rib bears two ovules, extending back toward the ovary wall. The individual ovules are thus separated from their neighbors (fig. 6-15,*c*), either by the thin attachment of the rib or by an additional radial flange of tissue issuing from the carpel wall. Each ovule, therefore, is in a separate pouch. The tenets of classical morphology would lead to the interpretation that the ovule-bearing wedges represent fused carpellary margins and that the radial flanges correspond to the median plane of the carpel.

Mature ovules have curved around in such a way

FIG. 6-14. *a, Pholisma depressum* (Morro Bay, California); *b, P. arenarium* (Japatul, California; courtesy Dr. Reid Moran); *c, Ammobroma sonorae* (preserved material without data, UC); *d, Pholisma depressum* showing linear seriation of inflorescences (Morro Bay, California).

that the micropyle faces the ovarian cavity. There is only a single integument, and the nucellus is no more than a single layer in thickness. These two ovular layers are almost entirely consumed by the expanding embryo sac. The behavior and development of the latter are insufficiently known, but the final product in *Pholisma* appears to be the usual eight-celled embryo sac.

Seed and Dissemination

Not having much left by way of young seed coat, the developing seed has taken recourse to the adjacent tissues of the carpel wall and its radial protrusions. A layer surrounding the ovular pouch differentiates into a sclerenchymatous tissue; the whole mature seed is thus neither a seed in the traditional sense nor a fruit. The mature fruit dehisces by means of an irregular circular slit; the top of the fruit wall with the attached style comes off as a calyptra-like structure to reveal a whorl of seeds (fig. 6-16,*b* and *c*). The seeds are more or less ovoid with a rough, dark brown or black surface (fig. 6-16,*a*). The interior of the seed is of simple construction (fig.

6-15,*b*). A considerable amount of endosperm surrounds the completely undifferentiated embryo.

There seems to be no well-developed dispersal mechanism. The old inflorescence dies without undergoing any morphological changes, bearing its seeds between the remnants of various floral structures. We may assume that the rather severe ecological conditions surrounding the plants contribute to the early destruction of the seed-bearing head, and that in this fashion some degree of seed movement is accomplished. Rabbits are active in the vicinity of the *Pholisma depressum* population which I have seen at Morro Bay, California, and may be similarly effective. I have not seen evidence of their interest even in the fleshy underground parts of the parasite, however, and their contribution may be incidental.

The seeds, under the influence of various natural agencies, may eventually be adjacent to the young roots of susceptible hosts. This may happen either by means of a downward movement of the seed into the soil, or by its burial through shifting sands in dunes which are subsequently invaded by new vegetation. It seems safe to predict an extended viability of the seeds under such conditions. Comparison with other parasites also leads me to suggest that the seed will be brought to germination only through the chemical stimulus inherent in host roots.

Systematic Position of Lennoaceae

A number of divergent suggestions have been made with regard to the natural affinities of Lennoaceae. The subject has been reviewed by Copeland (1935), from whom I extract the following summary. The original description of *Lennoa* (La Llave and Lexarza, 1824) suggested a possible relationship with Primulaceae, especially with *Lysimachia*. The parasitic habit seems to have led several botanists astray in proposing affinities with Orobanchaceae (Hooker, 1844), *Monotropa* (Torrey, 1867; Fournier, 1868), and even *Cuscuta* (Conzatti and Smith, 1910). Solms-Laubach (1870) accepted a position in Ericales (Bicornes). More recently, the position most commonly assigned has been in the vicinity of Hydrophyllaceae and Boraginaceae. It was first defended by Hallier (1923), and received the support of Suessenguth (1927) and most subsequent workers, including Drugg's palynological study (1962). The two known chromosome counts in the family, unfortunately, at present contribute little in this regard: *Ammobroma sonorae* and *Pholisma depressum*, both with a haploid number of 18 (Carlquist, 1953; Anonymous, 1962).

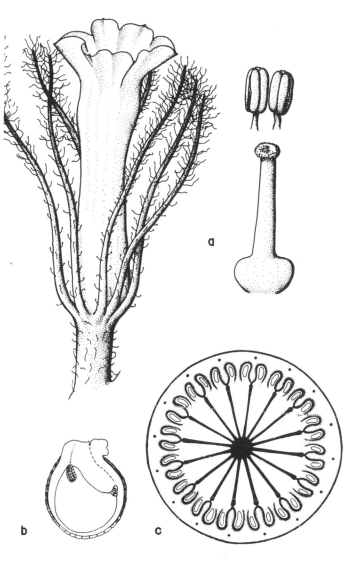

FIG. 6-15 (*above*) *a, Ammobroma sonorae*, flower and floral details (after preserved inflorescence of unknown origin, UC; × 10); *b, Lennoa*, longisection of seed; embryo and antipodal haustorium at opposite ends of endosperm (after Suessenguth, 1927; × 125); *c, Lennoa*, diagram of transection of ovary soon after fertilization (based on Suessenguth, 1927).

FIG. 6-16 (*Right*) *Pholisma depressum* (material from Morro Bay, California): *a*, seed (× 50); *b*, cap of fruit (× 12.5); *c*, fruit with seeds in natural position (× 25).

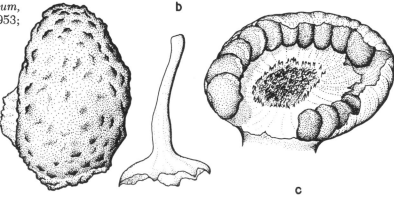

154

Cannon (1910), investigating the roots of various desert plants, made the unexpected discovery that the roots of *Krameria* were parasitic on neighboring plants. In all fairness it should be pointed out, however, that Edward Palmer, when collecting *Krameria* in Baja California in 1887, suspected it might be parasitic (Vasey and Rose, 1890). Cannon's report has had remarkably little currency, with the result that *Krameria* is ignored in almost every account of parasitic plants.

The genus has been known for a long time and is a fairly common shrub in certain localities. As in most parasites among the figworts and sandalwoods, the general appearance of *Krameria* gives no indication that it draws nutrients from other plants. Its habitat is much like that of many other desert shrubs (fig. 6-18). It is a small, dense bush, up to three or four feet in diameter, closely appressed to the rocky soil, profusely branched, with greatly reduced leaves and stiff branches, and of a blue-gray cast. In most of these features it is a typical desert shrub.

The genus has been partially monographed (Rose, 1906; Britton, 1930), and its geographic range can be summarized as follows. *Krameria* is mostly a Mexican genus, reaching north into Texas, Kansas, Oklahoma, Arizona, Nevada, southern California, Georgia, and Florida. It seems to occur throughout the Caribbean and is known from Guatemala. It reappears in South America, where records from Colombia, Venezuela, and Peru exist. The extent to which *Krameria* is a Mexican genus is shown in Rose's (1906) treatment for North America, where fourteen of the fifteen species recognized are represented in Mexico. While acknowledging its Caribbean occurrence, therefore, we can call *Krameria* an amphitropical genus with distinctly xerophytic preferences.

Habit

The root system of *K. grayi* was described by Cannon (1910) as consisting of a short taproot with half a dozen laterals arising from near the soil surface, extending in a fairly horizontal direction as far as 2 meters. The fact that no lateral root reached deeper than 18 cm indicated a remarkably shallow root system for a desert climate. Even the taproot was said to reach no farther than 20 cm. Cannon obtained the impression that the taproots had extended farther but had died back. His report dealt with plants in the Tucson region.

My own observations, superficial as they are, disagree with Cannon's. The same species in Joshua Tree National Monument has a root system extending much deeper (fig. 6-18). I was unable to find laterals of a horizontal tendency; instead, a number of laterals, taking their origin from just below the soil surface, reached obliquely downward in a fanlike manner. I can confirm the brittleness of roots mentioned by Cannon, and the relative lack of

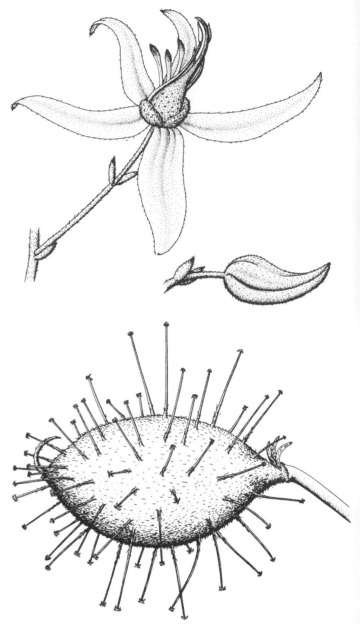

Fig. 6-17. *Krameria grayi*, flower, bud, and fruit (material from Joshua Tree National Monument, California; × 5).

branching. My observations were made in April. It is obvious that a comprehensive picture of the root system of *Krameria* is a prerequisite to the study of its parasitism.

Krameria stems are terete when young, bearing the leaves in alternate arrangement (fig. 6-18). Leaf shape is simple except in one Mexican species with small lateral pinnae at the base (*K. cytisoides;* Taubert, 1894). The leaves are up to 3 × 1 cm in *K. tomentosa*, but nearly nonexistent in *K. ramosissima*. The shape of the leaves ranges from elliptic-lanceolate in the former to almost linear in plants of *K. lanceolata* and *K. secundiflora*. An indumentum of white hairs covers all young vegetative parts of

Krameria except some Western species like *K. grayi,* which, at least superficially, is glabrous. The older stems, especially the woody base of suffrutescent species, shows a longitudinal furrowing due to cork formation.

Flower and Pollination

The papilionaceous appearance of the flower has led to the frequently accepted placement of the genus with Leguminosae (fig. 6-17). The calyx is four- or five-parted and forms the showy red or purple portion of the flower. The four to five petals are present in two extremely dimorphic groups. The lower two, symmetrically sheathing the ovary, are short fleshy scales with a silvery indumentum even at the time of anthesis. The remaining petals are slender thin organs curving upward somewhat, terminating in dark, nail-like sharp tips. The stamens are three to four, inserted freely below the ovary, with a basifixed anther which dehisces by means of a terminal pore (Taubert, 1894). The ovary is sessile, partly obscured by the petals, and extends in a long saber-like style. Within the ovary are two anatropous ovules, inserted at the base of the single ovarian cavity. I have not been able to ascertain the number of carpels.

We know nothing about pollination but may assume entomophily. The pollen gives us no clue, for it is quite smooth except for some very fine longi-

tudinal striations clustered around the poorly marked angles. The grain is nearly circular in polar view. My rather inadequate preparations suggest that only two pores are present, in the manner shown (fig. 6-19,*h* and *i*).

Fruit, Seed, and Germination

The enlargement of the ovary is accompanied by growth of the two fleshy petals which at maturity surround the entire fruit (fig. 6-17), with only the curved style protruding. This elaborate fruit envelope bears some of the most elegant spines known in the higher plants (fig. 6-19,*a* and *f*). Although I have not seen fruits of all species, there appear to be three general types of spines, linked by some intermediate forms. The first is a stout, nearly conical spine, its base clothed in white hairs, its apical portion occupied by many small, sharp retrorse barbs (*K. lanceolata, K. secundiflora*). The second type is very slender, and bears only about half a dozen slightly curved barbs along its upper half (*K. parvifolia, K. pauciflora, K. tomentosa*). Some white hairs are present lower down. In two species (*K. glandulosa, K. ixina*) I have been able to discover only a single recurved barb just below the tip of the spine. In the third type of spine, also very slender, the only barbs present are four large hyaline ones, slightly curved, at the apex of the spine, and extending sideways (*K. grayi, K. paucifolia*). There are some characters in these spines which may be very useful systematically. These elaborate structures must aid considerably in dispersal by animals.

The fruit enclosed in this spiny envelope is a nut

Fig. 6-18. *Krameria grayi,* branching habit of stems and roots (Joshua Tree National Monument, California).

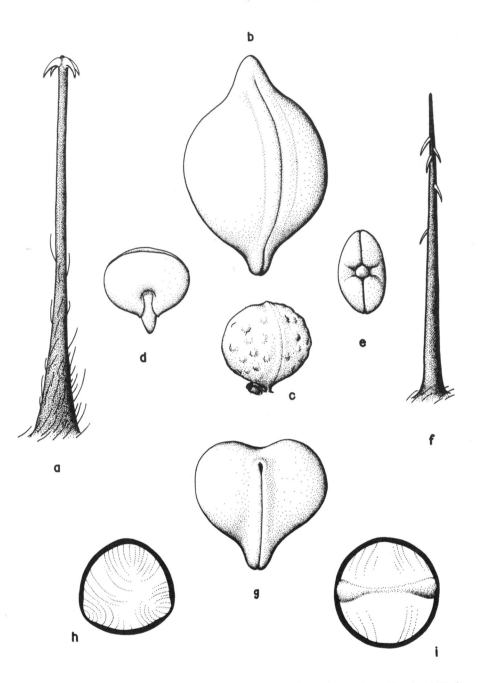

FIG. 6-19. *a, Krameria grayi,* trichome from fruit envelope (R. Wilson No. A 1087, from Joshua Tree National Monument; × 20); *b, K. grayi,* fruit (× 5); *c, K. grayi,* seed, with second, aborted seed visible below (× 5); *d* and *e, K. grayi,* embryo seen from side and below (× 5); *f, K. parvifolia* var. *glandulosa,* trichome from fruit envelope (R. Wilson No. A 1075, from Joshua Tree National Monument; × 20); *g,* fruit of same (× 5); *h, K. grayi,* polar view? of pollen grain (liquid-preserved material from same locality as above; × 550); *i,* same, equatorial view?.

in which only one seed has developed. The shape of the fruit in two species is illustrated in figure 6-19,*b* and *g*. The fruit wall is nearly smooth; the single longitudinal suture suggests a unicarpellate structure. The hard fruit wall is followed by a very thin membranous seed coat. There is no endosperm at maturity. The embryo is very large, filling the entire fruit, and is made up largely of two fleshy cotyledons. The outer surface of each cotyledon is nearly hemispherical, extending into two sagittate lobes below the cotyledonary node (fig. 6-19,*d* and *e*). The inner faces of the cotyledons are closely appressed to one another except at the apex, where they are parted in a liplike fashion. The epicotyl is small and shows a couple of minute leaves. The radicle is straight and stout, projecting almost as far as the cotyledonary spurs which flank it.

A remarkable and unexplained fact is that, at least in *K. grayi*, fully formed fruits frequently turn out to be sterile. Breaking open the seeds of such fruits, we find two shriveled ovules at the base of the empty cavity as proof that no seed ever developed. How common this type of sterility is, and what its meaning, I cannot say. Among about two dozen mature fruits of this species at Joshua Tree National Monument in 1965, I was unable to find a single seed; very few fruits showed evidence of insect damage.

We know virtually nothing about the process of germination. The structure of the embryo indicates a long period of autotrophic growth during which the nutrients stored in the cotyledons will be used. Palmer's field notes (Vasey and Rose, 1890) contain an interesting statement to the effect that the seeds of *K. grayi* in Baja California sprout upon the plant, "forming three rather fleshy leaves like the leaves of the plant and of a bronze color." These must have been true leaves, as the cotyledons could not look like foliage leaves; this, in turn, may mean that the cotyledons in reality stay within the fruit. The entire process of germination, about which Cannon (1910) unfortunately does not inform us, is unexplored. At what time parasitism makes its appearance, or whether indeed parasitism is a requisite of full development, are among the many questions which need to be answered in this interesting and isolated genus of parasites.

Systematic Position of Krameriaceae

The most obvious and common taxonomic assignment of *Krameria* is to Leguminosae. There was so little doubt in Taubert's (1894) mind in this respect that he assigned *Krameria* tribal status within the subfamily Caesalpinioideae. Yet, especially when we consider the calyx-like nature of the showy part of the flower, there is scarcely a feature which favors such an alliance, except possibly the rather undistinguished pollen grain (Erdtman, 1944). The most detailed study of the genus (Kunz, 1913) also pointed to the need for exclusion from Caesalpinioideae. The remarkably large chromosomes, comparable to those mentioned for *Phoradendron serotinum* and *Psittacanthus cuneifolius*, negate such affinities also (Turner, 1958). A tentative position within Polygalaceae was suggested by Bentham and Hooker (1862-1883); Hutchinson (1959) also used this alliance, placing Krameriaceae in Polygalales. The genus was made into a monotypic family by Chodat (1890), the position perhaps most generally accepted today. Many new facts need to be introduced before any systematic position, whether one of the above or a new one, can be adequately supported.

7

The Haustorium

THE TERM *haustorium* first appears in botanical literature in A. P. De Candolle's *Théorie élémentaire de la botanique* (1813), with special but not exclusive reference to *Cuscuta*. The term does not carry a morphologically uniform connotation in contemporary botanical thought. It has been applied to various cell types in the embryology of flowering plants (see under Santalaceae) and to absorptive structures of parasitic cryptogams. Not all aggressive haustorial organs in plants carry this term, as is evident from the foot of the moss sporophyte, which, though not usually called a haustorium, is nevertheless haustorial in action.

There is no *a priori* reason for all haustoria of parasitic phanerogams to be similar in origin, mode of attack, structural organization, or functional mechanism. We must be careful, in this most plastic of organs more than in any other, to be on the lookout for convergent evolution as a confounder of phylogenetic speculation.

The concept of haustorium as here utilized has considerable latitude. It is used to refer to the physiological bridge which joins host and parasite, or at least to the latter's contribution to this bridge. The haustorium is the specialized channel through which nutrients flow from one partner to the other. It is the organ which, in a certain sense, embodies the very idea of parasitism.

The haustorial organ often undergoes elaborate differentiation before it breaks into the host. This is true of many Santalalean haustoria, but applies also to some others like *Cassytha* and *Cuscuta*. The "efforts" of the haustorium at this stage are focused on penetration, and the establishment of the first effective conduit of nutrients must of necessity be postponed. This separation of functions is reflected to some extent in the architecture of the organs concerned, much as the young root of an ordinary plant shows differences from its later life when undergoing secondary growth. Because of this functional division of labor the term "prehaustorium" was proposed by Peirce (1893) to apply to the central superficial cells immediately in front of the meristematic tissue from which the mature haustorium later takes its inception. The prehaustorium of *Cuscuta* actually penetrates host tissues, but is quickly replaced by the true haustorium. In other haustoria, nothing comparable seems to take place except perhaps in *Striga* (Okonkwo, 1967). I feel that the haustorium is more usefully conceived as a single organ, even though it has functionally, and consequently structurally, different stages. The adhesive, intrusive, and conductive stages of these haustoria follow each other closely and, indeed,

overlap. These stages are manifested in separate tissues for purposes of adhesion to the host surface, often in a specialized haustorial portion or intrusive organ which first gains permanent entry into the host, and finally in a bridge of vascular tissue between the two partners.

One encounters other semantic difficulties in considering parasites whose absorptive organ, instead of being unified, has developed into a fragmented system of strands and filaments which permeate the host's tissues over a large area. To speak of a haustorium here, especially in Rafflesiaceae, where it constitutes the entire vegetative body, is not satisfactory; such expressions as "haustorial" or "absorptive system," "endophytic system" or simply "endophyte" gain in usefulness. The use of "mycelium" even in the most extreme cases of reduction runs counter to my own preference, and I hesitate to speak of the "hyphae" so typical of a complete dodder haustorium—a term which seems to have gained general acceptance.

In my account of this complex subject I shall deal first with the Santalalean haustorium and its many modifications. I draw heavily on my recent review (Kuijt, 1965*b*) and on Thoday's important series of papers. Next come the unique haustorium of *Cuscuta* and the haustoria of Scrophulariaceae, Orobanchaceae, Lennoaceae, and Rafflesiaceae. The extraordinary absorptive apparatus of Balanophoraceae concludes the section; the little information available for *Cassytha* and *Krameria* is presented in connection with the Santalalean haustorium. I am not prepared to defend this sequence of topics as the best possible. The great diversity of haustorial organization would certainly allow other arrangements. My principal aims here are economy and clarity.

To review the distinction between primary and secondary haustoria: a primary haustorium is the direct outgrowth of the apical meristem of a radicle and is therefore always terminal in origin; secondary haustoria, in contrast, are usually lateral organs, as in Santalales, Orobanchaceae, and Scrophulariaceae. It is in these groups that the distinction is most significant, and has interesting phylogenetic implications. In other parasitic groups the distinction is more questionable, either because of the absence of secondary haustoria (many Balanophoraceae, Rafflesiaceae) or of primary ones (*Cuscuta* and *Cassytha*), or because of our ignorance of these aspects of the parasite's mode of life (Lennoaceae, Krameriaceae, and Hydnoraceae). In *Pholisma* (Lennoaceae) small, modified roots grow directly into the host organ: we may still call such haustoria secondary, since they are produced by roots other than the radicle.

In duration, haustoria are extremely variable. In most perennial parasites at least one original haustorial union persists until the parasite dies. This may not be true of the few parasites that perennate and reproduce vegetatively. A few dodders apparently do so, and thus the individual outlives its haustoria, which may be replaced continuously. The literature of Santalaceae and Scrophulariaceae has many references to the ephemeral nature of haustoria, but the actual record shows little precise information. We are certain of longer duration only in haustoria in which secondary growth normally takes place, as in *Lathraea* of Scrophulariaceae (Heinricher, 1895) and *Buckleya* of Santalaceae, where haustoria may remain active for fifteen years or more (Kusano, 1902), but these may be the only convincing examples. In the other perennial root parasites of these families it seems to be a matter of renewed search for suitable host roots each growing season. This is probably true even of such trees as *Santalum album*, *Nuytsia floribunda*, and *Gaiadendron punctatum*, where large haustoria leading to hosts other than themselves are not known.

SANTALALES AND CASSYTHA

The basic type of Santalalean haustorium, as represented in many terrestrial parasites of this order, is the logical starting point for discussion. In some of the more advanced types, which have established themselves as parasites on the branches of trees, a number of radical evolutionary changes have taken place which will be discussed later. Since all parasites in Santalales have chlorophyll, they are hemiparasites.

The basic type is represented in either primary or secondary haustoria (figs. 7-1, 7-2) or both, as in mistletoes with truly epicortical roots. Thus the haustorium may terminate the radicle or it may have a subterminal position on a lateral root. If the former, it will mould itself into a conical or hemispherical mass of tissue closely appressed to the host surface. If the latter, a conical outgrowth becomes visible just below the region where one might expect root hairs. The outgrowth continues to swell, reaching a spherical or oblong shape by the time the apex makes contact with the host organ. The surface layers of the mother root and the haustorium are at all times continuous and show no break. No wonder so many authors have regarded this organ as exogenous.

The margin (mantle) of the haustorium now extends sideways, applying itself to the contours of the host surface. The precision of this "fit" is sometimes amazing. The stomatal crypts of *Nerium* leaves may be invaded and filled by the haustorial organ without disturbing the epidermal cell of the leaf (see fig. 7-7,*c*). The mantle, especially when the host organ has a small diameter, may grow around the host more than in its longitudinal direction, and the haustorium assumes the shape of a saddle (see fig. 7-4). In some instances the two main

lobes nearly meet at the other side of the host root, forming an effective clasp (fig. 7-1). In *Ximenia* the lobes actually do meet upon occasion (Barber, 1907*c*).

The situation in *Nuytsia floribunda* needs further study. According to the only account available (Herbert, 1918-1919), the two opposite lobes of the mantle seem to continue growth until they meet, and then fuse to form an unbroken ring. From the inside of this haustorial ring, several intrusive processes are said to penetrate the imprisoned organ. This extraordinary mode of attack deserves investigation. Almost any plant root in the vicinity of a *Nuytsia* tree may become thus constricted. Indeed, even underground electrical cables are not safe from attack (fig. 7-3,*a*).

Bilaterality may express itself not perpendicular to the host organ but parallel to it. The haustoria of *Phthirusa* (see fig. 7-6) and *Struthanthus* are elongated in the direction of the mother root. As the latter's course usually follows that of the host branch, the haustoria are longitudinally oriented with regard to both organs.

The mantle effectively hides the act of entry into the host. Sometimes, however, especially in terrestrial forms, the initial mantle is followed by a second one which is interpolated between the first mantle and the host, lifting the first mantle while forming a clasplike hold (fig. 7-1,*b* and *e*). The second, in turn, may be similarly replaced by a third, and so forth. This repeated mantle formation results in a compound haustorium. It is to be expected in the secondary haustoria in most Santalalean parasites, but not in all: *Antidaphne* and *Olax* appear to have only simple haustoria (Barber, 1907*c*; Kuijt, 1965*b*).

In order to understand the action of the haustorium we must inspect its internal organization. The great mass of cells making up this organ is parenchymatous. Before penetration of the host has begun, the haustorial cells forming the contact surface have been modified into columnar cells which presumably secrete an adhesive. Reaching from the margin of the mantle partway up the flanks of the haustorium is a thin layer of parenchyma cells which have apparently been crushed. This "collapsed zone" is perhaps the most distinctive feature of the Santalalean haustorium.

The interpretation of the haustorial core is more difficult. I have recently attempted to harmonize the available information in order to understand how entry into the host is effected (Kuijt, 1965*b*), and from that account I reconstruct a tentative sequence of events. The young parenchymatous core at an early stage begins to divide actively in planes parallel to the host surface. The core thus assumes a tiered appearance: it consists of a series of plates of parenchyma cells, stacked up like coins. In the axial region of the haustorium a further differentiation takes place. Two layers of cells, cutting right across the tiers which have been formed, assume a columnar appearance. Their constituent cells elongate, and their protoplasm becomes extremely dense

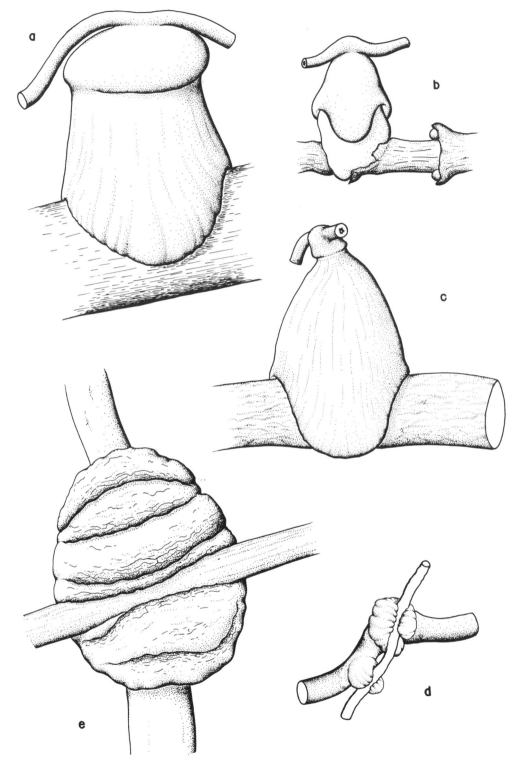

Fig. 7-1. *a-d, Geocaulon lividum,* haustoria in various stages of development; *b,* on rhizome of *Linnaea borealis,* others on roots of *Vaccinium* sp. (Manning Park, B. C.; × 12.5); *e, Comandra umbellata* subsp. *pallida,* compound haustorium (Oliver, B.C.; × 50).

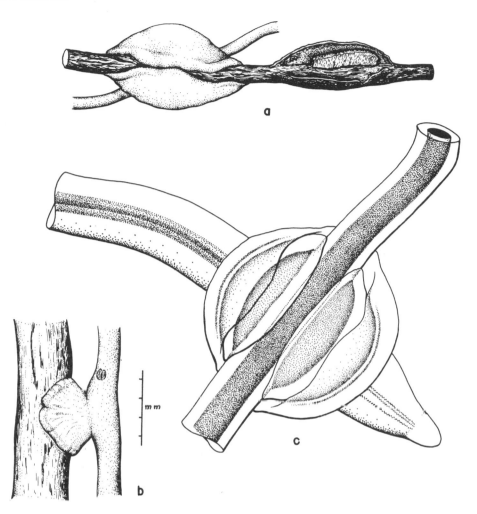

FIG. 7-2. *Gaiadendron punctatum* (*a* and *b*, after Kuijt, 1963a; *c*, after Kuijt, 1965a):
a, functional haustorium on root of *Vaccinium consanguineum;* scar left by older haustorium
on right (× 5); *b*, self-parasitic haustorium; *c*, partially cleared haustorium on fern
root, showing three successive mantles (× 50).

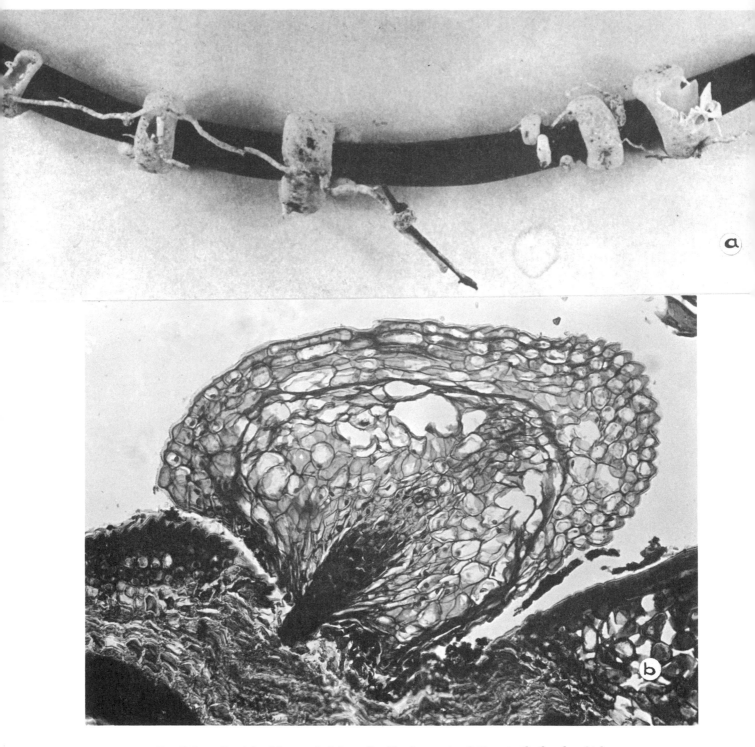

FIG. 7-3. *a*, Coaxial cable attacked by collar-like haustoria of *Nuytsia floribunda* which
had penetrated insulating polymer to produce short circuits; cable about 2 cm in diameter
(courtesy Forests Dept. Government of Australia); *b*, haustorial penetration by *Arceuthobium
campylopodum;* note collapsed zone, and clearly endogenous intrusive organ (after Scharpf,
1963; courtesy of author).

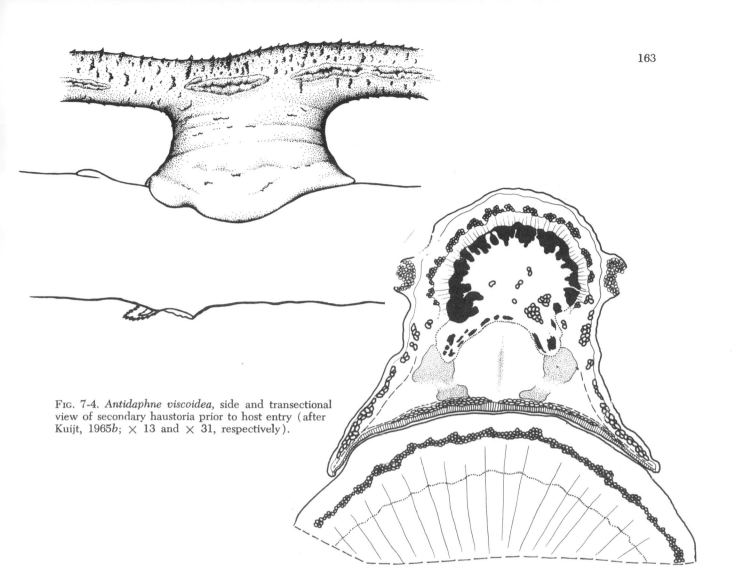

FIG. 7-4. *Antidaphne viscoidea*, side and transectional view of secondary haustoria prior to host entry (after Kuijt, 1965*b*; × 13 and × 31, respectively).

FIG. 7-5. *Antidaphne viscoidea*: *a*, epicortical roots of older plant, demonstrating flaring margins of secondary haustoria; *b*, same, showing distal portions of epicortical roots dying off, and those nearest primary haustorium (upper left) growing rapidly (both after Kuijt, 1964*c*).

and granular. Some cells belonging to these two median plates and nearest the mother root now appear to withdraw their cytoplasm from the median plane which divides the two plates. This curious behavior seems to be followed by a type of auto-lysis of the cells. The action proceeds toward the host surface. By the time the latter is within reach, a cavity, or gland, is present, presumably containing enzymatic materials which are released onto the host and aid in penetration.

The origin of the intrusive organ has not been adequately dealt with in the literature. At least in *Phthirusa* this organ seems to originate in a small group of meristematic cells just above the internal cavity. The intrusive organ grows through the cavity and any intervening haustorial tissues to reach the host's surface. Whether the same situation prevails in other Santalalean haustoria is a matter of speculation.

What, then, is the significance of the so-called collapsed layer? Its very name suggests that it is compressed or crushed, perhaps by the expanding core. Such an explanation has been assumed by most students of Santalalean haustoria. But it is difficult to explain how parenchyma cells even in the flanges of the mantle would collapse as a result of this pressure. The suggestion that this layer, instead of passively collapsing, actually contracts and thus reinforces the thrust of the intrusive portion of the haustorium should be considered with care in future work. Against this notion is the fact that at least in *Antidaphne* haustoria the collapsed layers (there are two separate ones in this genus) are present even before the intrusive organ is recognizable (fig. 7-4).

The actual mechanism of host entry is a matter of conjecture. Most recent workers have assumed a combination of mechanical and enzymatic action. The cementing contact layer might contribute to the former in two ways. First, it provides a counter-force for the intrusive organ. In some mistletoes, for example *Phthirusa* and *Struthanthus*, both the mantle and the haustorial contact layer are much reduced, but now the root itself adheres to the host surface along its entire length. Thus the seat of the adhesive counterforce has been partially transferred to the epicortical root. Adhesion by the haustorium, however, serves a second function as well. At the moment that the intrusive organ forces its way through the columnar layer, the latter is forced aside, too. In this way the subjacent host cells may also be forced aside in opposite directions, causing a split where the parasite is about to enter.

Enzymatic action simultaneously with this physical action has been assumed, but has not been conclusively demonstrated. It is strongly suggested by the fact that the haustorial wedge of *Phthirusa* can slice neatly through the leaf of the host, emerging on the other side, apparently without causing much damage to contiguous tissues (see fig. 7-8,*c*). The existence of a glandlike haustorial tissue also tends to support this idea. It is clear, however, that the columnar adhesive layer of the haustorium is not active enzymatically; this role seems to be limited to the intrusive organ which attacks the host tissues.

One of the chief difficulties in the study of these early developments is the great rapidity with which host entry seems to be accomplished. At least in highly differentiated haustoria, all signs point to a slow build-up, an elaborate preparation, followed by a sudden thrust into the host. Such rapidity would explain the difficulty in finding haustoria in the process of entering. Only Barber (1907*b*) has come near to capturing the crucial moment of entry. Even he complains of the same difficulty: of 130 *Olax* haustoria he inspected, none were in the act of penetration. As Barber says, a great and sudden elongation of the cells of the core probably coincides with a split in the host bark. We must be careful, however, not to assume that these events are the same in all haustoria in this large order of parasites.

The formation of compound haustoria would seem to be based on repeated efforts to gain entry into the host. When the first meristematic wedge fails to puncture the host surface it flattens out against it and becomes a replica of the first mantle. It is supplied with the usual collapsed layer and columnar contact surface. It may be that a complete new gland is formed internally. Renewed efforts and repeated failures result in a series of superimposed mantles with a complex of often interrelated collapsed layers, together representing a haustorium of great complexity. Such a series of later meristematic crests, each emerging from the inner tissues of the previous one, is illustrated for *Phthirusa* (fig. 7-6).

The young haustorium is sometimes a great deal more complex, anatomically, than I have indicated. Some writers speak of "phloeotracheids," a cell type intermediate between sieve elements and tracheids, and presumably combining their functions. Immediately surrounding the young haustorial core are some puzzling tissues in which starch may be stored; the function of these tissues may be to supply nutrients to the intrusive organ. In a few parasites of the order a large amount of sclerenchyma is formed subdermally, both along the flanks and along the contact surface of the haustorium (fig. 7-4). Further details are given in the article already cited (Kuijt, 1965*b*).

The vascular structure of the haustorium shows much variation. Normal Santalalean haustoria, so far as I know, are without differentiated phloem. The xylem seems to be composed of vessel members in a matrix of parenchyma; the vessel members are short cells with simple perforation plates. The secondary wall ranges from reticulate to scalariform or pitted. Arrangements resembling protosteles and siphonosteles have both been reported. In the latter the shape of the stele is usually somewhat bulbous in the area of the haustorial core. Haustorial vasculature is at a minimum before host entry.

The haustorium of *Cassytha* is mentioned in this context not because of systematic affinity but because it resembles the simple Santalalean haustorium

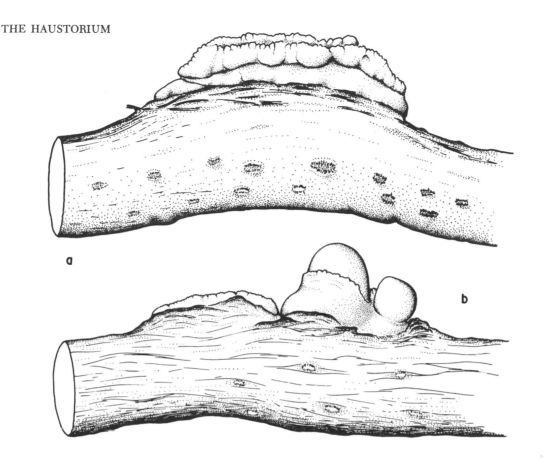

FIG. 7-6. *Phthirusa pyrifolia*: *a*, side view of free
secondary haustorium; *b*, side view of one (or two?)
free secondary haustoria, two endogenous lateral roots
having emerged from interior of haustorium at right
(after Kuijt, 1965*b*; × 23).

more than any other (fig. 7-9). It has a mantle, an
adhesive disk of columnar cells, and even a collapsed layer (Cartellieri, 1928). Nothing resembling a gland has been reported, but the intrusive
organ is clearly of endogenous origin, and is in the
shape of a wedge. The core shows a collection of
vessel members which are reminiscent of the tracheal
head of Rhinanthoideae. The intrusive organ reaches
the host stele, but little fragmentation occurs. A
direct xylem bridge leads from the host stem to
that of the parasite. The haustorium of *Cassytha*, as
far as studied, offers nothing unique, but combines
features known from other parasitic plants.

We know virtually nothing about the initial behavior of the haustorium in the early stages of its
endophytic existence. The external part of the haustorium has accomplished its mission—establishment
of an endophyte. If the physiological bridge is successful, the haustorium will now undergo secondary
growth, unless it is one of the more evanescent haustoria of many terrestrial Santalales. The mantle is
normally pushed up, or relaxes, and disappears from
sight in larger haustoria. The presumption is that
the intrusive wedge grows directly toward the vascular tissues to establish a physiological conduit.

The discussion of the mature, fully formed haustorial organ is drawn mainly from my survey of
mistletoe parasitism in the New World (Kuijt,
1964c) and from Thoday's excellent series of exploratory articles cited in the bibliography.

In the terrestrial parasites of Santalales the haustorium undergoes little modification after the host
stele is reached. It often happens that the most
advanced part of the endophyte spreads against the
xylem body of the host, forming two lobes. But usually a stout wedge is inserted into the host with very
little lateral expansion. Within this wedge a system
of vessel members links the xylem of the mother root
with that of the attacked organ. In *Exocarpos* a
number of endophytic strands emerge from the initial wedge and make their way laterally into other
tissues of the host (Fineran, 1963*f*). From these
longitudinal strands sinkers (radial strands) occasionally extend into the host wood. Although this
haustorial system does not achieve the complexity
of that of *Viscum* or *Phoradendron,* it nevertheless
gives us the first glimpse of a degree of endophytic
dimorphism, a tendency which is fully exploited in
Viscaceae.

In those Santalalean parasites which parasitize
the branches of trees, two basic trends in the evolution of the haustorial organ may be recognized. As
these trends characterize the two families of mistletoes, we may speak of the Loranthacean and Viscacean types of haustoria, although some overlap
occurs.

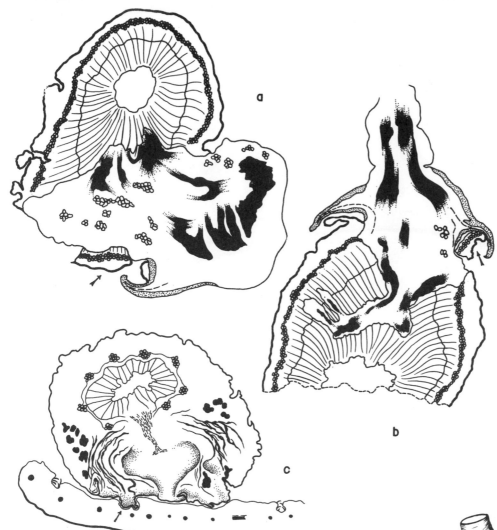

Fig. 7-7. *a* and *b*, *Antidaphne viscoidea*, sectional views of young primary haustoria; *a*, is somewhat older than *b*, and shows elimination of host tissues (arrow) due to action of radial shafts of endophyte visible in *b* (× 20); *c*, *Phthirusa pyrifolia*, young secondary haustorium formed on *Nerium* leaf (arrow indicates invasion of stomatal crypt of host by haustorium) (× 30; all after Kuijt, 1965*b*).

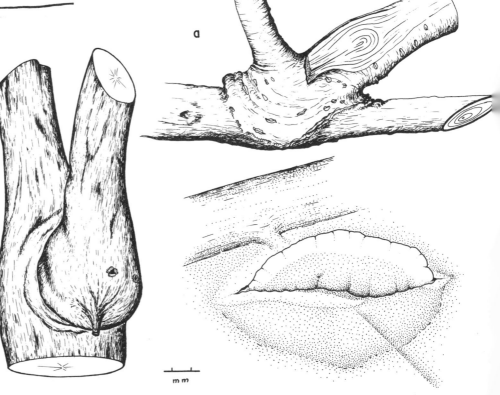

Fig. 7-8. *a*, *Lepidoceras kingii*, haustorial attachment on Myrtaceae (West No. 4879, Chile, UC; ca. × 2); *b*, *Gaiadendron punctatum*, self-parasitic haustorium (after Kuijt, 1963*a*); *c*, *Phthirusa pyrifolia*, crest of secondary haustorium emerging from upper surface of *Nerium* leaf (after Kuijt, 1965*b*; × 62).

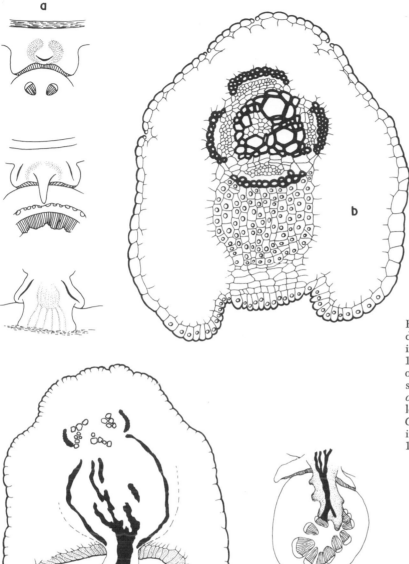

FIG. 7-9. *a, Cassytha pubescens,* three diagrams showing haustorial entry into host stem (after Cartellieri, 1928); *b, C. glabella,* transection of young haustorium and adjacent stem (after McLuckie, 1924; × 125); *c,* same, older haustorium entering leaf of *Xanthorrhoea* sp. (× 75); *d, C. melantha,* transection of haustorium in stem of *Ficus* sp. (after McLuckie, 1924; × 47).

The Loranthacean endophyte is a massive, unified body with little or no lateral intrusion into adjacent host tissues. It may, through its increase in size, displace host tissues, but it does not invade them and become fragmented. One of the most interesting features of this haustorium is the local host response: the adjacent host cambium increases in activity with the age of the haustorium (see figs. 7-11,*a* and *b*; 7-12). This change is visible in the amount of wood deposited. The enlarged host surface assumes the character of a placenta the shape of which is determined by the species combination. The host wood, by means of fluted columns, may elevate the parasite's haustorium and root (fig. 7-10,*b*). In *Myzodendron* a smooth, regular cup has been described (Hooker, 1846). Something of the sort is noticeable also in the early stages of *Antidaphne* parasitism, but later the "placenta" has the shape of a saddle, with fine ridges of wood radiating in all directions.

The best known of such objects are the curious

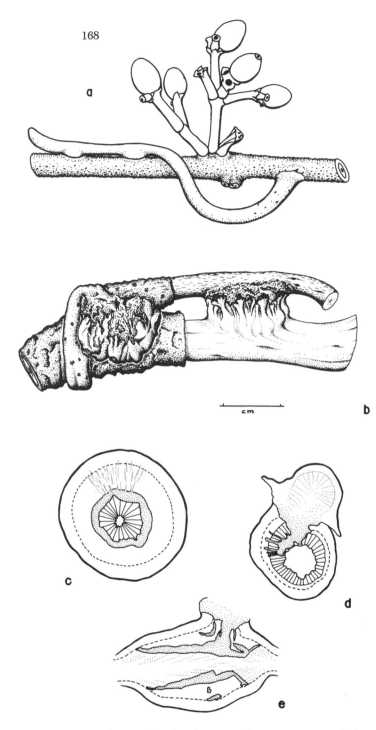

woodroses known as *rosas de palo*, *flores de palo*, or *rosas de madera* in Mexico and Guatemala (fig. 7-12). Many tourists buy these decorative pieces, but few have any idea of their true nature. In fact, the development of woodroses has never been studied in detail. The host and the parasite, a species of *Psittacanthus*, at the contact surface seem to compete in lateral expansion, and the haustorium as a whole increases in depth as well as in width (fig. 7-11,*a* and *b*). The greatest activity is doubtless found near the margins, where the haustorial surface develops ridges which become increasingly convoluted with age. Eventually a system of delicate radial folds and dichotomizing ridges may, in exceptional cases, grow to a foot or more in diameter. The woodrose, however, is host tissue only. It is the tree's placenta-like outgrowth which once supported the massive, domelike absorptive organ of the mistletoe. Upon dying, the woody tissues of the mistletoe, which have a great deal of parenchyma, decay rapidly and fall away from the host. The woodrose, therefore, is a three-dimensional negative —a cast, as it were—of the parasite's haustorial surface.

The body of this massive absorbing organ is by no means solid wood. It is a dense mesh of many hundreds of small vascular strands which fuse with neighboring ones and branch again, gradually leading to the host. Close inspection of the fluted surface of a woodrose may show many small tubercles representing the tips of vascular strands, severed from the rest of the mistletoe plant. These strands, which in larger haustoria must number in the thousands, thus represent the actual xylem bridge, although details of continuity still need to be elucidated. The haustorial matrix is parenchymatous, with numerous sclereid groups. Sclerenchyma in this form is very common in the haustoria of many mistletoes at the point of entry, but has not been reported in the internal parts of the endophyte.

There is a regrettable lack of information with regard to the mature haustorium of other Loranthaceae, but from Thoday's papers and others the following points may be made. A radically different type of endophytic organization prevails in *Phrygilanthus aphyllus* and *Viscum minimum* (Reiche, 1904; Engler and Krause, 1908). The former, a member of Loranthaceae, is parasitic on certain columnar cacti in Chile; the latter, belonging to Viscaceae, on columnar *Euphorbia* in South Africa. The leaves of each species have been reduced to mere scales, and other external parts of the parasite have diminished to very short internodes bearing flowers. Within the tissues of their succulent hosts the two mistletoes have an extensive system of strands leading in all directions. The extent to which the course of these strands is influenced by the conductive tissues of the host is not known. We can only presume that there are vascular contacts, for none have been clearly demonstrated, although the convergence of endophytic strands below and the production of flowers from host areoles seem to indicate such a

FIG. 7-10. *a, Struthanthus oerstedii*, autoparasitism (after Kuijt, 1964*c*; × 2); *b, S. oerstedii*, showing part of epicortical root with two old haustoria on a branch of peach; bark of one haustorium has been stripped away to reveal woody core (after Kuijt, 1964*c*); *c, Korthalsella opuntia* on *Altingia excelsa*, transection of continuous endophytic cylinder from which shafts (only a few indicated) radiate into host tissues (parasite shaded) (after Thoday, 1958*a*; × 2.2); *d, Macrosolen cochinchinensis* on *Achras sapota*, haustorial connection with radial shafts (after Thoday, 1961; courtesy The Royal Society; × 3.5); *e,* same as *c,* longisection (× 1.5).

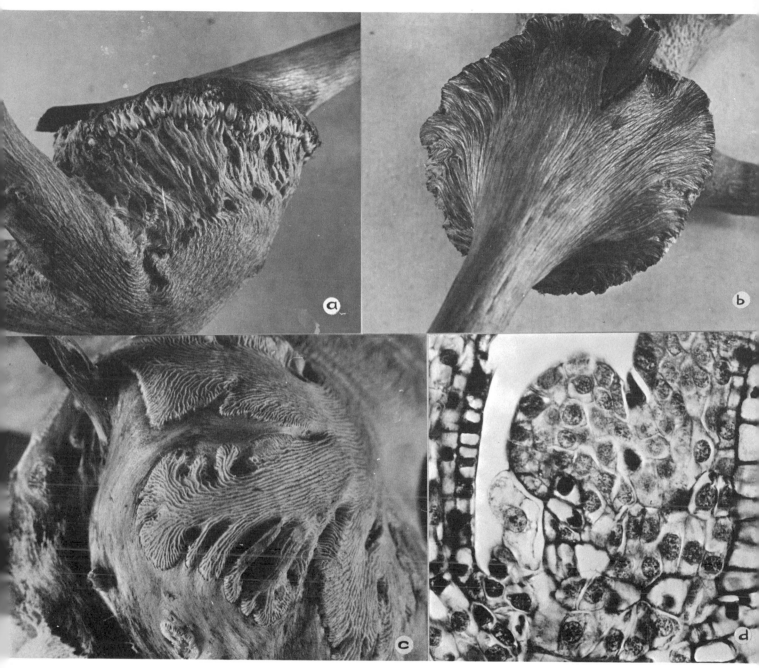

Fɪɢ. 7-11. *a* and *b, Psittacanthus schiedeanus,* young haustorial organ on *Drimys* sp.,
in Costa Rica; plant has developed on lower surface of host branch; all bark has been
stripped from both partners; *b,* dorsal view of young woodrose which is being formed
(placenta 30 mm in diameter); *c, Phoradendron densum,* part of expanding haustorial
system on *Cupressus sargentii,* California, with all bark removed; host wood (smooth)
has become a clavate supporting organ; distal portion of host branch (upper left) is dead,
3 mm in thickness; *d, Arceuthobium douglasii,* emergence of apical cell of uniseriate filament
from apex of branch of *Pseudotsuga menziesii* broom; emergence has occurred in axil
of young bud scale, visible on left; leaf primordium of host is present above apical cell, the
latter having a very large nucleus (× 280) (*a-c,* after Kuijt, 1964*c; d,* after Kuijt, 1960).

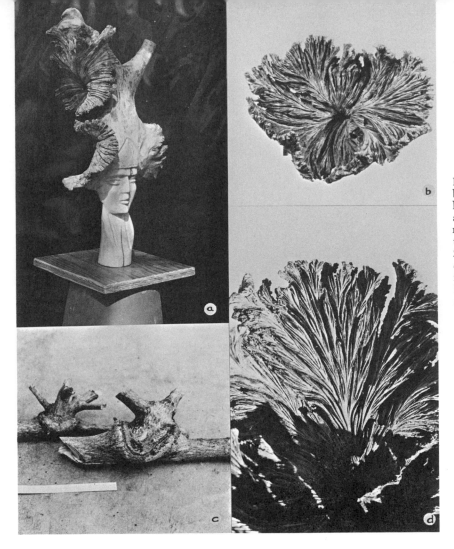

Fig. 7-12. *a*, Carved head on branch bearing three woodroses, Mexico locality unknown; parasite probably a species of *Psittacanthus*; *b*, woodrose from Valle de Bravo, Mexico, said tob e caused by *P. calyculatus*; width 23 cm. (courtesy Olivia Converse); *c*, attachment of *P. calyculatus* to its host, *Spondias purpurea*, Mata Limon, Costa Rica; *d*, detail of *b* (all after Kuijt, 1964c).

condition in *P. aphyllus*. These absorptive systems do not fit comfortably in either of the two haustorial categories: while they are greatly fragmented and diffuse, there seems to be nothing comparable to a division into a sinker system and a longitudinal system (see the discussion of Viscacean haustoria). The degree of convergence between the two mistletoes is indeed remarkable.

The endophyte of *Loranthus europaeus* is broken up into a number of strands. The relations between the strands of this mistletoe and its host are flexible, and depend on the relative size and vigor of both partners. The inner margin of the strands is closely applied to the host wood, but often shows a stepwise progression giving the appearance of a periodic lifting of the tip of the strand by additional host cells which differentiate into xylem. Many features of this variable interaction remain unexplained (Thoday, 1961).

Another interesting development in mistletoes such as *Korthalsella* and *Antidaphne* (Viscaceae) and *Macrosolen cochinchinensis* (Loranthaceae) on *Achras sapota* (fig. 7-10,c-e) is the centrifugal growth of shafts from the deepest part of the endophyte, radially outward, reaching the host bark. In

this way sectors of the host wood may become isolated to a degree, and even eliminated, at least in *Antidaphne*.

In at least two genera of Loranthaceae (*Ileostyles* and *Struthanthus*) haustorial lobes which have spread tangentially against the host wood produce slender brushlike structures from their ventral surface. These organs consist of hairlike cells that have apparently penetrated into the host wood and occupy a small cavity which contains the remains of some host cells.

For the details of these and other variations, consult the articles by Thoday in the bibliography. The variety of structural interactions is indeed infinite when we consider how small a percentage of the possible combinations has been studied.

In the Viscacean type of haustorial system, as exemplified in many species of *Viscum* and *Phoradendron*, we are struck by the great number of cortical strands which grow out from the place of mistletoe insertion. The strands, which actually are surrounded by secondary phloem of the host rather than by cortex, ramify and fuse on all sides, extending especially in the longitudinal direction of the host branch. Their mature organization is dorsiventral, not radial, in *V. album,* and perhaps in others as well (fig. 7-13, *a* and *b*). The apex of each longitudinal strand is capped with a number of elongated cells (Melchior, 1921; fig. 7-13,*c*). Just behind them is a small group of meristematic cells. Thus the over-all organization of the apex appears to be rootlike, even though the details are different.

Radial sinkers are produced from the inward side of the longitudinal strands. It is not clear whether these sinkers penetrate beyond the cambial layer or whether they are passively embedded by wood once they have grown to the cambium. The end result, nevertheless, is an encapsuled condition. The fundamental characteristic of the Viscacean haustorial system thus is a division of labor, a dimorphic organization into longitudinal strands extending laterally outside the host wood, and a system of radial sinkers extending in the direction of the host pith. The absorptive organ, seen in its entirety, is diffuse and greatly fragmented, lacking the solidarity of the Loranthacean equivalent.

A recent comprehensive study of the endophyte of *Phoradendron* permits a better account of the sinkers in the genus (Calvin, 1966). These organs are flat wedges attached to the cortical strands. The part of the sinker which is surrounded by the host phloem is largely parenchymatous. At the level of the host cambium the sinker of *Phoradendron,* as that of *Arceuthobium,* has an intercalary meristem which allows the apical portion of the sinker to be stationary throughout the host's cambial activity. The distal part of the sinker is completely embedded in the host wood and consists of mostly tracheary elements (perhaps all vessel members) and some parenchyma. The sinkers in *Phoradendron,* it can be seen, are not essentially different from those of *Arceuthobium* (Kuijt, 1960; Srivastava and Esau, 1961*a*). Calvin has also demonstrated the presence of sieve tube members in cortical strands, and in the extraxylary parts of the sinker of *Phoradendron*—features which are said to be absent in *Viscum album* (Melchior, 1921).

The next point of reference in the haustorial variation of mistletoes is the parasitism of *P. densum* on species of *Cupressus* in California. The young plant at first produces a small number of lateral outgrowths into the cortex or phloem of the host. Each strand curves toward the host cambium and eventually terminates in a sinker which anchors the pilot strand. The foot of the pilot strand expands laterally just outside the host cambium. Many pilot strands are branching, buttress-like formations. The subsequent growth in thickness frequently allows different branches to anastomose, after which a lobed, common front advances on the host. It is interesting to see the differing behavior of the haustorial lobes in a single individual: some parts form branching pilot strands, clasping the host branch firmly; others develop a flat, even front, interlocking harmoniously with the xylem of the host branch (fig. 7-11,*c*).

In another California mistletoe, *P. juniperinum* on incense cedar (*Libocedrus decurrens*), the cortical strands (fig. 7-14,*b*) pursue a very regular course parallel to the grain of the host tissues, giving off sinkers at measured intervals. Here the sinkers are undoubtedly *lateral* productions; the apex of the strand continues in its undeviating longitudinal path.

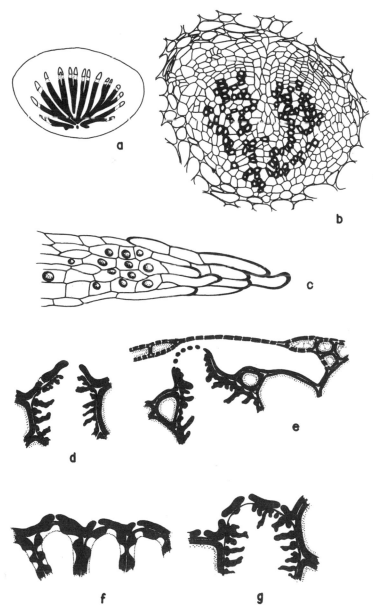

Fig. 7-13. *Viscum album,* details of endophytic system: *a* and *b,* transections of cortical strands, showing typical acentric vascular organization (× 8 and × 50); *c,* apex of cortical strand (magnification not known); *d,* haustorial tracheary cell pushing into fiber tracheid (above) of *Pinus silvestris* (magnification not known); *e,* a similar cell bulging into vessel member of *Pyrus malus* (magnification not known); *f,* several tracheary elements of haustorium abutting on a fiber tracheid of *Abies balsamea* (magnification not known); *g.* similar cell of a sinker bulging into a fiber tracheid of *Abies balsamea* (magnification not known) (all after Melchoir, 1921).

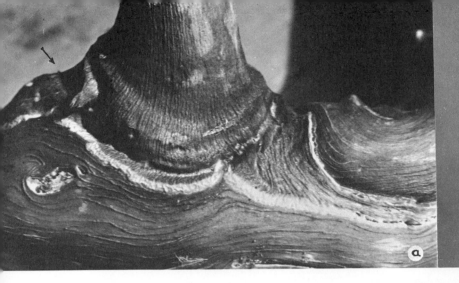

FIG. 7-14. a, *Phoradendron* sp. on unknown euphorbiaceous tree, Costa Rica, illustrating main features of haustorial system; photograph taken in fresh condition, showing white membranous margin of endophyte; arrow indicates a more "emergent" endophytic strand; b, *P. juniperinum* in wood of *Libocedrus decurrens,* showing pattern of sinkers; mistletoe base has been at lower right, toward which sinker series converge (after Kuijt, 1964c).

This feature represents a fundamental difference from that described for *P. densum,* where the longitudinal strand seems to reach the host cambium with its apex.

Such accounts are based on the xylem only. The bark has been stripped away, and with it go the more delicate tips of the haustorial system, and any phloem that might belong to the parasite.

The great phenotypic variability of mistletoe haustoria is indicated by the parasitism of one species of mistletoe on another. *Phoradendron* is generally characterized by a highly fragmented endophyte. When a species of this genus is attached to, say, a *Dendrophthora* (fig. 7-15,b), no separate cortical strands or sinkers are formed. The haustorium of the hyperparasite is, in fact, scarcely distinguishable from the wood of its host, so completely has it fused with the latter's tissues. Even with the aid of a dissecting microscope the line between the two systems is practically imperceptible. The fusion thus accomplished seems to become more complete and harmonious with age.

The haustorial features associated with different species of mistletoes and hosts have been so poorly explored in detail that only a few modifications of the Viscacean type can be mentioned. One difficulty is that we know virtually nothing about the comparative behavior of a single mistletoe species on different hosts, or the structural changes with age on a single host. When the trend toward a dimorphic endophytic system is reported to be poorly expressed on some trees (Thoday, 1957) and much better on others, we do not know which variable we are dealing with.

The parasitism of *Phoradendron perottettii* on *Protium insigne* is very similar to that of *Loranthus europaeus* (Engler and Krause, 1935). From the base of the mistletoe a number of straight, unbranched strands proceed in a longitudinal direction down the host branch. These strands are sharply pointed, but expand in width and eventually coalesce. A sharp, edged "keel," perhaps running the length of the strand, seems to take the place of sinkers. A similar endophytic system, but even sparser, characterizes a *Phoradendron* on Euphorbiaceae in Costa Rica. An additional feature here is a membranous white seam lining all the endophytic strands of this plant (fig. 7-14,a).

A periodic and alternating overtaking of the contiguous tissues of both partners seems to be the basis of the *Korthalsella* haustorium. In the simplest situations the endophyte spreads along the wood and lifts the cambial layers, which do, however, remain alive. This process, when continued, leads to a haustorial girdle completely surrounding the host axis (fig. 7-10,c). Meanwhile, the host cambium deposits more wood against the outside of these tangential lobes, which, simultaneously, appear to produce outwardly growing parenchymatous shafts. After some time, another tangential lobe originates from either the shafts or the original haustorial axis, and again separates the cambium from the host wood. No complete girdles are formed after the first one. The haustorial organ appears to insert its lobes tangentially into the host wood. The encapsuled portions, all of which may be presumed to have a degree of physiological contact with each other, are composed of vessel members and an abundance of parenchyma. The *Korthalsella* haustorium would seem to represent a special development within the Viscacean type where, in a curiously competitive fashion, the host and parasite attempt to surround one another's tissues. There seem to be no sinkers; they may have been replaced by centrifugal shafts.

The genus *Phacellaria* of Santalaceae deserves mention here, although no one has yet studied its absorptive system. The emergence patterns of its shoots from host branches (fig. 3-7,a) leaves no doubt of the existence of a very diffuse endophyte. Comparison with such mistletoes as *Arceuthobium* and *Phoradendron juniperinum,* which it greatly re-

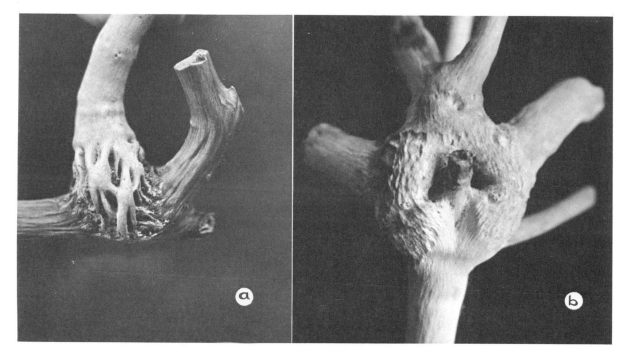

FIG. 7-15. *a, Dendrophthora ambigua* on *Weinmannia pinnata*, showing main features of haustorial system; bark of both partners removed; note fusions of some large endophytic strands; finer, more peripheral parts of endophyte have been torn away; mistletoe stem ca. 10 mm. in diameter; *b, Phoradendron* sp. on *Dendrophthora ambigua*, with all bark removed, showing hyperparasitism: parasite (above) clasps host branch (smooth, below), distal portion of which has died; rough surface texture corresponds to haustorial organ, which is 35 mm in diameter, in striking contrast to that in *a* (both after Kuijt, 1964c).

sembles in general aspect, would lead one to suspect a well-developed system of sinkers and longitudinal strands. The description of this haustorial organ is one of the most tantalizing, unexplored areas of the kind.

The endophytic system of *Arceuthobium* represents the extreme in the tendency of lateral development into extraxylary host tissues (fig. 7-16, 7-17, 7-18; Srivastava and Esau, 1961b). It may assume one of two distinct forms. The first is a localized system, associated with a rather severe swelling on the part of the host. The internal parts of the parasite are limited to this hypertrophy. A central cluster of aerial shoots emerges from the cortical strands corresponding in position to the original court of entry. The shoots are subtended by a plexus of cortical strands ramifying in all directions and probably forming anastomoses. A feature very reminiscent of Rafflesiaceae is the termination of the cortical strands in uniseriate filaments and therefore in a single apical cell, presumably situated near the periphery of the host swelling. The sinkers, which extend radially into the host wood, issue from the longitudinal strands.

The second endophytic type in *Arceuthobium* is of a much more diffuse sort, associated with the witches'-brooms which account for much of the notoriety of the genus in North America. These brooms are indeed remarkable, and place a distinctive stamp on the landscape in a number of areas of western North America (fig. 7-19; 8-5). Based on the host response and the distribution of endophyte and mistletoe shoots, two types of *Arceuthobium* brooms may be distinguished. In the first type there is a great deal of hypertrophy in the central part of the broom; the swollen branches taper off rapidly toward the periphery. The mistletoe shoots, and probably also the endophyte, are restricted to the regions of hypertrophy, and the shoots show no regularity in emergence.

The second type of broom is more interesting, for it represents a very advanced sort of adaptation of the parasite to the host. The branches of such brooms are slender, often stiffly erect at first but pendulous in older brooms, with a negligible amount of hypertrophy. On these brooms the mistletoe shoots emerge in predictable, annually repeated patterns (fig. 7-19, *a*; 7-20). The ultimate filaments have achieved an intimate relationship with the apical meristem of the host shoot. The permanent presence of the apical cells of the endophytic filaments in the apical meristem of the host brings about a synchronous rhythm in their growth activities. This apparently allows the mistletoe to lay down shoot primordia in certain places, for example in the axis of host leaves, and the elements of the regularity of shoot emergence are thus present. Shoot production in these brooms is limited to a few years. The mistletoe's reproductive zone is just within the periphery of the broom, but

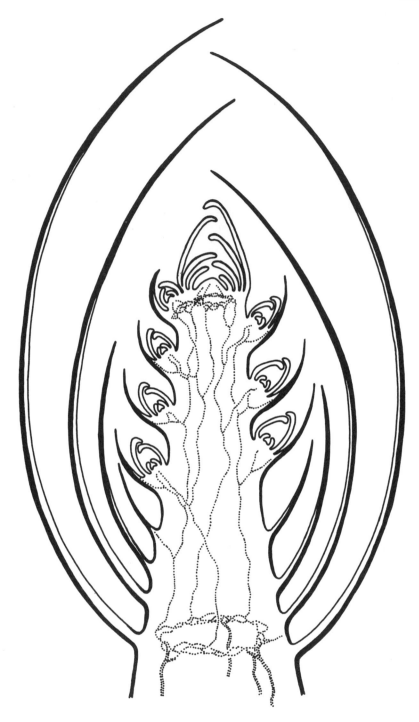

Fig. 7-16. Diagrammatic representation of distribution
of endophyte of *Arceuthobium americanum* in dormant
bud of broom induced by this parasite on *Pinus
contorta;* six primordial dwarf shoots are drawn and
one young lateral bud (upper left) (after Kuijt, 1960).

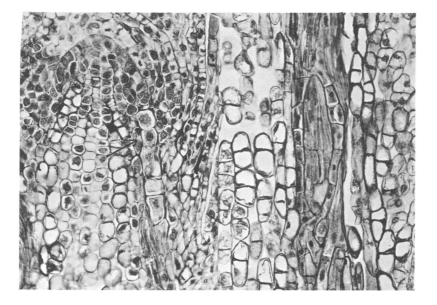

Fig. 7-17. *Arceuthobium americanum* on
Pinus contorta, showing uniseriate stage of endophytic
system: *a*, longisection of shoot apex of broom; filament
(at right of host pith; see arrow) of parasite extends
to within a few cells of surface of host apex; one
subterminal cell of filament has divided longitudinally
at level just below arrow; *b*, largely uniseriate filament
in procambial area of dormant bud from a broom; host
pith at right (both after Kuijt, 1960; × 220).

Fig. 7-18. *Arceuthobium americanum*, system
of radial sinkers connecting longitudinal strands in
host pith with those in the cortex; area illustrated was
just below a dormant terminal bud of a broom on
Pinus contorta, *a*, longitudinal filament, partly in pith
and partly in protoxylem of host; *b*, longitudinal strands
in cortex of host; *c*, — *e*, connecting sinkers, partly
flanked by ray parenchyma of host; *f*, young sinker
with single apical cell; *g*, host cambium (all after
Kuijt, 1960; × 250).

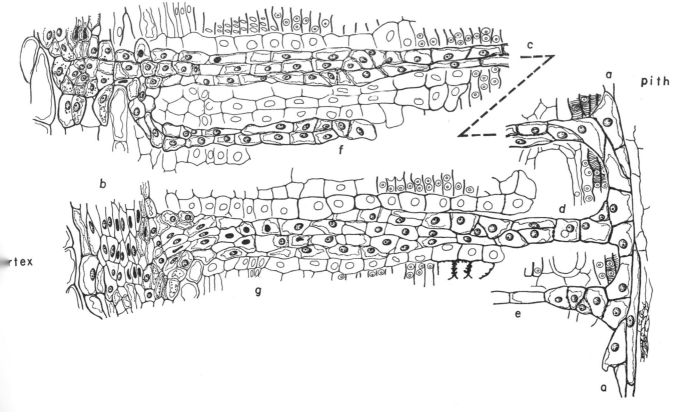

Fig. 7-19. *Arceuthobium americanum* parasitic on *Pinus contorta*, Manning Park, B. C.; *a*, close-up of branch from a broom, with all but previous season's host leaves removed to show distribution of mistletoe shoots; *b*, young broom with pronounced vertical growth habit (both after Kuijt, 1960).

sufficiently far out to allow for its pollination and seed dispersal. The uniqueness of the absorptive system lies in its extreme fragmentation and its remarkable extension in the tissues of the host; even leaves may be invaded, and the apex of the filament may emerge from the host epidermis. This extraordinary genus, more effectively than any other mistletoe, has adapted its structure and life cycle to the internal environment and growth rhythm of its host. Among vascular parasites, it is only in *Pilostyles* and possibly in *Mitrastemon* of Rafflesiaceae that we find a similar parasitism.

These various manifestations of dwarf mistletoes are determined by both the host species and the parasite species. *Arceuthobium douglasii* on *Pseudotsuga menziesii* induces the second ("regular") type of broom, which I have called "isophasic" (Kuijt, 1960). Yet when the same parasite grows on *Abies grandis* the broom is one of the first kind (anisophasic). In a parallel fashion, *A. americanum* differs in its behavior and host response on *Pinus contorta* and on *Picea glauca*. The host's contribution to this is indicated by the fact that, in contrast to the last example, *A. campylopodum* on *Pinus contorta* results in no brooming, whereas the eastern species *A. pusillum* on *Picea* is again isophasic. The symptoms of infection are thus associated with a particular host-parasite combination, not with either organism alone.

Before leaving the Santalalean haustorium we shall review the available information on the precise interlocking of host and parasite. What is the nature of the vascular contact at the cellular level? From the time of Solms-Laubach (1875) it has been known or suspected that, while the general tendency is certainly toward a xylem-to-xylem contact, the haustorial vessels do not invariably lead to the host's

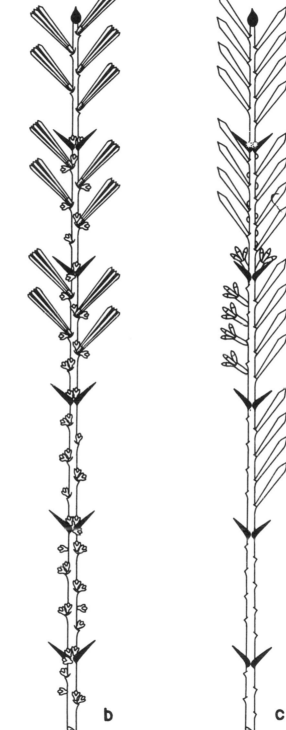

Fig. 7-20. Diagrammatic representation of shoot emergence from brooms: *a, Arceuthobium douglasii* on *Pseudotsuga menziesii; b, A. minutissimum* on *Pinus excelsa; c, A. pusillum* on *Picea glauca*. Host leaves have been removed from one side of third and fourth segments of *a* and *c*. All branches represent dormant condition; bud scales drawn in solid black (after Kuijt, 1960).

tracheal elements. There seems to be little or no selection in this regard for the various tissues of the host (Kuijt, 1965*b*). If a precise correspondence of pitting of the two organisms is found, however, it seems to be no more than a chance occurrence. Indeed, it is conceivable that morphogenetic factors emanate from the host xylem, causing tracheal differentiation in that direction, as is known in *Cuscuta* haustoria (see below).

Melchior (1921) has provided some interesting anatomical details which are germane at this point (fig. 7-13,*d-g*). Pit correspondence is prevalent here. In apple, the wall of a host vessel may be partially resorbed by an adjacent parasitic parenchyma cell, leaving only a scalariform area to separate the two cells. In *Pinus silvestris,* the pit membrane of a tracheid may be resorbed; in *Tilia* a large perforation is produced. It is not uncommon for the parasitic cell, before laying down its secondary wall, to bulge into a host element, a condition reported in the haustorial system of *Phoradendron* also (Calvin, 1966).

CUSCUTA

As we have seen (chap. 6), the dodders (*Cuscuta*) are twining parasites which manufacture very little chlorophyll. Contact with the soil is lost soon after germination when the parasite begins to wrap itself around the stem and leaves of the host.

The haustorium of the dodder was the first to receive special study (Guettard, 1744). (In fact, the term "haustorium" was first applied to *Cuscuta* by De Candolle in 1813.) A number of excellent accounts have since been added; so we can now give a reasonably complete description of this organ.

The small superficial protuberances which mark the sites of future haustoria are usually limited to the contact surfaces of the tight, so-called haustorial coils. They point toward the host and may elongate considerably if the host organ recedes (fig. 7-21,*a*). It has been said that dodder, like leaf hoppers, places its absorptive structures in such a way as to hit the vascular bundles (Lackey, 1953). Since this would involve a perception of host bundles even before the haustorium is initiated, it is difficult to conceive of an applicable mechanism.

The dodder haustorium has nothing which can be called a mantle, and also lacks collapsed layers. Nevertheless, the contact surface is bordered by a columnar tissue filling the crevices of the host (fig. 7-21,*b, c;* 7-22). The haustorial coil, in addition, provides a strong counterforce; so the intrusive organ is well braced.

Fig. 7-21. *a, Cuscuta salina* branch parasitic on *Salicornia,* showing free haustoria not associated with haustorial coil (Saanichton Beach, B. C.; x 5); *b, C. europaea,* section of young haustorium (after Zender, 1924; magnification not known); *c, C. reflexa,* section of prehaustorium and young haustorial meristem; prehaustorium has entered a stick of pith through center of which a nutrient solution was siphoned (after Thomson, 1925; x 75).

Before the actual haustorium makes its appearance a curious ephemeral organ may be formed which since the time of Peirce (1893) has been called the "prehaustorium." A number of cells in the center of the contact surface elongate greatly and seem to sink into the host tissues below, forming a brushlike structure (fig. 7-21,*c*). Simultaneously, just above it, a small group of meristematic cells becomes prominent. This cushion-like tissue, by means of elongating some of its constituent cells, pushes itself along the course followed earlier by the prehaustorium, destroying the latter on its way. Thus a peglike or wedgelike organ, the true haustorium, is inserted into the host. As the prehaustorium is rarely mentioned in the scientific literature, it may not be a general feature.

The haustorium of *Cuscuta* stops its intrusive growth at a certain depth. Its superficial cells, especially those near the apex, then begin to grow into separate filaments in various directions. They do so by dissolving a small opening into the partitions between adjacent parenchyma cells of the host. The tip of the filament, or "hypha," as it is called, bulges into the punctured cell and grows toward the cell wall at the far end, where the process is repeated. Cell divisions within the hypha take place during these successive invasions, resulting in a uniseriate filament. Some of the hyphae reaching out from the haustorium may eventually contact a tracheal element of the host, or insert themselves between several adjacent ones (fig. 7-24,*c*). Such a hypha now

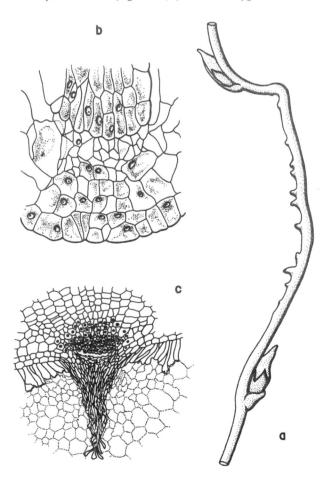

swells up so that it fills all the parenchyma cells which it has traversed. The cells making up a hypha now differentiate into tracheal elements, starting with the cell abutting the host vessel member or tracheid (fig. 7-23). Xylem differentiation proceeds basipetally until the uniseriate vascular strand reaches the simple axial strand of xylem which meanwhile has been formed in the base of the haustorium proper. The xylem bridge is now complete. From a single haustorium several vascularized hyphae may reach out to the host bundles in the vicinity. In cleared preparations such a haustorium reminds one of a small, stalked medusa (fig. 7-24,a), but such preparations show nothing but tracheal tissues. It is, in fact, in nonlignified hyphae that we may encounter the most fascinating features of the dodder haustorium.

The hyphae making contact with functional phloem undergo a radical change in behavior. The first indications of this came from the work of Zender (1924), who described how entry of a hypha into a sieve tube was followed by a "rhizopodial" lobing of what he thought was the parasite's naked protoplast. It is easy to see in these fine protoplasmic branchlets a specialized absorptive apparatus adapted to the host phloem, from which one expects most of the nutrients to come for the nearly holoparasitic dodders. Later work, however, has cast doubt upon the existence of a naked protoplast

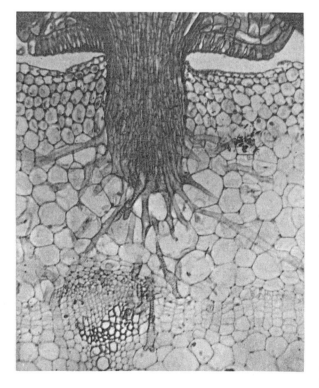

Fig. 7-22. *Cuscuta campestris*, haustorium in stem tissues of *Impatiens* sp.; a number of hyphae are vasculated (after MacLeod, 1961, courtesy of author; x 125).

(Kindermann, 1928; Schumacher, 1934). The rhizopodial lobes appear to be surrounded by a cell wall, after all, even if the wall is exceedingly thin. While some lobes are said to enter into host cells, others insert themselves in an attenuated form between the host cells, or grow along the surface of sieve elements, clasping them in a digitate fashion (fig. 7-24,d-g).

There has been much controversy over the occurrence and position of phloem. The matter has been summarized by Bennett (1944) and Truscott (1958). Several workers, including the latter, have described cells resembling sieve tubes in the haustorium of *Cuscuta*. At least one student has maintained that there is a direct phloem bridge, with the sieve elements of the parasite joining those of the host—in fact, sharing sieve plates. Others have categorically denied the occurrence of haustorial sieve elements. But virtually all workers agree on one thing: there is no true phloem bridge. Parenchyma cells separate the sieve elements of parasite and host.

It is probably better to keep an open mind with respect to all these facts. Many statements which appear to be at variance may yet be reconciled when we consider the possibility of structural variability. We know that the organization and differentiation of dodder haustoria can be influenced by the host species attacked. Kindermann (1928) has studied these facets in some detail. Haustorial penetration, by itself, does not necessarily prove the acceptability of a host. The main haustorial body is apparently limited in the extent of its penetration, and may be out of reach of host bundles. True hosts, in general, are only those for which hyphal, tracheary bridges are completed. Thus, in addition to normal haustoria, there are some which, because of distance or sclerenchyma barriers, fail to reach a vascular supply. A third type of haustorium lacks both hyphal development and axial vasculature. Some of the smaller dodder species may still grow and reproduce with imperfect haustoria, but not vigorously so.

Another aspect of haustorial variation concerns the rhizopodial processes themselves (fig. 7-24,f and g). These curious lobes have been demonstrated only in *Cuscuta europaea* (Zender, 1924; Kindermann, 1928) and *C. odorata* (Schumacher, 1934). Could they be present in some species, and absent in others, or is this another example of phenotypic variation? Another question pertains to the conjoining walls of host and parasite. Considering the pattern of vasculation in the hyphae, it would not be surprising to find a significant degree of pit coincidence at the contact surface. The hyphal walls are perforated by numerous plasmodesmata (Schumacher and Halbsguth, 1938) which are strangely lacking where the contact is with sieve tubes. Much work remains to be done here; especially exciting is the prospect of electron microscopy of the hyphae in their various forms, as shown by some exploratory work on young hyphal cells (Bonneville and Voeller, 1963). Even after two centuries of study the dodder haustorium is still a fascinating object of research.

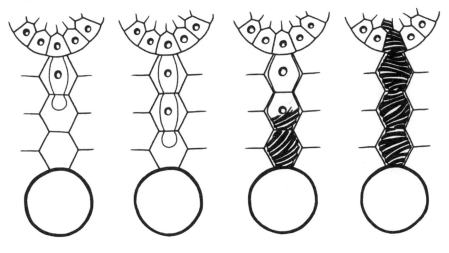

Fig. 7-23. Diagrams showing establishment of a xylem bridge from haustorium of *Cuscuta* (above) through intervening host parenchyma to a host vessel member (below) (after Thomson, 1925).

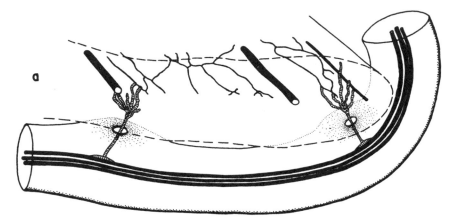

Fig. 7-24. *a, Cuscuta salina* on *Trichlogin maritima* leaf, a cleared preparation showing nature of haustorial xylem (cross-hatched) and its connection to mother stem and host vasculature (Saanichton Beach, B. C.; × 25); *b, C. relfexa*, haustorial hyphae entering host parenchyma without evidence of distortion (after Thomson, 1925; × 100); *c, C. reflexa*, nature of xylem bridge in transection of *Salvia* stem (after Thomson, 1925; × 100); *d, C. europaea*, formation of secondary wall on hypha which has entered a vessel member of *Lathyrus heterophyllus* (after Zender, 1924; magnification not known); *e, C. odorata*, hyphal tip growing in parenchyma cell (after Schumacher, 1934; × 700); *f, C. odorata*, showing beginning of rhizopodial formation in contact with sieve tube of host (× 700); *g, C. odorata*, another view of rhizopodia (× 700).

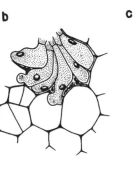

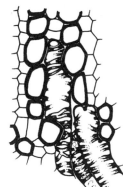

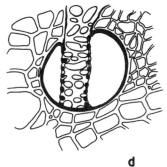

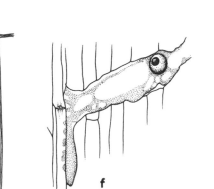

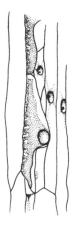

SCROPHULARIACEAE AND OROBANCHACEAE

In dealing with the haustoria of the alliance of
Scrophulariaceae and Orobanchaceae we have root
parasites to consider. Primary haustoria, resulting
from the direct transformation of the radicular apex
after it reaches the host, are known for all Orobanchaceae, and for *Striga* and possibly a few other
figworts such as *Harveya* and *Hyobanche*. Later
roots are usually part of the subterranean equipment of these parasites, and bear small secondary
haustoria which are lateral in position. Most parasitic
figworts have secondary haustoria only.

Since few illustrations of the haustoria of this
alliance have been published, I am supplying new
illustrations of a few of the most common ones,
demonstrating considerable variation. *Rhinanthus*
seems to have particularly large haustoria (figs. 7-25,
7-26,*b*), approaching some of those seen in Santala-
laceae. They are conical or even spherical, with two
massive lobes clasping the host root. Curiously, they
seem to be preceded by a number of elongated epidermal hairs comparable to root hairs. These root
hairs are probably pushed aside by the haustorium
developing in their midst, for long hairs can often
be seen at the base of older haustoria. Similar hairs
are observable in *Castilleja miniata* haustoria, which
are also spherical at maturity but much smaller than
those of *Rhinanthus*. Piehl (1962*a*) has suggested
that such hairs in *Melampyrum* may hold haustoria
to host roots.

In *Castilleja* we can scarcely speak of a clamp, or
mantle, as the margin of the haustorium seems to
advance around the inner tissues of the captured

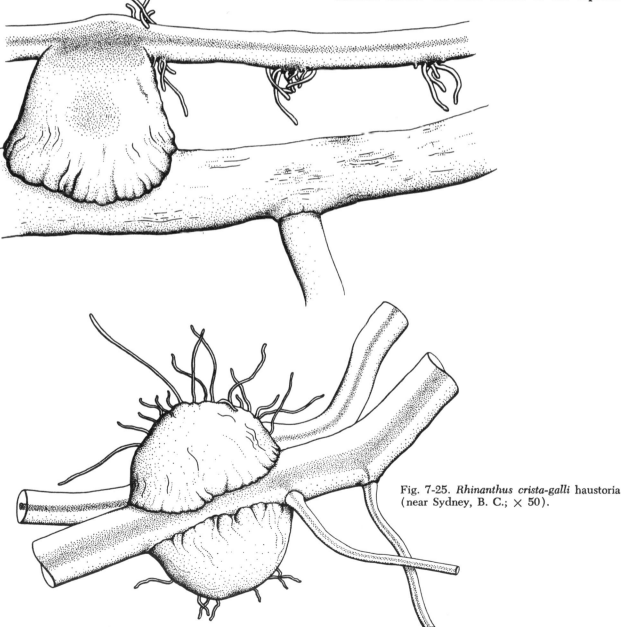

Fig. 7-25. *Rhinanthus crista-galli* haustoria
(near Sydney, B. C.; × 50).

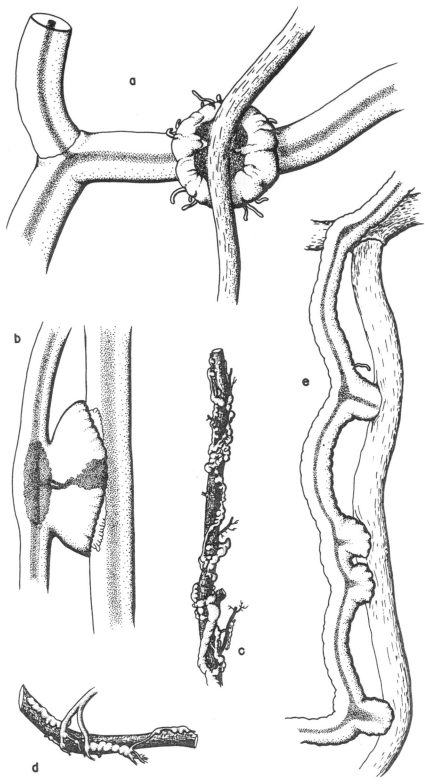

FIG. 7-26. Haustoria of several Rhinanthoideae: *a, Castilleja miniata* (Kaleden, B.C.; × 50);
b, Rhinanthus crista-galli, cleared (near Sydney, B.C.; × 50); *c, Lathraea clandestina*
on root of *Salix* (after Heinricher, 1893; × ½); *d, L. squamaria* on root of *Alnus* (after
Heinricher, 1893; × ½); *e, Euphrasia arctica* var. *disjuncta* (Vancouver; × 50).

root, pushing up the outer tissues (fig. 7-26,*a*). The haustoria of *Pedicularis lanceolata* are much like those of *Rhinanthus* (Piehl, 1965*a*). In *Euphrasia* the haustoria are very inconspicuous, lack anything resembling a mantle, and appear to enter the host root directly (fig. 7-26,*e*). This is especially true in *Orobanche*, where a conical emergence is simply inserted into the attacked root (fig. 7-27). Later, a swelling in the broomrape root forms two liplike bulges around the point of contact (fig. 4-17,*g*).

Although careful anatomical studies of most genera are yet to be carried out, the haustoria appear to be very simple structurally. The main body consists of undifferentiated parenchyma. An adhesive layer, cementing the haustorium to the outer surface of the host, is at best poorly delimited. There is no collapsed layer in young haustoria, although such a layer apparently forms in *Striga* after the xylem bridge with the host is functioning (Okonkwo, 1967). Aside from the xylem, the only differentiation is in the size of cells. The haustorial core of *Rhinanthus* is composed of very small cells; those forming the mantle, or clamps, are much enlarged, probably mainly around the attacked root (fig. 7-28,*a*). In *Euphrasia* even such differences are absent (fig. 7-28,*b*).

The vascular supply of a functioning haustorium is very simply constructed. There is no sign of phloem. The xylem forms a slender strand, sometimes a little irregular or even double (fig. 7-27), connecting the xylem of the partners. On either end a disklike body of xylem reinforces the continuity.

Fig. 7-28 (*Below*) *a, Rhinanthus minor*, haustorium on grass root, transection (after Solms-Laubach, 1867-1868; × 40); *b, Euphrasia* haustorium, transection (after Wettstein, 1896; magnification not known); *c*, face view of same, early stage (magnification not known); *d, Lathraea clandestina*, haustorial xylem abutting on xylem of dicotyledonous host root (after Solms-Laubach, 1867-1868; × 119); *e, Orobanche ramosa*, portion of endophyte parasitic on *Cannabis sativa* root, in region of host phloem (after Schumacher and Halbsguth, 1938; × 200).

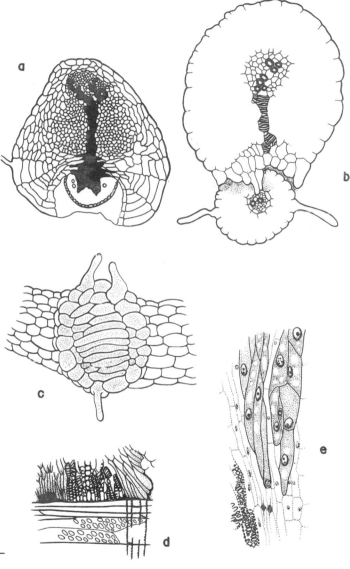

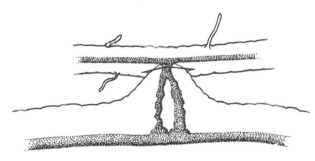

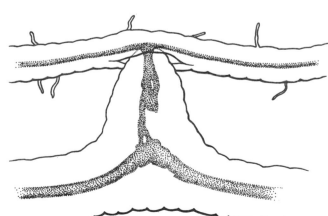

Fig. 7-27. (*left*) *Orobanche uniflora*, haustoria parasitic on roots of *Sedum album*, cleared to show vascular connection (Vancouver; × 50).

The more complex haustorium of *Lathraea* has led to an interesting concept of its action which might, to a lesser degree, apply also to haustoria of *Rhinanthus* and others (Ziegler, 1955). The central tracheal strand widens to a tracheal "head" just below the mother root. This tracheal head is interspersed with an abundance of parenchyma, the core parenchyma or hyaline body. As its name implies, the tissue, in contrast to surrounding cells, lacks starch. It is rich in protoplasm and has exceptionally large nuclei. The tracheal head was earlier thought to represent a water-storage tissue in the initial stages of the haustorium (Sperlich, 1925). Ziegler, instead, suggests that the combination of a large surface area within the fragmented tracheal head, and the specialized core parenchyma forming a matrix, point to a mechanism allowing the latter to resorb organic materials passing through the xylem. *Lathraea*, being a holoparasite, must somehow obtain organic materials from the host. It would be interesting to know to what extent the other, simpler haustoria of Scrophulariaceae and Orobanchaceae employ such a mechanism. Anatomically the intrusive portion of the small haustoria can often be seen to spread against the host xylem in both directions. This parenchymatous front consists of elongate, digitate cells (fig. 7-26,*b*). In two instances (*Lathraea* and *Striga*) actual penetration of host cells has been reported (Heinricher, 1895; Okonkwo, 1967). In its median area, vascularization spreads outward to form a disklike or conelike xylem body. There is nothing in the haustorium which may be called phloem.

The xylem elements of the connecting strand have been variously referred to as vessels (Piehl, 1962*a*, 1963; Sprague, 1962*a*; Cantlon *et al.*, 1963) or tracheids (Stephens, 1912; Rogers and Nelson, 1962; Sprague, 1962*a*; Cantlon *et al.*, 1963). A careful recent study of *Orthocarpus* shows that the haustorium of that parasitic figwort contains no tracheids, only vessel members (Thurman, 1965). It is difficult to believe that such radical differences exist in the very similar haustoria of closely related plants. The controversy brings to mind a similar one in Santalaceae which has been decided in favor of vessel members (Kuijt, 1965*b*). In Scrophulariaceae and Orobanchaceae, also, the xylem bridge may always consist of vessel members. The small size and perforations may have produced the alternate notion.

The anatomical structure of the primary haustorium has received little attention. Where present, it is considerably larger than secondary haustoria. In *Orobanche* the primary haustorial body usually has several wedgelike processes terminating in long slender cells (Schumacher and Halbsguth, 1938; fig. 7-28,*e*). Many of these ultimate cells are said to be multinucleate, which is a rather surprising phenomenon. Expansion of the endophyte is only by means of interpolation between host cells, not by their penetration as in *Cuscuta*. The terminal haustorial cells often become exceedingly attenuated.

Rarely, their tips become somewhat comblike or clamplike, but no elaborate structures comparable to those of the dodder haustorium are known.

It is not altogether clear whether the intrusive organ is endogenous or exogenous. The first cells to enter are modified superficial ones (see fig. 7-28,*b*). However, this is true also of *Cuscuta*, where the haustorium is undoubtedly endogenous. A careful developmental study of the maturing haustorium is yet to be made. Meanwhile, it must not be thought that, if the endophyte turns out to be exogenous, we must automatically deny it the status of a modified root. The roots of *Orobanche*, also, are exogenous, but they are unmistakably roots.

LENNOACEAE

The haustorial organ of *Lennoaceae* had not been studied until a brief account of *Pholisma* appeared (Kuijt, 1967*b*), except for a few lines nearly a hundred years earlier (Solms-Laubach, 1870). Little is known even now about the early stages of host entry and establishment, except for what has been discussed in chapter 6.

The mature endophytic organ of *Pholisma* is an irregularly lobed organ from which endophytic processes extend into adjacent host tissues. These parenchymatous lobes interlock tightly with the host and taper to margins no thicker than one cell. The margins probably expand and, in so doing, part and eventually crush the host tissues. When intervening host areas have been crushed, adjacent haustorial lobes may fuse again so that the absorptive body does not become more fragmented. The network of crushed host cells can be seen even in older haustoria.

With regard to vasculature of the endophyte, the absence of phloem-like cells must be emphasized. Strands of vessel members meander through the parasitic tissue in various directions, sometimes returning upon themselves in a circular fashion. A solid xylem body is lacking, and single vessels predominate.

There is nothing comparable to a radial sinker system in *Pholisma*. The nearest parasitic strands are invariably parallel to the host xylem. Vascularization in such strands begins in cells adjacent to the host xylem; so we find the side walls of tracheary elements of the two organisms in intimate contact. Were it not for the distinctive, often reticulate secondary walls of the parasite, it would be very difficult to delimit the endophyte precisely. The continued activity of the host cambium leads to an encasement of parasitic strands which, however, seem to retain a degree of continuity with extraxylary portions. Thus encapsuled strands may function both as anchors and as absorptive structures. The youngest strand appears to be uniseriate, having a single apical cell as have *Arceuthobium* and Rafflesiaceae. No specialized contact with the host phloem is in evidence, although the path of the longitudinally expanding endophyte seems to lie through the phloem area. The mature physiological contact, as in many other parasites, is a bridge of xylem, not of phloem.

RAFFLESIACEAE

Rafflesiaceae are holoparasites in most of which no more than individual flowers emerge from the infected host tissues. In this family, also, the mature endophyte must be taken as the point of departure, since nothing is known about the way in which these parasites enter the host.

The nature of the endophyte of Rafflesiaceae, constituting the entire vegetative body, defies description. The uniseriate filaments which form the youngest portion leave scarcely a tissue or an organ of the host unexplored. The endophyte has frequently been compared to a fungus mycelium, ramifying and anastomosing throughout the host. The periphery of the endophyte is known in only one or two cases. We do not know whether the flowers of *Sapria* and *Rafflesia* represent separate individuals, or whether the parasite is systemic through part of the host, giving rise to several flowers. In *Cytinus* the haustorial system produces a continuous sheath around a certain length of host stele, beyond which it is broken up (fig. 7-29,*f*). The longitudinal strands at both poles terminate in extremely attenuate filaments, possibly uniseriate at first, which anastomose later with adjacent filaments to extend the sheath. The sheath appears to lift up the host cambium which continues its activity, thus encapsuling part of the endophyte in the xylem.

In *Rhizanthes* the endophyte consists mostly of uniseriate filaments (Cartellieri, 1926), as indeed had been known since Solms-Laubach's classic paper (1875). Small unistratose plates of parasitic cells also occur, and rarely a few larger strands, usually about the base of a flower. The diffuse mycelial nature of the endophyte is not, as has been thought (Peirce, 1893), a result of secondary growth on the part of the host, even though some such disruption seems to occur. In this genus, and possibly in others, it is now believed that most parasitic growth occurs in the cambial region of the host and conjoining parts of the phloem. No host tissue is truly safe from infection, however, even if certain preferences exist. Cambial activity embeds some filaments and produces convolutions in others.

We can scarcely speak of sinkers in Rafflesiaceae, although there are occasional connecting strands between the pith region and the secondary phloem (fig. 7-29,*e*), especially just below developing flowers. Radial filaments can be seen in various developmental stages (fig. 7-29,*a* and *b*), and the same is true of the plates (fig. 7-29,*c*) of parasitic tissue. As such radial connections are frequently oblique and are not a regular feature, the term "sinker" is not a useful one in the family.

Although the endophyte may be encountered in nearly all tissues, most of it is enclosed within host phloem. The main haustorial organ may be visu-

alized as a hollow cylindrical body made up of a complex reticulum of parasitic cells which tends to become more massive and unified with advancing age. The situation is, therefore, not radically different from that in *Cytinus*.

Nothing resembling either phloem or xylem has ever, to my knowledge, been reported in the endophyte proper, since the entire vegetative body is reduced to parenchyma. It is only immediately below flowers that filaments of xylem vessels are seen, but sieve elements apparently do not differentiate even in the emergent flower (Schaar, 1898; Endriss, 1902).

The behavior of the endophyte and the consequent pattern of flower emergence from the host in *Pilostyles* and *Mitrastemon* have been indicated in the section on Rafflesiaceae (chap. 5). These parasites are essentially isophasic in that the flowers emerge in predictable places along the host root or stem, and the ultimate filaments are lodged within the apical meristem of the host or nearly so. When the host apex is pushed upward by its own rib meristem it carries with it some cells of the parasite, which, through transverse divisions, maintain continuity with older portions. The host's meristem is thus permanently infected. The parallel with certain situations in *Arceuthobium* parasitism is indeed remarkable. In *Mitrastemon* we even encounter a brooming response on the part of the host. This has not been reported for *Pilostyles haussknechtii*; infected *Astragalus* branches are said to be usually sterile and stunted (Bornmüller, 1909, 1913).

The extent of the endophytic system in other species of *Pilostyles* and possibly in *Apodanthes* is not well known. Herbarium material frequently gives the impression of regularity, of a restriction of flowers to host wood of a certain age. Here, too, a reproductive zone seems to exist in the parasite, maintaining a certain distance from the host apex (see, e.g., Smith, 1951). The endophyte of these species may be as extensive as that in *P. haussknechtii*. We know that filaments of *P. ulei* reach into petioles and capitula, and, indeed, to the vicinity of the apical meristem of the host (Endriss, 1902). Unfortunately, the ultimate filaments are difficult to distinguish in meristematic regions.

The usual assumption in the literature is that filaments traverse host tissues with the aid of enzymatic action and intercellular growth. In isophasic behavior neither process is, of course, required; the parasite is simply borne along within the meristematic region. Even in this type of parasitism, nevertheless, host tissues must have been traversed initially in order to reach this privileged position. From the observations made on *Rafflesia* and *Pilostyles* it appears that intercellular extension involving the dissolution of the host's middle lamella is a favorite method of progress (Schaar, 1898; Endriss, 1902). *Pilostyles* occasionally penetrates primary cell walls in the manner of *Cuscuta* (fig. 7-29,*d*), but this is probably exceptional. Of particular interest are the filaments which penetrate the differ-

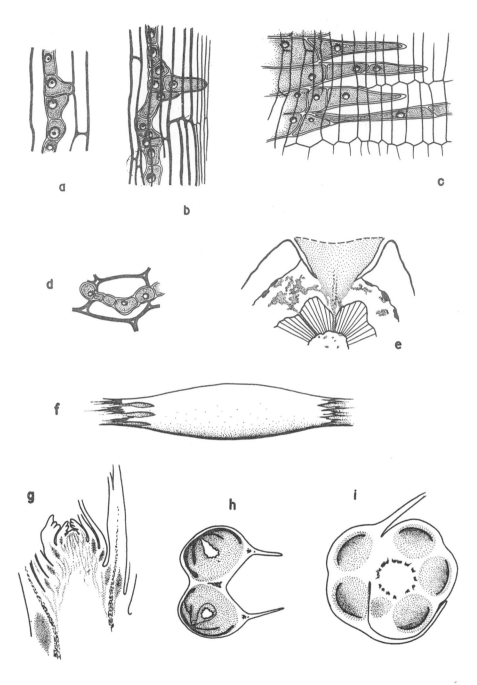

Fig. 7-29. *a, Rhizanthes,* longitudinal filament of endophyte, and initiation of radial filament (after Cartellieri, 1926; ca. × 100); *b,* same, a later stage; *c,* same, radial plate of parenchyma of endophyte in region between medullary ray and wood; *d, Pilostyles ulei,* uniseriate filament penetrating pith cells of *Mimosa* stem (after Endriss, 1902; no magnification given); *e, P. ulei,* transection of infected *Mimosa* stem, showing distribution of endophyte (stippled), largely in host phloem (after Endriss, 1902; no magnification given); *f, Cytinus hypocistis,* diagrammatic view of endophyte; external host tissues have been removed (after Solms-Laubach, 1867-1868; no magnification given); *g–i, P. haussknechtii* on *Astragalus* sp. (after Solms-Laubach, 1874*a*): longisection of infected shoot apex showing distribution of endophyte (stippled) and young floral cushions; *h,* transection of base of petiole of host with two young flowers of parasite; *i,* transection of infected host stem with three pairs of floral cushions corresponding to three successive host leaves.

entiating vessel members of the host and subsequently are surrounded by a pitted jacket-like secondary wall belonging to the host (Cartellieri, 1926). Penetration of host cells, so far as I am aware, has elsewhere been reported only for *Cuscuta, Lathraea,* and *Striga,* and doubtfully for *Geocaulon* (Moss, 1926) and *Atkinsonia* (Menzies and McKee, 1959).

BALANOPHORACEAE

In Balanophoraceae the absorptive equipment is so extraordinary that we must leave behind all we have learned about haustoria. A detailed haustorial study has been made only in *Balanophora* (Heinricher, 1907; Strigl, 1908; Fagerlind, 1948*b*) and *Thonningia* (Mangenot, 1947). The haustoria of other genera have remained largely unexplored. All members of Balanophoraceae are holoparasitic, and attack only the roots of the host.

Even when *Balanophora* was suspected of fungus affinities, the nature of its tuber was correctly diagnosed (Blume, 1827). The tuber is a "corpus intermedium," as Blume called it, an organ belonging neither wholly to the parasite nor wholly to the host. Each partner makes its peculiar contribution to it. There is no point in comparing such a dual organ to a hypocotyl (Fagerlind, 1948*b*). It is a chimeral system the like of which is not known elsewhere in the world of higher plants.

In Balanophoraceae, as in Rafflesiaceae, we have no precise information on the extent of the endophyte. The apparent gregariousness of inflorescences in a few genera has prompted some workers to suggest that the endophyte, instead of being restricted to the tuber, ramifies through adjacent host tissues (Heinricher, 1907). Except for Beccari's (1869) early findings, which certainly bear checking, this is an inference only, and not really a necessary one (Strigl, 1908). If endophytic portions are indeed present in host tissues at some distance from the tubers, such organs must be totally different, with respect to structure, behavior, and host response, from those inhabiting the tuber.

The development of the *Balanophora* tuber is based on a reversal of the usual pattern. The host root, when first contacted by the parasite, is stimulated to produce bundle-like outgrowths from its stelar surface. These processes grow actively into the host, dichotomizing repeatedly. Cambial stimulation on the part of the host is common in the immediate vicinity of mistletoe infections, and sometimes results in elaborate "placental" productions. Only in the *Balanophora* type of tuber, however, does the parasite induce the formation of aggressive individual strands which, furthermore, are exploited by the parasite in a remarkable manner.

Not until the first decade of the present century were the more intricate details of the *Balanophora* tuber revealed (Heinricher, 1907; Strigl, 1908). Even Solms-Laubach (1867-1868) had noted nests of peculiar vesicular cells along the entire length of the cords entering the tuber from the host. These vesicular cells were very large, with correspondingly large nuclei, and were embedded in a matrix of small cells. Although in section they sometimes seemed to be isolated from one another, close scrutiny often established a continuity with others. By means of elongated *Balanophora* cells the vesicular cells were continuous also with the outer parenchyma cells of the tuber (fig. 7-30), which certainly belong to *Balanophora*.

What are we to make of this extraordinary dual structure? I suggest that the vesicular cells are the main absorptive cells of *Balanophora*—cells which are in intimate contact with the host bundle leading from the attacked root. The vascular cords which enter from the host root themselves are dual structures, containing both vascular elements of the host and vesicular cells of the parasite. Nutrients may be transferred from the host xylem and adjacent host parenchyma to the vesicular cells of the parasite and from there, via the upper part of the filaments, to the massive parenchymatous tissues of *Balanophora*. These filaments may well be comparable to the commissural veinlets described for *Thonningia* below, and possibly even to the radial sinkers of some mistletoe haustoria. No vascular tissues of *Balanophora* seem to be present near the tuber strands. The continuity of bundles from one partner to the other, as mentioned by several early workers, appears to be only an illusion in *Balanophora*. Everywhere there seems to be parenchyma intervening between the tracheary elements of the two organisms. There are bundles, truly belonging to *Balanophora* and leading to the vasculature of the inflorescence, which reach into the top of the tuber and between the fanlike system of dual bundles. The *Balanophora* bundles themselves are easily identified, for they lack vesicular cells.

At the apex of each tuber bundle is a meristematic region which is really a chimeral meristem, since it includes meristematic cells of each partner (fig. 7-30,*b*). Fagerlind (1948*b*) has elucidated this unique situation. The matrix of the meristem consists of small cells of the host. It is surrounded by a mantle of cells of *Balanophora*. The innermost cells of the mantle are papillar, and many divide transversely to form uniseriate filaments which reach down into the axial region of the tuber strand. Such filaments are initiated at the summit, but subsequent events displace the tops of the filaments sideways. The mantle's meristematic region probably also adds to the tuber parenchyma above.

The organization sketched for the *Balanophora* tuber prevails generally in *Thonningia* (Mangenot, 1947). We have already noted the peculiar gall-like organs with which *Thonningia* roots eventually surround the captured host roots. A section of such a root in its original condition is represented in figure 7-31,*a*. In the stelar region we encounter a rather unusual diarch condition. The cortex contains a large number of vascular bundles, the smaller ones near the surface of the root. There are also a number of fiber bundles.

In *Thonningia*, as in *Balanophora*, a large number
of host bundles penetrate the parasite, branching
profusely. The host bundles occupy the mid-region
of the tuber parenchyma (fig. 7-31,b). The bundles
are very similar in structure to those in the *Balanophora* tuber: they are made up of cells of host and
parasite in the same arrangement (fig. 7-31,c).
Again the core of the tuber strand consists of host
xylem in which filaments of vesicular cells are embedded. An addition, however, seems to be the differentiation of commissural veinlets in the outer
region of the tuber strands. These veinlets are parallel to the host bundle for some distance, but eventually reach across to the tuber parenchyma and to
one of the cortical bundles.

Many aspects of this curious interaction remain
unknown. How, for example, is the tuber initiated?
Do the commissural veinlets represent a true xylem
link between the tracheary tissues of the two partners? Is it possible that the vesicular cells inserted
into the growing tuber strand from the apex are involved in the strand's ramification?

One of the interesting facts which Mangenot
(1947) has been able to establish is that the same
general organization is maintained in the tuber

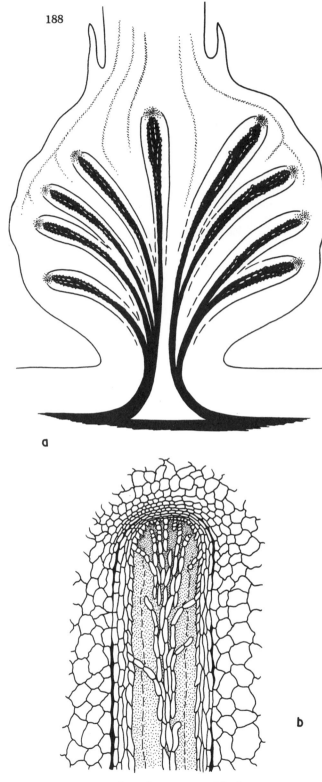

FIG. 7-30. *Balanophora*; *a*, simplified
view of organization of tuber, seen
in longisection (as interpreted by
Strigl, 1908; Heinricher, 1907; and
Fagerlind, 1948b);*b*, diagrammatic
view of tip of tuber strand, seen
in longisection (after Fagerlind,
1948b).

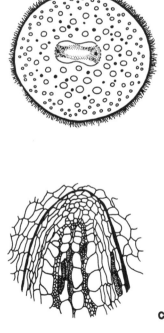

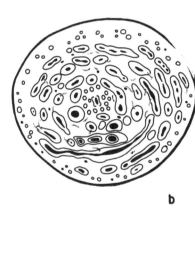

FIG. 7-31. *Thonningia sanguinea* (after Mangenot, 1947);
a, transection of root with vascular bundles (open
circles) and fiber bundles (black dots) ($\times$ 15); *b*,
transection of tuber, cut at some distance from host
root ($\times$ 5); *c*, longisection of apex of tuber strand,
showing filaments of large vesicular cells of parasite
(center) embedded in smaller host cells, some of
which have differentiated into tracheary elements; apex
of tuber strand, as in *Balanophora*, may actually be
continuous with surrounding tuber parenchyma ($\times$ 70).

strand from host species to host species. Always there are vesicular cells and associated cells, even if the host contribution is derived from a different species.

We might, indeed, be tempted to speculate on the structural variations which still lie unexplored in the haustoria of other genera. It is fortunate that the only two genera studied so far show appreciable differences, giving us an incentive to study other Balanophoraceae. There are also obvious similarities between *Thonningia* and *Balanophora*. In *Cynomorium* (Weddell, 1858-1861) and *Lophophytum* the complexities described above are not present. In *Dactylanthus,* where woodroses occur (Hill, 1926), and perhaps in the related *Hachettea,* we may expect a very simple interlocking of parasite and host. Further knowledge of the haustorial organs in other genera may show types of interactions yet unknown, as well as intermediate situations, and may thus contribute significantly to a better understanding of affinities within the family.

SOME GENERAL COMMENTS

The most durable impression we may carry with us from the previous account is the evolutionary plasticity of the haustorial organ. That of Balanophoraceae, for instance, has nothing discernible in common with that of Santalaceae and each is fundamentally different from that of *Cuscuta*.

One cytological feature may well be standard in all parasitic angiosperms: the nuclei of the haustorial organ are very large in comparison with those of the adjacent host cells. This fact is apparent or has been commented on frequently in Loranthaceae (Kummerow and Matte, 1960) and Viscaceae (Kuijt, 1960), Rafflesiaceae (Schaar, 1898; Cartellieri, 1926), *Orobanche* (Schumacher and Halbsguth, 1938; Kadry and Tewfic, 1956), *Balanophora* (Strigl, 1907), *Dactylanthus* (Moore, 1940), and *Pholisma* (Kuijt, 1967b). From this sampling we can probably conclude that a similar discrepancy exists between partners in each case of parasitism. It is interesting to find very large nuclei in cells which are presumably charged with the transfer of nutrients and often extensive intrusion and enzymatic activities. The large size of these nuclei makes it much easier for the anatomist to distinguish them from host cells.

The most impressive fact about the haustorium is that in both holoparasitic and hemiparasitic forms the emphasis is not on phloem continuity with the host but rather on xylem continuity. There is not a single reliable record of the sieve elements of one partner having contact with those of the other. The links between the two phloic systems consist at most of somewhat specialized parenchyma. In *Cuscuta* there is much specialization, but perhaps only in certain species. Even in dodders, however, the xylem bridge is very obvious. The existence of xylem continuity has been described in figworts and broomrapes, in many Santalales (e.g., Kuijt, 1965b; Calvin, 1966), in nearly every paper on *Cuscuta*,

in *Cassytha* (Cartellieri, 1928), *Pholisma* (Kuijt, 1967b), and *Krameria* (Cannon, 1910). Its extent in Rafflesiaceae would appear to be very limited, and in Balanophoraceae it may be lacking.

Can any generalization be made with regard to the mechanism regulating the initiation of haustoria? What determines where and when a haustorium appears in those parasites where haustoria are distinct organs? I have briefly touched upon this topic for *Cuscuta*. It remains to be said that, while haustorial formation and haustorial coils are developmentally linked, we are by no means certain what triggers the former. Our ignorance is underlined by a recent study demonstrating the haustoriogenetic influence of infra-red radiation on *Cuscuta* (Zimmerman, 1962).

In terrestrial root formation we cannot simply say that haustoria are chemically induced by neighboring host roots. Chemical induction must play a role, for if there were no such directive we could not anticipate the numerous successful haustoria separated by lengths of normal root of the parasite. However, there must be other stimulating factors in the subterranean environment. There are many reports of haustoria having formed on nonliving objects such as dead roots, pebbles, or sand grains in Rhinanthoideae (Sablon, 1887; Koch, 1889, 1891; Sperlich, 1925; Yeo, 1961; Piehl, 1962a; Sprague, 1962a; Cantlon et al., 1963; Malcolm, 1966), Loranthaceae (Kuijt, 1963a), and Santalaceae (Barber, 1906). The Australian mistletoe *Nuytsia floribunda* has attacked underground electrical cables, producing short circuits (fig. 7-3,a). Where haustoria are attached to dead roots it may be claimed that the root died following attack, but sand grains and pebbles are in a different category. It is possibly the stimulus of pressure exerted by such unyielding objects upon the growing root which makes a haustorium appear. After all, a similar stimulus seems to be effective when a mistletoe radicle pushes against the host surface and transforms itself into a primary haustorium.

The epicortical roots of mistletoes invite further comments on the nature of haustorial initiation. Whatever the mechanism is, it must be directed by the subjacent host branch, as haustoria invariably point in that direction. It seems too extreme to suppose that there is a chemical stimulus from the host, although the possibility cannot be excluded. It is more likely a light stimulus, possibly a suppression of haustoria by light, limiting them to the dark side of the root.

But what determines the periodicity of haustorial formation along such an epicortical root? I did not succeed in discovering a pattern in distribution in several mature plants of *Phthirusa pyrifolia* by measuring distances from the primary haustorium (Kuijt, 1964c). One of the difficulties in that species is the very inconspicuous nature of young haustoria; another difficulty is the frequent abortion of root tips.

It now appears that I may, literally, have been

working from the wrong end. In a collection of *Eremolepis* from Peru (fig. 7-32) scrutiny reveals a regular pattern, even though only a few roots are present. The two epicortical roots branching from a single mother root bear haustoria in perfectly matching pairs (one haustorium is invisible, behind the branch). Starting from the tips of these two roots, haustoria occur in precisely the same positions. It is possible that, if root apices in *Phthirusa* and others are taken as starting points, we shall be able to discern a similar pattern.

What conclusions can be drawn from this pattern? The main one would seem to be that haustorial initiation in epicortical roots is not directly dependent on the substrate. Instead, owing to some unseen physiological condition of the entire plant, a stimulus is experienced in all the subterminal root portions where haustoria can be formed. We get the impression here of an endogenous direction of haustorial formation, simultaneously throughout the root system.

Such a mechanism can scarcely play a role in terrestrial parasites, however. It could not explain the usual restriction of haustoria to certain places, whether at points of contact with other roots or with stones. What all secondary haustoria do seem to have in common in this respect is that they are initiated only in a zone approximately corresponding to the root-hair zone of normal plants. Only there can undifferentiated root tissues translate the relevant stimuli into the sequence of events culminating in haustorial action. This is manifested also by the acropetal differentiation of haustoria: the youngest ones are always found nearest the root tip.

What, then, is the haustorium, in terms of contemporary morphological categories? Students of parasitic flowering plants have racked their brains trying to decide whether these variable organs are rootlike, shootlike, or neither. The resulting opinions are as variable as the haustoria themselves. The difficulty is that we try to judge very extraordinary organs by means of the ordinary canons of morpholo-

gy. The criteria used in reaching a decision, of course, are based on stelar structure and ontogeny of ordinary roots and stems. The latter organs in these same parasites are not necessarily equipped, and do not necessarily develop, in the same way as those of more normal angiosperms. Elsewhere I attempt to bring together some deviations of this sort (chap. 9).

Some botanists, in the morphological equivalent of a fit of pique, have resorted to the concept "organon sui generis." According to this concept the haustorium is neither rootlike nor stemlike, but a separate organ spontaneously arisen. My objection to this theory is that it gives the impression of being a philosophical cul-de-sac: it would put a stop to further morphological thought. Perhaps more importantly, when we survey haustoria in angiospermic parasites we find an entire spectrum of forms. At one end of the spectrum are the haustoria of dodders and of Lennoaceae which few people would call anything but rootlike in origin. Farther along we encounter the haustoria of Santalales in their manifold manifestations. There are no real barriers to regarding these organs as the evolutionary derivations of roots (Kuijt, 1965*b*). At the far end are the extraordinary absorptive systems of Rafflesiaceae and Balanophoraceae. The haustorial system of Rafflesiaceae is not sufficiently different from that of some mistletoes to cause any serious problems in interpretation. Those of Balanophoraceae, notwithstanding their very unusual features, are also probably morphologically comparable with roots.

Taking this over-all view of haustoria, and bolstered by a strong faith in the evolutionary plasticity of plants, I find no difficulty in accepting the view that these organs are evolutionarily what they seem to be functionally, namely highly modified roots. That haustoria in many ways clash with our traditional concepts about the roots of vascular plants should not lead us astray. *The haustorium of parasitic angiosperms represents a root in function and evolutionary origin.*

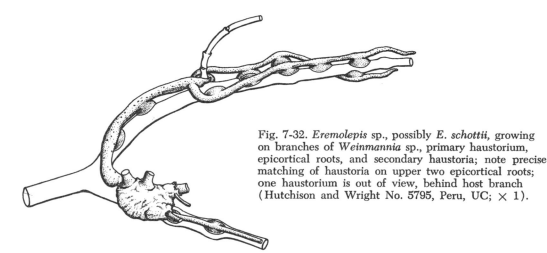

Fig. 7-32. *Eremolepis* sp., possibly *E. schottii,* growing on branches of *Weinmannia* sp., primary haustorium, epicortical roots, and secondary haustoria; note precise matching of haustoria on upper two epicortical roots; one haustorium is out of view, behind host branch (Hutchison and Wright No. 5795, Peru, UC; × 1).

8

Physiological Aspects of Parasitism

GERMINATION

The seedling of a parasitic flowering plant, when germinating, demands a great deal more from its environment than its nonparasitic fellows: in order to survive it needs to find, as quickly as possible, a suitable host plant. There seem to be three different ways to make physiological contact with the host more likely. In a number of parasites the seed has sufficient food reserves to allow the radicle to grow extensively while it is seeking a suitable host plant. The seeds or disseminules of such parasites are relatively large, as exemplified by *Cuscuta, Cassytha, Rhinanthus, Melampyrum,* and all root-parasitic Santalales. The germination requirements of these plants are not very specialized.

Another mechanism ensuring high frequency of establishment relies on birds. A precise mode of dispersal has evolved here which tends to deposit the seed on the host branch, in actual contact with it. The examples are epiphytic mistletoes, *Phacellaria,* and possibly other epiphytic Santalaceae. This mechanism requires a seed of considerable size to allow the formation of a rather large haustorium. The active photosynthesis in the endosperm of some of these plants presumably provides an additional source of nutrients.

In both these mechanisms, through the increase in the size of the seeds, their numbers have been reduced, usually to one seed per flower.

The third method represents a radical innovation which may have had a strong bearing on the evolutionary history of these plants. It involves the need for a biochemical exudate, produced by the host root, to initiate the germinative sequence. This was first demonstrated by Vaucher (1823) for *Orobanche* seeds.

If this is to be a successful mechanism, vast numbers of seeds must be produced, for their chances of becoming lodged in the immediate vicinity of a suitable root are slim. This in turn implies a drastic decrease in seed size and greater longevity of seeds. Another prerequisite is a chemotropic growth of the radicle toward the host root, as exemplified in *Striga* (Rogers and Nelson, 1962) and Orobanchaceae. There is a good chance that, in addition to all Orobanchaceae and a few genera of Scrophulariaceae, most Balanophoraceae, Rafflesiaceae, Hydnoraceae, and Lennoaceae fit this pattern.

These three different ways of coping with the seedling's most immediate requirement, a host, may show the following evolutionary connections. In all probability, the first method is the most ancient, and has separately produced the other two. (See chaps. 2 and 4 for the intermediate stages in this evolution.)

In considering the germination of parasites, we can omit Lennoaceae, Krameriaceae, Hydnoraceae, Rafflesiaceae, and Balanophoraceae, not because they have no interest but simply because we know virtually nothing about their germination. (For a possible exception in each of the last two families, see the relevant chapters.) In *Cassytha* germination seems to have no specialized features (Mirande, 1905a), and in *Cuscuta* and the mistletoes the important facts have already been stated in their own chapters. I shall therefore limit the following discussion to Scrophulariaceae and Orobanchaceae.

The seeds of parasitic figworts and broomrapes show great variation in structure and size (see chap. 4) and in the number produced per capsule. In *Melampyrum* and *Rhinanthus* each ovary produces only two or four rather large seeds. At the other extreme is *Aeginetia,* which may have as many as 70,000 seeds per capsule. No wonder the fresh weight of the seed of a broomrape such as *Orobanche minor* is no more than 5 mg, the average volume 5×10^{-3} mm^3, even when fully imbibed (Sunderland, 1960b).

The literature on germination of parasitic figworts contains many misleading statements. It is certain that germination may take place in the absence of host plants, as in *Castilleja* (Heckard, 1962), *Euphrasia, Orthantha, Odontites, Rhinanthus, Bartsia,* and *Melampyrum* (Heinricher 1898a, 1898b, 1908a). Yet Sunderland (1960b) states: "It is now well known that usually the seeds of the parasite do not germinate even under the most favorable conditions of temperature and moisture unless stimulated to do so by a substance which escapes from the root of the host." This statement takes little account of our knowledge of germination of all Santalalean root parasites and many genera of parasitic Scrophulariaceae, or of our almost total ignorance of germination in five other families of root parasites.

As the acquisition of the need for a germination stimulant must be an advanced feature, we shall turn first to those parasitic Scrophulariaceae which are known to germinate under natural conditions without a stimulant. Most genera fall in this category; the exceptions are discussed below. A few genera in the present group have already been mentioned. In *Rhinanthus* germination can be induced simply by exposing the seeds to moisture at 2°C (Vallance, 1952). *Euphrasia* also requires exposure to moist winter cold (Yeo, 1961), but great variation exists even within a single species: some samples require cold and others do not, and the percentage of seeds germinating is variable. *Orthocarpus faucibarbatus* seeds (Thurman, 1965) and those of several species of *Castilleja* (Heckard, 1962) germinate readily. Species of *Castilleja* from high elevations, however, failed to germinate, suggesting the need for stratification. Recent work with *Melampyrum lineare* (Curtis and Cantlon, 1965) has revealed a somewhat more complex system, in which germination requires activation during storage at 20° C followed by a long period of cold. The effect of storage can be duplicated by gibberellic acid. There appears to be a close parallel between the ability to germinate and the activity of endogenous gibberellin-like materials (Curtis and Cantlon, 1968). The dormancy mechanism of this species, unlike many other plants, does not involve synthesis or activation of starch-digesting enzymes in the embryo, but rather primarily a hydrolysis of the thick, mannan-containing cell walls of the endosperm. Both the variation in germination response between samples in nature, and the complex germination pattern tend to extend the effectiveness of a single seed crop over several seasons. A similar variation in germination totals has been mentioned for *Striga* (Kust, 1963) and *Cuscuta* (Dawson, 1965).

This enumeration must not blind us to the possibility that the chemicals emanating from host roots may nevertheless stimulate germination of the seeds. Although these seeds will germinate without chemical stimulus, the possibility remains that the latter may speed up germination or increase the percentage of resultant seedlings. Indications of this sort have been reported for *Melampyrum* (Sperlich, 1908; Heinricher, 1908a), and a similar response may be possible in other parasitic figworts.

In addition to Orobanchaceae there are four genera whose germination depends on host stimulants: *Lathraea* (Heinricher, 1930), *Tozzia* (Heinricher, 1901a), *Striga* (Sunderland, 1960a, 1960c), and *Alectra* (Botha, 1948). It would not surprise me if *Hyobanche* and *Harveya* were eventually added to this list.

Experimentally it is known that a moisture treatment is required in *Alectra* and *Striga*, in the nature of sensitization, before root exudates have an appreciable effect on germination (Botha, 1948; Williams, 1961a). In *Alectra*, where there is no evidence of the leaching of an inhibitor from the seed (Botha, 1950a, 1950b), no relation seems to exist between the moisture treatment and the seed coat's permeability to the stimulus. Whatever change is brought about is sensitive to temperature and does not occur in the absence of CO_2. These various

phenomena suggest that a substance needed for germination is synthesized during pretreatment.

The information on *Alectra* does not agree in all details with that on *Striga*. In the latter genus it has been demonstrated that germination inhibitors and growth inhibitors are leached out of the seed during pretreatment (Williams, 1961b). Changes in the respiratory quotient indicate changes in the respiratory substrate as well (Vallance, 1951). Such facts are difficult to reconcile with reversibility.

Not one but a complex of substances is involved in the stimulation of germination in *Striga* and *Orobanche*. Some chemical characterization of the active substances has been carried out, but none has yet been identified. Several features suggest affinities to certain coumarin-like derivations or related compounds (Worsham *et al.*, 1964). Within the complex of stimulants produced by roots of *Zea mays,* certain substances stimulate germination of *Orobanche minor* but not of *Striga hermontica*. Linseed, sorghum, and maize all produce a similar complex of stimulants, differing only in the proportions of the different constituents (Williams, 1961a). A number of known compounds have shown some germination activity (Worsham *et al.*, 1964).

This already complicated situation may be further confounded by the presence, among the complex of host exudates, of natural inhibitors. When some early workers (Pearson, 1912; Saunders, 1933) were unable to stimulate germination by means of expressed host juices, natural inhibitors may have been dominant. Dilution of the host juices might have had radically different results.

The production of root exudates which may act as germination stimulants is probably general in higher plants, whether or not they are parasitized. That many organic materials are given off in this fashion to the soil is gradually becoming better known (Winter, 1961). In tomato roots it has been demonstrated that a mixture of seventeen amino acids is released (Butcher and Street, 1964). In an evolutionary sense, the host-dependent seeds have "learned" to be guided by root exudates as indicators of where establishment is likely to occur and often in what direction to grow to find the host.

Much work has been done in pinpointing the source of the stimulant. In maize, production is most active in the region from 3 to 6 mm behind the root apex, but this maximal area differs slightly between *Striga* and *Orobanche*. Within a few hours after the maize has begun to germinate, or with the differentiation of the first root hairs, the stimulants make their first appearance (Worsham *et al.*, 1964). The germination stimulants present in maize cells are used up during extension growth.

Apparently the stimulants from different host roots may have a synergistic effect upon one another (Sunderland, 1960b). Separate, comparable extracts of maize root and of linseed root proved effective in only 4 and 5 percent, respectively, of the *Striga* seeds tested. Yet, when the two extracts were mixed, their effectiveness increased to 51 percent. Since the same has been shown with *Orobanche hederae,* and with sorghum root diffusate, such synergistic effects may predominate. This is not a function merely of concentration, for increases in maize-root extract have no such effect.

This aspect of parasitism has aroused great interest lately, but the nature of the stimulating action is not known. Its rapidity is spectacular: immersion for thirty seconds may bring about 60 percent germination in broomrapes (Chabrolin, 1939). In *Alectra* seeds the duration of immersion and the concentration of the stimulus complement one another (Botha, 1951a). Sunderland (1960b) has suggested that the stimulant removes an inhibitor which stands in the way of a correct balance of metabolites essential for growth. Brown *et al.* (1949), on the contrary, have suggested that the stimulant is an extension growth hormone. Yet another student has shown that the root exudates actually lower the rate of extension growth of the radicle of *Striga* (Williams, 1961b). It has long been known that in *Striga* the stimulants do not act on the isolated embryo unless endosperm is added (Saunders, 1933).

An entirely different facet of host-root exudates has recently begun to be explored. These substances, at least in *Striga,* not only initiate germination but also have morphogenetic effects on the seedling (Williams, 1961b). They guide the direction of growth toward the host root. This chemotropic response is not perfect, but nevertheless ensures host contact for many seedlings. No less than five other morphogenetic effects have been ascribed to the root exudates: increased diameter of the radicle; production of root hairs; cortical proliferation; ramification resembling dichotomous branching; and lowered rate of extension growth. Many of the intimate physiological interrelationships between the more advanced parasites and their hosts still await discovery.

The need for a host stimulant in germination implies a chemotropic response of the parasite's radicle toward the host root. Whether this is true in all cases is not certain. In the species whose germination depends on host stimulants, we should like to know whether the chemotropism of the radicle results in the formation of a primary haustorium. Does the apex of the radicle become transformed into a haustorium? The answer seems to be yes, except for *Lathraea* and *Tozzia.* Significantly, these genera have larger seeds, by far the largest in the plants here discussed, a condition related to dispersal by ants. The seedlings are thus able to live independently much longer than others and do not need, or have not yet evolved, the growth stimulus leading them directly to the host.

Surveying the known facts about germination of parasitic figworts and broomrapes, we can perhaps postulate that the pretreatment necessary in the more advanced groups before host exudates can take effect, is comparable to the totality of germination requirements of seeds which can germinate without the benefit of a host. We should not expect complete coincidence in detail, for there is much variation among both groups. But we can be reasonably certain that the required stimulation by host-root exudates is an evolutionary *addition* to the germination mechanism of Rhinanthoideae, and has not *replaced* the already existing mechanism.

NUTRITION AND WATER ECONOMY

The popular conception of a parasite as a plant lacking chlorophyll is very misleading as a generality. The holoparasites are the following: Lennoaceae, Balanophoraceae, Rafflesiaceae, Orobanchaceae, and, among Scrophulariaceae, *Lathraea* and possibly *Hyobanche. Tozzia,* also in the latter family, spends nine-tenths of its existence as a subterranean holoparasite, after which it bolts and develops chlorophyll for its brief flowering and fruiting period.

One earlier worker (Wiesner, 1872) claimed to detect traces of chlorophyll in *Orobanche,* but no one else seems to have bothered to check on the presence of chlorophyll in the holoparasites. The refined methods of modern physiology and biochemistry would make this a rewarding undertaking. The electron microscope could also contribute much if chlorophyll turns out to be present in marginal amounts, as such work might permit the identification of individual chloroplasts. Recent studies by Laudi (1964) on *Cuscuta* and Menke and Wolfersdorf (1968) on the achlorophyllous saprophyte *Neottia nidus-avis* hold much interest in this regard.

As early as 1883 (Temme, 1883) it was shown that photosynthesis occurred in *Cuscuta europaea.* Several subsequent studies have confirmed the existence of chlorophyll in these parasites (MacKinney, 1935; Walzel, 1952a; MacLeod, 1963; Bertossi, *et al.,* 1964; Pattee *et al.,* 1965; Baccarini, 1967). Both chlorophyll *a* and *b* are present, in normal proportions but in very low concentrations. The small quantities of chlorophyll involved seem to be due to the fact that chloroplasts are composed of no more than two coupled discs, somewhat like the prochloroplasts of other plants (Laudi, 1964). The nutritive condition of *Cuscuta* determines to some extent the concentration of chlorophyll, as shown in sterile cultures of the plant (Loo, 1946; Walzel, 1952a) and by the fact that the tips of starving seedlings become greenish. Chlorophyll is normally most conspicuous in the seed and ovary. Notwithstanding all these indications, even in recent literature there are occasional references to the total absence of chlorophyll in *Cuscuta* (Fritsché, *et al.,* 1958).

Some other parasites are occasionally cited as holoparasites, but this is erroneous. Among the most reduced squamate mistletoes *Arceuthobium* and *Phrygilanthus aphyllus* are known to have chlorophyll (Follmann, 1963; Hull and Leonard, 1964b); other squamate types such as *Korthalsella* and *Ixocactus* are also green in appearance. Proof of the existence of chlorophyll in leafy mistletoes is scarcely necessary, but has been provided for *Phoradendron serotinum* (Freeland, 1943). *Striga,* also, has functional chlorophyll (Rogers and Nelson, 1962).

As far as I know, the transfer of organic materials between the two partners has been demonstrated only in chlorophyll-bearing plants. We can, of course, take this process for granted in truly holo-parasitic plants, but even there the details deserve attention. Modern techniques with radioactive tracers have made it relatively easy to demonstrate the passage of organic materials from host to parasite. This has been done for *Odontites* (Govier et al., 1967), *Striga* (Rogers and Nelson, 1962), *Arceu-thobium* and *Phoradendron* (Hull and Leonard, 1964a and 1964b), *Melampyrum* (Cantlon et al., 1963), *Orthocarpus* (Atsatt, 1966), and *Cuscuta* (Littlefield et al., 1966). Using a vital stain, eosin Y, passage has been observed from hosts to *Castilleja coccinea* (Malcolm, 1962) and *Pedicularis canadensis* (Piehl, 1963). Other observations on the transfer of organic materials to semiparasites have been made by Govier and Harper (1965) and Rogers and Nelson (1962). In the last instance it was demonstrated that *Striga asiatica* can grow to maturity and set viable seed in total darkness, thus obtaining all organic materials from the host. This, of course, is not the normal situation; other green Rhinanthoideae are said to photosynthesize as actively as the nonparasitic Scrophulariaceae (Kostytschew, 1924), contradicting older statements to the opposite effect (Bonnier, 1893). Another conclusive demonstration of transfer of organic materials, although it never seems to have been looked upon in that light, is through the use of *Cuscuta* as a virus bridge (see chap. 1). Some of these studies tell us that earlier and more subjective conclusions about different degrees of parasitism were essentially correct: plants such as *Arceuthobium* and *Striga* straddle the dividing line between holoparasitism and semiparasitism. They have some chlorophyll, but obtain substantial amounts of organic materials from the host. The rate of CO_2 fixation in *Arceuthobium*, at least, is very low (Hull and Leonard, 1964b).

The haustorial organ is apparently not completely unidirectional in action. De Candolle (1801) was able to show the passage of the dye cochineal in aqueous solution from a mistletoe into the host tree. Pitra (1861) came to opposite conclusions in this regard. The work of Rediske and Shea (1961) and Hull and Leonard (1964a) on mistletoes and that of Atsatt (1966) on *Orthocarpus* indicate a similar "leakage" into the host. I have earlier (Kuijt, 1964c) discussed field observations showing the continued growth of trees which are without foliage except that of mistletoe. If this is a situation of long standing we may justifiably speak of an inversion of parasitism, the host receiving a significant amount of organic nutrients from the parasite. In the dodders, also, we can see the transfer of organic materials (viruses) from the parasite to the host when the former is used as a virus bridge. We must, of course, not overemphasize this apparent inversion of parasitism. It is likely to be temporary or insignificant. Yet it must be accounted for if we wish to

understand the complex phenomena of parasitism.

The study of the physiology of parasitism has not yet reached the stage at which we know the actual materials transferred, or the exact pathway they take through the haustorial organ. Recent work on *Odontites* (Govier et al., 1967) points the way to more precise work of this type. It includes the important demonstration that different host species contribute significantly varying sets of organic compounds to the parasite. Fragments of information indicate that parasitism is sometimes involved in the hormonal economy of the two partners. I have elsewhere mentioned relevant observations on mistletoes (Kuijt, 1964c), but more precise information has come from other sources. A number of workers have wondered whether the flowering periods of host and parasite might not coincide, but this is most unlikely in the majority of flowering parasites. The work of Holdsworth and Nutman (1947) on *Orobanche*, however, does seem to imply a transfer of materials affecting flowering. On red clover exposed to a long-day treatment, inflorescences of *O. minor* appeared, but not so where the host was under a short-day regime, even though subterranean broomrape plants were present. Thus it appears that in this particular partnership the parasite does not initiate flowers unless the host has reached the flowering condition. To what extent other broomrapes follow this pattern is not known. Either the specific flowering hormone of the host or its precursor might be transferred to the tissues of the broomrape in which flowering is induced. The possibility of nonhormonal explanations should not, however, be neglected. The later work of Kribben (1951) shows how little extrapolation can be done here. Kribben was able to obtain flowering in several species of *Orobanche* on hosts which were in purely vegetative condition. I might add that *O. uniflora* in the Pacific Northwest usually flowers and often sets seed before the host flowers open. It would be helpful to have similar field observations extended as a guide to more analytical work.

More recent work has shed new light on hormonal interactions in parasitism. Fratianne (1965), extending earlier work by von Denffer (1948) on *Cuscuta gronovii*, has shown that flowering inhibitors may have a dominant role. Again, we must keep in mind the parameters of such work, as flowering dodder is often reported on vegetative hosts (Denffer, 1948; Fritsché et al., 1958), and in pure culture behaves as a short-day plant (Baldev; 1959; Jacob, 1966). Fratianne (1965) states that *C. campestris* flowers only when the host itself is flowering, even though the flowering pattern of the host remains unaffected by the parasite. When two hosts, linked by dodder bridges, were given different photoperiodic treatments, however, no transmission of flowering hormone from one host to the other could be demonstrated. Instead, the various experimental results support the notion that a flowering hormone inhibitor is translocated from the noninduced host to the induced one via the dodder bridge. Dodder may become a promising tool in the study of floral induction, and from such studies we cannot fail to learn more about its parasitism.

Even in recent literature there are casual references to a supposed saprophytism of green parasitic figworts (Yeo, 1961). This notion would seem to take its origin from the work of Sablon (1887) and Koch (1887a), and appears to be based merely on the observation, often repeated since that time, that haustoria may be attached to dead organic matter. The early report of a very low photosynthetic capacity in green parasitic figworts seemed to agree with the idea (Bonnier, 1891), but was later contradicted by Kostytschew (1924). Recent data on *Castilleja coccinea* are similar to the former (Malcolm, 1962).

I must agree with Heinricher (1908a) that the evidence for saprophytism is non-existent. Even today, half a century later, we know too little to allow a reasonable interpretation of the stimuli leading to haustorial formation. Heinricher regarded the formation of haustoria on dead organic materials as a sign of starvation. He also ascertained that such haustoria are anatomically imperfect. The case for saprophytism in these plants is very weak and, indeed, can probably be dismissed. This is not to say that some organic materials may not enter the root from the soil, but this scarcely sets these plants apart from nonparasitic, nonsaprophytic angiosperms (Winter, 1961). Saprophytism in the sense accepted by most botanists implies a mycorrhizal association which is totally lacking in parasitic flowering plants.

There are few topics in the realm of parasitism so intangible as the degree of dependence upon the host. We encounter here the concepts "facultative" and "obligate parasitism," which I have not had occasion to use elsewhere in this book. This in itself is perhaps significant. These two concepts may actually have little meaning when applied to parasites in their natural environment. A facultative parasite would refer to one which may or may not parasitize a host but does not require haustorial contact to complete its life cycle. We know of parasites to which this may apply: *Melampyrum lineare* (Curtis and Cantlon, 1965), *Rhinanthus minor* (Hambler, 1958), *Odontites verna* (Govier and Harper, 1965), *Orthocarpus faucibarbatus* (Thurman, 1965), and some species of *Castilleja* (Heckard, 1962). But the fact remains that in nature these plants are always encountered in the parasitic state. As Thurman (1965) has pointed out, some parasitic figworts may be facultative under cultivation, but these optimum conditions are never encountered in the field. In the context of their natural environment even these plants must be regarded as obligate parasites. This is true of *Cuscuta* also, which no one would call facultative, even though it can complete its full life cycle on aseptic media (Baldev, 1959). The progress of embryo culture of various parasitic angiosperms might lead one to expect similar results in future (Maheshwari and Baldev, 1961; Johri and Bajaj, 1962, 1964, 1965; Rangaswamy, 1963; Rangan, 1965; Rao, 1965; Bajaj, 1966; Rangaswamy and

Rangan, 1966). The division into facultative and obligate parasites, at the level of the angiosperms at least, only reflects the state of contemporary ignorance. Because of the flexibility these plants exhibit in nature, it is impossible to draw a dividing line between the two categories.

Beyond the implications of anatomy, we know nothing about the tissues active in the transfer of organic compounds. Probably in all haustoria the greatest contact surface is occupied by parenchyma, and it is by means of these apparently undifferentiated tissues that organic materials must be led. We should not believe, however, that the xylem of the host contains only water and minerals. The relevant work here has been done mostly on trees, where a variety of organic materials, including sugars, nitrogenous compounds, and enzymes, are known to be present in the xylem exudate, although in greatly fluctuating amounts (Kramer and Kozlowski, 1960). An appreciable movement of carbohydrate, at least in spring and late summer, has also been recorded. Considering the prominence of the xylem bridge in most haustoria, we must keep such facts in mind.

In their water economy, parasitic plants have high transpiration rates. It is common knowledge to plant collectors that semiparasites wilt almost as soon as they are cut. This phenomenon is not limited to Rhinanthoideae, as is evident in *Comandra* (Piehl, 1965c) and *Epifagus* (Cooke and Schively, 1904). In the more fleshy holoparasites the rigidity of the stem and the absence of foliage leaves may hide a similar phenomenon. Even in mistletoes, transpiration is very high (Kamerling, 1914) although subject to great, unexplained fluctuations (Härtel, 1956).

The strength of the pull on the water-conductive system of the host is shown by the fact that, under conditions of moisture stress, the host wilts first (Thurman, 1965). Another indication is found in the great increase in the parasite's transpiration when its cut stem is placed in water. The consequent water loss in *Rhinanthus* may range from two to ten times the normal loss (Kostytschew, 1924). Similar reports come from *Euphrasia* (Heinricher, 1917b; Seeger, 1910), *Pedicularis* (Härtel, 1959; Sprague, 1962a), and *Odontites* (Seeger, 1910).

Thurman (1965) has provided a simple demonstration of the fact that all water and minerals of an established plant of *Orthocarpus faucibarbatus* can be supplied by the host (fig. 8-1). An individual of this parasite was grown in a pot together with a plant of *Agoseris grandiflora*. A large root of the host was allowed to grow out of the bottom of the pot and into another pot immediately below it, where it ramified profusely. When the parasite began to grow rapidly—a sure sign of functional haustoria—the upper pot was no longer watered, but the lower one was kept moist. This treatment resulted in fully developed *Orthocarpus* and demonstrated that at least in later growth the host is able to supply all the water and minerals it requires.

This is more than a hypothetical situation: it may frequently be duplicated in nature where the only

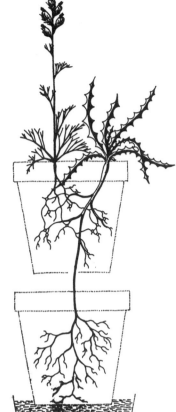

Fig. 8-1. *Orthocarpus faucibarbatus* on *Agoseris grandiflora* an experiment to demonstrate host's capacity to supply virtually all water and inorganic materials to a mature root parasite: (haustoria indicated by small circles) (after Thurman, 1965).

californicum on creosote bush, *Larrea divaricata,* and always in *Cuscuta salina* on *Allenrolfea occidentalis. Cuscuta* on *Impatiens,* on the contrary, had an osmotic gradient leading in the direction of the parasite. Another worker has established the same expected result in *Orobanche crenata* growing on *Vicia faba* (Bergdolt, 1927).

It seems, therefore, that we can form no unified concept of the role of osmotic concentrations in the parasitic partnership. When we consider the complexity of osmotic gradients in the normal growing plants we might wonder whether the above-cited studies have really contributed much to a general understanding of parasitism. Except for Bergdolt, none of these students refer to the haustorial organ, which is, after all, the active site of parasitism. Furthermore, mistletoes are not the only plants growing on trees and having relatively high osmotic concentrations (Senn, 1913). To what extent do high osmotic values reflect the xerophytic environment of epiphytes in general? The instances of an apparently inverted gradient are difficult to account for. A variable which seems to have been neglected in earlier discussions is the possible effect of edaphic fluctuations, which, at least in *Striga,* apparently influence the discrepancy between host and parasite in the matter of osmotic pressure (Solomon, 1952). The flexibility of parasites in these respects must not be underestimated, as is also shown by the phenomenon of hyperparasitism.

Botanists have looked for specific structural features in some Rhinanthoideae (Goebel, 1897; Heinricher, 1901*a*), which might represent adaptations to high transpiration rates. The lower leaf surface of these plants, because of these structures, is believed to be the seat of active transpiration, at least in *Euphrasia, Odontites,* and *Orthantha* (Seeger 1910). *Cuscuta* stomata may also function as hydathodes (Renner, 1934).

More precise knowledge of the parasite's leaf vasculature may give us further clues. In *Pedicularis canadensis* leaves the veins in the tips of each lobe are expanded massively (fig. 8-2). This sort of vascular development points to active guttation. In the leaves of many mistletoes, similarly, the ultimate veinlets bear so-called storage tracheids which may be similar in function. It remains to be explored how common such features are in foliaceous parasites.

available hosts have taproots reaching beyond the desiccated surface soil confining the parasite's root system. Thurman reports field observations where the most successful *Orthocarpus faucibarbatus* were found in close proximity to deeply rooted plants such as *Plantago* and *Hypochoeris.*

About the precise mechanism of water transfer we know nothing, but the usual xylem continuity between host and parasite suggests that the parasite may be regarded as an excessively demanding graft in this regard. The mechanism of movement of water and minerals from host to parasite is likely to be the same as that from the main stem of a plant to its secondary branches. Besides this passive movement via the transpiration pull, a certain amount of water probably enters the parasite through osmotic action. Both mechanisms presuppose an osmotic gradient between the two partners (Härtel, 1956). Moreover, the host's root pressure may be of assistance (Ziegler, 1955).

It is not always clear that such an osmotic gradient actually exists, or what its significance is in the water economy of host and parasite. Harris and coworkers gave much attention to osmotic comparisons of hosts and parasites (Harris and Lawrence, 1916; Harris, 1918, 1924; Harris and Valentine, 1921; Harris *et al.,* 1930; Harris *et al.,* 1930-1931). From these studies they concluded that osmotic concentrations of parasitic chlorophyllaceous tissues tend to be higher than those of the host leaves. Nevertheless, some puzzling exceptions were brought to light. The osmotic gradient seemed to be in the opposite direction in the parasitism of *Phoradendron*

HOST RANGES

In considering what hosts might be attacked by a certain parasite, we must take into account the variable susceptibility of hosts, degrees of virulence of the parasite, the immunity of certain plants to attack, and related topics. In trying to ascertain whether a certain parasite can thrive on a certain host, we are dealing with two systems of genetic variation simultaneously: one belonging to the host, the other to the parasite. It would be incautious, indeed, to draw inferences from one level while ignoring the other. Superimposed on this genetic complexity is the system of environmental influences which might, in certain cases, also determine the success of a parasite.

When we review the numerous relevant field observations, we are haunted by the difficulty of interpreting negatives. The fact that a tree does not have mistletoe plants growing on it, even though neighboring trees do, may or may not indicate a kind of resistance. It might be advisable, therefore, to summarize the causes of apparent resistance in such situations.

In the branch-inhabiting parasites of Santalalean affinities it is possible that the absence of a parasite from certain trees actually has nothing to do with the tree itself, but is due to the habits of the birds which disseminate the seeds. Not only are there clear habitat preferences for isolated trees, or open parkland, or the margin of a forest; it has even been shown that some birds are ecologically limited to definite parts of the crowns of trees (MacArthur, 1958). It should not surprise us, therefore, if the apparent preference of some mistletoes for open-grown or marginal trees is actually dependent on the behavior of birds. It is conceivable that in *Dicaeum*, a bird requiring a complex maneuver to void its mistletoe seeds, preference might be exhibited for the branches of certain forest trees. Nothing of the sort has been proved, but such possibilities must be considered before passing judgment on the resistance of trees free of the parasite.

Some parasites are very intolerant of shade, and this predisposition may place out of reach many otherwise acceptable hosts which normally live under shaded conditions. Mistletoes lacking extensive vegetative reproduction can be shaded out through later growth of the crown of the host. In a similar way terrestrial green parasites such as many Rhinanthoideae and herbaceous Santalaceae have eliminated from their natural host range many plants which are found in shady places. Here we see a great evolutionary advantage for the chlorophyll-less parasite. A dark station has become acceptable to such a plant. Most Rafflesiaceae and Balanophoraceae are typical inhabitants of shaded forests. Curiously, the broomrapes have not followed this pattern (although *Boschniakia* is equally at home in exposed or shaded places), and neither have Lennoaceae.

The mechanical properties of the host organ to be attacked may have an influence also. Again, the most obvious examples are seen in the mistletoes. An early and extensive cork formation on young branches will go far in stopping the penetration by the young haustorium. It has been said that the striking exfoliation of bark in such trees as *Eucalyptus, Platanus, Arctostaphylos* and *Arbutus* accounts for the absence of mistletoes from such trees, but this notion does not survive closer scrutiny. *Eucalyptus* is a favorite host for many mistletoes in Australia (Blakely, 1922-1928), and there are four other reports of parasitism on this host: one in Ceylon (Weeraratna, 1960), one in South America (Hoehne, 1931), a third in India (Patvardhan, 1924), the

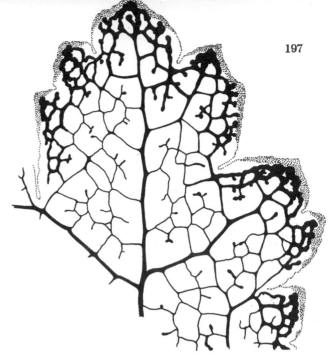

Fig. 8-2. *Pedicularis canadensis,* cleared portion of leaf (Lundell No. 11127, Texas, UC; × 25).

fourth in Costa Rica (Kuijt, 1964*d*). *Platanus* and *Arbutus* are heavily parasitized by various species of *Phoradendron* in Mexico and the United States (Trelease, 1916). These types of exfoliation occur only on older branches where, in other trees, bark formation of a more permanent nature might be at least as good a defense. The apparent local immunity of such trees must therefore be explained in other ways.

In *Cuscuta* mechanical barriers may be very important in halting the haustorial advance. The branches of *Quercus* and *Digitalis* and the leaves of *Pinus* and *Picea* possess an effective sclerenchyma barrier (Gertz, 1915). In some cases sclerenchyma seems to be especially differentiated as a mechanism of defense (Gertz, 1918). Excessive cork formation may make mistletoe infection impossible (Scharpf, 1963). Wound cork sometimes surrounds *Cuscuta* haustoria, but appears to be of limited significance (Thomson, 1925). Other instances of mechanical tissues being similarly effective in attack by Santalalean haustoria have been given by Barber (1906, 1907*a*). The resistance of some sorghum varieties may be of this type (Williams, 1959). Such a defense, however, could be completely successful only in weak haustoria utilizing mechanical force alone, and it is doubtful whether such haustoria exist.

Other causes of apparent immunity may be ascribed, usually by inference, to biochemical incompatibility of host and parasite. Here we can place a great number of host-parasite combinations which we ordinarily fail to consider because they have not yet been reported. We know that *Arceuthobium* does not grow on angiosperm trees, even though in nature it very often has this opportunity, and it may be assumed that a biochemical incompatibility is involved. The high acidity of the tissues of *Begonia* and *Oxalis* is believed to confer immunity with respect to *Cuscuta*, and the same may be true of alkaloids and other toxic substances in *Euphorbia, Papaver, Datura,* and other genera (Gertz, 1915). However, some plants containing known toxic substances have been reported as hosts to the same parasitic genus (Gaertner, 1950). We cannot take it for granted that all materials present in the host are transferable to the parasite, as is shown by the failure of nicotine to do so (Walzel, 1952a).

As we have noted, haustorial penetration in itself does not prove the host to be a suitable one. This uncertainty haunts us in nearly all host records of dodders. Sometimes it results in contradictions. In the extensive lists provided by Gaertner (1950) we find, for example, that *Cuscuta europaea* parasitizes *Rhinanthus crista-galli,* but elsewhere it is said that this host is unable to support the same dodder. *Beta vulgaris* and *Satureja vulgaris* are sometimes susceptible, and sometimes virtually resistant to *C. gronovii.* According to another report, in each of the genera *Solanum* and *Amaranthus* one species is highly susceptible and the other is apparently resistant to dodder. In all records of this sort in *Cuscuta,* unlike other parasites, we need to examine the exact circumstances of the observations. Many names have been entered as host plants whereas they actually serve as little more than support.

From total immunity an entire series of intermediate steps takes us to conditions of optimum development. Several years ago I reported on a curious infection of larch by hemlock mistletoe (Kuijt 1964e). Large swellings were visible on the larch, but the endophyte had not produced mistletoe shoots. The larch is susceptible to this mistletoe in the sense that establishment occurred, but not susceptible in a sense relevant to the parasite's ability to reproduce. On rare hosts we may find very small mistletoe plants and/or very slow growth and, consequently, a greatly reduced reproductive capacity, as in the *Phoradendron* reported on a cactus by Moran (1962). The amount of hypertrophy in mistletoe parasitism sometimes serves as a guide to the degree of physiological incompatibility: extensive hypertrophy indicates a high degree of incompatibility.

In some instances, mechanisms which are not really based on mechanical barriers or biochemical incompatibility allow a host to escape infection. A Russian report tells of sunflowers being attacked by broomrapes less when subjected to a long-day regime (Zakharov, 1940). The suggestion here is that the photoperiodic treatment determines the amount of germination stimulant secreted by the sunflower roots. There is also a variety of sorghum which, in contrast to other varieties, is unable to stimulate germination of *Striga* seeds (Williams, 1959).

One must avoid being misled by both the negative and the positive reports in host records of parasites. A parasite reported to be restricted to, or greatly preferring, a certain host species may in fact be confined, through other ecological or historical factors, to a vegetational zone where the "favorite host" happens to be dominant. Thus *Psittacanthus schiedeanus* in Central America, which appears to be limited to oaks, in reality is not, but may, instead, be limited to the *oak zone* (Kuijt, 1964c). To cite another hypothetical case: *Cuscuta salina* may not actually be confined to saline plants, but its survival may be determined by other factors in an environment where such shrubs happen to be abundant.

The phenomenon of hypersensitivity as known in fungus phytopathology (Gäumann, 1950) has scarcely been explored in parasitic angiosperms. The parasitism of *Viscum album* on pear may be a genuine example of hypersensitivity (Paine, 1950). Either the seed or the radicle of the mistletoe when on certain pears sets off a reaction that kills the adjacent bark, making it impossible for the mistletoe to become established. The variables in this situation are so diverse, however, that it is difficult to reach any general conclusions (Scholl, 1957). I have attempted to explain the absence of self-parasitism in *Psittacanthus* in the same fashion. The excessive formation of cork or gummy exudates by the host in the area of infection can repel the entering haustorial organ (Cowles, 1936; Scharpf, 1963); and may represent a type of hypersensitivity.

The known host ranges of various parasitic groups will be discussed in the light of some of the qualifications given above. The general trend in the mistletoes is toward a very low degree of host-specificity (Abbiatti, 1946; Docters v. Leeuwen, 1954; Kuijt, 1964d). *Dendrophthoe falcata* has no less than 343 recorded host species (Narasimha and Rabindranath, 1964). The phenomenon is by no means restricted to the tropics, as shown by studies on *Viscum album* (Tubeuf, 1923) and *Phoradendron serotinum* (Baldwin and Speese, 1957). In certain mistletoes the host restrictions are especially striking. For example, *V. cruciatum* may occur only on olive trees (Stocken, 1964), and *Arceuthobium minutissimum* has been found only on *Pinus griffithii.* Elsewhere in this genus, *A. oxycedri* in nature grows on *Juniperus* only, and *A. pusillum, A. douglasii,* and *A. americanum* are virtually limited to *Picea, Pseudotsuga,* and two closely related species of *Pinus,* respectively (Kuijt, 1955). *Phrygilanthus aphyllus* is known from eight genera of cacti only (Follmann and Mahu, 1964); *Viscum minimum* is limited to a few species of *Euphorbia* (Engler and Krause, 1908), *Korthalsella dacrydii* to *Podocarpus* and *Dacrydium.* Beyond these are the curious examples of *Phacellaria* in Santalaceae (Danser, 1939), *Viscum capitellatum* (Weeraratna, 1960), *Dendrophthora epiviscum* (Kuijt, 1961c), and *Ixocactus* (Kuijt, 1967a), all of which appear to be parasitic on mistletoes or other epiphytic Santalaceae only. I have suggested (Kuijt, 1964c), that the feeding habit of birds may account for much of this hyperparasitism, i.e., parasitism of one parasite on another. A spectacular case of hyperparasitism is the three-storied parasitism of *Loranthus ferrugineus* on *Viscum articulatum,* and the latter on *Elytranthe bar-*

FIG. 8-3. *a, Arceuthobium* spp. on branch of *Pinus contorta; A. campylopodum* (left) and *A. americanum* (right), both plants male (near Kimberley, B. C.; courtesy Canada Dept. Forestry, Calgary); *b, Wallrothiella arceuthobii,* remarkable fungus which parasitizes only flowers of female plant of certain spring-flowering species of *Arceuthobium* in North America; fungus, visible as black stroma crowning nearly every flower of *A. americanum* here parasitic, in turn, on *Pinus contorta* (Kootenay National Park, B. C.; after Kuijt, 1955).

nesii parasitizing the durian, *Durio zebethinus* (Sands, 1924). Particularly inexplicable is the apparently double host-specificity of *Phoradendron bolleanum* subsp. *bolleanum* on *Arbutus* and *Juniperus* (Wiens, 1964b; Hawksworth and Wiens, 1966).

In root parasites generally, we find little specialization with regard to hosts. *Aeginetia* seems to be specialized with regard to monocotyledons (Beck v. Mannagetta, 1930), although Gleicheniaceae and Ericaceae are also recorded. *Orobanche,* in contrast, is not known to parasitize monocotyledons, ferns, or gymnosperms (Hegi, 1907-1931). *Conopholis* is perhaps restricted to red oak (*Quercus borealis*) (Percival, 1931). For other data on Orobanchaceae, Beck von Mannagetta (1890, 1930) is the best source. Two louseworts (*Pedicularis*) possibly restricted to *Pinus ponderosa* have been reported by Sprague (1962a).

It is tempting to think of a high degree of host-specificity as an advanced feature. Perhaps this notion is applicable only to root parasites which have a single haustorial attachment. At any rate, recent work with *Odontites verna* (Govier *et al.*, 1967) suggests that for parasites with many haustoria there may be a real advantage in attacking different host species simultaneously. If the host of *O. verna* is barley, nonnitrogenous compounds (e.g., monosaccharides) predominate in the organic materials transferred to the parasite. If the host is clover, nitro-

genous compounds are far more important. It may well be that the optimum growing conditions of such a root parasite include haustorial contacts with several different host species simultaneously.

Of special interest in connection with host ranges are the extremely advanced parasites Rafflesiaceae, Hydnoraceae, and Balanophoraceae. Although information here is scanty, host specialization, at least in the last two families, is apparently rare. In the first family, *Cytinus* may be restricted to *Cistus, Mitrastemon* to Fagaceae, and the members of the *Pilostyles-Apodanthes* complex to Leguminosae, whereas *Rafflesia* prefers Vitaceae. But in Hydnoraceae only *Prosopanche americana* seems to show any preference, for it is reported only on Leguminosae, especially *Prosopis. Prosopanche bonacinae,* in contrast, may parasitize Chenopodiaceae, Aquifoliaceae, Anacardiaceae, Rhamnaceae, Malvaceae, Umbelliferae, Solanaceae, and Compositae (Cocucci, 1965). In *Hydnora* the following are given as hosts: *Acacia, Euphorbia, Cotyledon, Zygophyllum, Albizzia, Adansonia* and *Kigelia* (Vaccaneo, 1934). In the third family, Balanophoraceae, few authentic host records exist. *Cynomorium* can parasitize *Obione (Halimus), Salsola, Inula, Tamarix, Melilotus, Lepturus,* and *Statice* (Weddell, 1858-1861); *Rhopalocnemis phalloides* attacks *Ficus, Quercus, Macaranga, Schima,* and *Salix* (v. Steenis, 1931). *Thonningia coccinea* is said to have no specificity (Mangenot, 1947), and the same may well be true of *Dactylanthus taylori* (Moore, 1940).

In the four unrelated parasitic groups treated in chapter 6 (*Cuscuta, Cassytha, Lennoaceae* and *Krameriaceae*), the situation is similar. The best treatment of *Cuscuta* host ranges is given in Gaertner (1950). It is difficult to interpret the earlier host records in the genus. Many species of *Cuscuta* when first described were thought to be restricted as to hosts, and thus we find such misleading epithets as *epithymum, epilinum, cephalanthi, coryli,* and *polygonorum.* When it was discovered that more than one host species could be attacked, concepts swung, like a pendulum, to the opposite extreme and these parasites were assumed to be polyphagous. It should be clear by this time that neither notion reflects reality. Some species seem to show a high degree of host-specificity, even though none is known to grow on one host to the exclusion of others. *Cuscuta epilinum*, for example, may be virtually limited to *Linum* in nature, but it will mature on *Impatiens* also. The total host list of *Cuscuta*, in spite of its inherent fallacies, is an impressive one. It includes (notwithstanding some published remarks to the contrary) Characeae, Polypodiaceae (five genera), Marsileaceae, Equisetaceae, Isoetaceae, Pinaceae, and a vast number of monocotyledons and dicotyledons. Some of the latter are succulents, others climbers, plants with latex or alkaloids, parasites such as *Euphrasia* and *Rhinanthus*, and aquatics.

Host records in *Cassytha* are rather inadequate, but perhaps we may assume the situation to be essentially the same as in *Cuscuta*. The hosts of *C. filiformis*, in any event, include both monocotyledons and dicotyledons, showing no particular preferences within these groups (Nayar and Nayar, 1952). In *Krameria*, *K. parvifolia* has been found attached only to *Cercidium microphyllum*, but *K. grayi* to eleven genera representing six unrelated families of dicotyledons plus a gymnosperm, *Ephedra* (Cannon, 1910). No data have been added here, however, since the original discovery of parasitism; so the list is far from complete. In Lennoaceae, work has not advanced sufficiently to make possible a meaningful statement except that *Pholisma* seems to show little host-specificity.

Host-specificity can evolve only where there is an abundance of a single host species. For this reason the narrowest of host ranges usually have the temperate zone as their setting. Tropical vegetation is generally too heterogeneous, and relatively pure stands of one species are very rare. It is the temperate zone which is characterized by extensive stands of a few species. In the host-specific parasites listed above, the temperate zone habitat clearly predominates. Mistletoes inhabiting equatorial forests could not survive if they were too exclusive with regard to their hosts. It might be countered that *Phacellaria* and other habitual hyperparasites are tropical, but their specificity, if real, would seem to be a special situation brought about by dispersal agents (birds) which are selective in the plants they frequent. *Aeginetia*, also tropical, is restricted mostly to grasses, but these hosts are perhaps the most likely of all tropical plants to occur in more or less pure stands over considerable areas.

We are now in a position to pass judgment on two other notions: that holoparasites are more specific as to hosts than semiparasites, and that closely related parasites attack closely related hosts (Gibbs, 1954). The latter idea can be dismissed at once, for there is no evidence to support it in angiosperm parasites; the commonly wide host latitude renders it indefensible. The first notion, the parallel condition of host specialization and evolutionary advancement, would seem logical. In Santalalean families several of the above-mentioned examples of host-specificity might apply. *Arceuthobium*, *Phrygilanthus aphyllus*, and *Viscum minimum* are certainly very advanced among the mistletoes, and the same is true of *Ixocactus*, and of *Phacellaria* among Santalaceae. How much caution is required, however, is illustrated by two very closely related species of *Viscum*, *V. album* and *V. cruciatum*: the former is extremely polyphagous, but the latter is known from only a single species, *Olea europaea*.

Outside Santalales the idea scarcely survives. Even if we accept the support which Rafflesiaceae might provide, the information from Hydnoraceae and Balanophoraceae destroys it as a notion of general validity. It would be difficult to find parasites further evolved than the members of these two families; yet they show no sign of host specialization. The hypothesis that more advanced parasites attack fewer host species—an idea probably zoological in derivation—can never have much significance in relation to parasitic flowering plants.

EFFECTS OF PARASITISM ON THE HOST

The effects of the parasite's interference in the physiology of the host vary all the way from spectacular malformations to absence of recognizable symptoms of disease or discomfort. In the former category are the witches'-brooms induced by *Arceuthobium* on various coniferous hosts (fig. 8-5). These brooms exhibit a variety of changes in growth habit, depending on age, species of tree and parasite, and perhaps other factors. A radical change in tropisms is sometimes evident, and a variable amount of hypertrophy may be expected (Kuijt, 1960). The genus *Mitrastemon* has a similar brooming effect on the roots of its host trees (Watanabe, 1936-1937). *Phrygilanthus tetrandrus* on poplar in Chile shows an irregular type of brooming (Reiche, 1904), and a few other mistletoes, when parasitizing uncongenial hosts, may have comparable effects (Hawksworth and Wiens, 1966).

An effect distinct from brooming, although sometimes leading to it, is the revitalization of the dormant shoot apex of dwarf shoots or short shoots in *Pinus* and *Larix* (Kuijt, 1960; fig. 8-4,*d*). I have observed something of the sort also where *Viscum album* plants become established near a short shoot which suddenly elongates. All such reports almost certainly involve interference in the hormonal economy of the host, although the factor of nutrition cannot be ignored. I suspect also that the presence of the haustorial system of *Arceuthobium* and *Viscum* may, upon occasion, inhibit the differentiation of the host's dormant bud, with the tip of the branch dying during the cold season.

When part of a host is heavily or systemically attacked by a parasite, the former's reproductive potential drops or disappears. There is no record of cones being formed on any brooms caused by

Arceuthobium. In *Astragalus* attacked by *Pilostyles haussknechtii* the host is invariably sterile in the infected portion (Bornmüller, 1909); *P. ulei* allows at least the formation of flowers on the host (Endriss, 1902), but this does not guarantee seed-set.

A common feature in the attack by parasites is the death of that part of the host which extends beyond the point of infection. Mistletoes frequently appear to be terminal on the host branch for this reason. The effect is not expressed so clearly in parasites with a more diffuse, endophytic development. Another peculiarity I have noted (Kuijt, 1964c), which may be related to haustorial differentiation, is that none of the squamate mistletoes seem to have a comparable effect.

In all but the most diffuse endophytes we may expect a certain amount of hypertrophy in the host (fig. 8-4, a-c). The parasite's tissues account for some of the swelling, but much of it is due to an unusual activity of the host cambium. This is clear as soon as the parasite is removed from the host branch. In mistletoe parasitism we thus observe the elaborate placental structures known as woodroses, or club-shaped malformations of the host wood riddled by the parasite's sinkers (see under Haustorium).

Gall-like swellings of the host often result from the presence of dodder haustoria (Dean, 1937b). We do not know what host tissues are involved in this hypertrophy. Thomson (1925) speaks of an irritation set up by haustorial entry stimulating neighboring cambial tissues to increased activity, in addition to the formation of wound cork initiated around the shaft of the haustorium. In hosts like *Cucurbita maxima* both inter- and intrafascicular cambium may be stimulated. Gertz (1918) had emphasized the general lack of formation of wound cork, speculating that an inhibitor substance might emanate from the haustorium.

That the presence of the endophyte of *Arceuthobium* may bring about striking anatomical changes in the host has recently been demonstrated in detail (Srivastava and Esau, 1961b). Host rays which contain parasitic filaments are frequently fused marginally, and the number of rays may be increased. It has also been suggested that sinkers, upon making contact with cambial cells, can actually induce the latter to initiate new rays. Even the individual axial tracheids of the host, when in the vicinity of infected rays, often become deflected and compressed or otherwise deformed, and tend to be shorter than normal. It must be remembered that *Arceuthobium* is a parasite with an unusually advanced endophyte, and that few of these details may apply to other parasites. Certainly in the case of evanescent haustoria we cannot expect significant changes in the structure of individual host cells.

A curious response to a nearby haustorial organ is the formation of tyloses in tracheal elements of the host. I have reported this phenomenon in *Antidaphne* parasitism (Kuijt, 1965b), but Barber in his various articles on *Santalum, Olax,* and *Cansjera*

had spoken of the same situation. It is not impossible that the pitted walls of tyloses have misled Moss (1926) and Menzies and McKee (1959) to speak of penetration of a vessel by a haustorial element, although this sort of penetration has been demonstrated in *Cuscuta* and *Pilostyles* (see under Haustorium). Tyloses under similar conditions have been described by Weeraratna (1960) and Piehl (1963) in hosts attacked by *Dendrophthoe falcata* and *Pedicularis,* respectively.

In summarizing the parasitic angiosperms which are of economic importance as pests, or are causal agents of diseases, I am fortunate to be able to draw on a recent book in which a chapter deals with this topic (King, 1966, q.v.).

Outside the mistletoes, no Santalales are known to do significant damage to economic plants through their parasitism. The occasional attachment of *Osyris alba* to the grapevine in southern Europe is no more than a curiosity, although *Thesium humile* may do some damage to crops in Spain.

Mistletoes cause enormous economic loss in many parts of the world. The matter has been admirably summarized by Gill and Hawksworth (1961). If most tropical mistletoes have not yet received the stigma of "pest," this may be because the cultivation of trees is usually at a very primitive level in the tropics. In Costa Rica, for example, few people are concerned about the species of *Struthanthus, Oryctanthus,* and *Phoradendron* growing in *Citrus* and cacao plantations, although considerable damage must be done. Mistletoes of several genera are serious forest pests in many parts of Australia (May, 1941), and in Chile eradication of *Phrygilanthus tetrandrus* is required by government decree (Ortiz, 1936). In Europe, *Viscum album* may injure a variety of trees, for example silver fir (Klepac, 1956) and orchard pears. *Phoradendron* in North America is troublesome in walnut groves, plantations of pear and pecan, and native stands of true fir, *Abies,* and incense cedar, *Libocedrus,* in California. Man's attitude to the "Christmas type" of mistletoe through the years has been an ambivalent one. As protector of trees he must regard mistletoes as the cause of serious losses. Yet what plant is more safely "rooted" in Western folklore? Mistletoe eradication programs must never take popular support for granted.

Much has been written about the economic importance of dwarf mistletoes (*Arceuthobium*). Outside North America the damage done by this genus does not seem to be significant. The exact ranking of *Arceuthobium* with respect to other major forest diseases in the American and Canadian West varies, depending somewhat on the experience and leanings of the observers. The trees perhaps most seriously affected are Ponderosa pine and Western hemlock, although nearly all members of Pinaceae may be attacked. Species such as *Pinus torreyana* and *Abies bracteata* are free of mistletoe probably because of their geographic isolation. It seems reasonably certain that the damage from *Arceuthobium* on Ponderosa pine exceeds the combined estimates for all

Fig. 8-4. *a*, Swelling (diameter 10 cm) caused by *Phoradendron "macrophyllum"* on *Robinia pseudacacia*, southern California; *b*, effect of infection of *Arceuthobium campylopodum* on trunk of *Abies* sp., California; no mistletoe shoots visible; mistletoe, in fact, may not survive; original area of infection completely decayed; *c*, hypertrophy caused by *Phoradendron "macrophyllum"* on *Populus fremontii* in California; *d*, very young infection of *Arceuthobium campylopodum* on *Larix europaea*, showing conversion of short shoot to long shoot at site of infection (*a and c*, after Kuijt, 1964*c*; *b*, after Kuijt, 1960; *d*, after Kuijt, 1964*e*).

Fig. 8-5. Witches'-brooms caused by two species of *Arceuthobium; a, A. americanum* on *Pinus banksiana,* Calling Lake, Alberta; *b, A. douglasii* on *Pseudotsuga menziesii,* Kootenay Lake, B. C. (after Kuijt, 1960).

other diseases of this tree (Anonymous, 1955). The situation in Western hemlock has not been worked out in as much detail, but may be comparable. The dwarf mistletoes pose a major silvicultural challenge which will keep generations of foresters and botanists occupied. The immediate prospects for adequate control are not very good.

It is difficult to make a precise estimate of the damage done by these parasites, for one is attempting to measure a negative quantity: for example, the volume of wood which might have materialized in Ponderosa pine in a certain stand if *Arceuthobium* had not been there. An accurate estimate would require careful comparison with a nearly identical stand elsewhere which is free from *Arceuthobium,* but identical stands of trees are notoriously difficult to find. We must rely heavily on rather subjective estimates made by local personnel. Rarely can the results derived from one stand be applied to another.

Control of economically important mistletoes is very difficult. Removal of aerial shoots is quite ineffective, for the endophyte normally regenerates shoots rapidly. Pruning must be done with knowledge of the extent of the haustorial system, or it also is without effect. Where large limbs are parasitized in fruit trees or where the main stem is infected, pruning is not practicable. No truly effective and reliable hormonal spray is known against mistletoes. Removal of heavily infected trees often does no more than give the minor unseen infections on surrounding trees an impetus to develop into major infections. In forests infected with mistletoes the most effective control will be one not superimposed on, but carefully integrated with, a silvicultural program of management. In the coniferous forests of the Pacific Northwest this stage seems far in the future.

Balanophoraceae, Rafflesiaceae, Hydnoraceae, Lennoaceae, and Krameriaceae do not constitute economically important parasites, although *Prosopanche bonacinae* has been reported from cultivated cotton (Burkart, 1963). In *Cassytha* we are concerned not so much with a debilitating effect on host plants as with a great nuisance. *Cassytha* is an ungainly plant which covers masses of otherwise attractive native or ornamental vegetation. It seems to be a real menace only in Puerto Rico, where it has actually been called "Public Enemy No. 1." On the whole, nevertheless, *Cassytha* seems to be more easily kept in check than some species of *Cuscuta*. The dodders (*Cuscuta*), according to King (1966), are considered noxious weeds in forty-seven states of the contiguous United States. By no means all species, or even most species, are of economic significance. But when certain species, often because of the planting of impure seed, become established on leguminous crops like *Lespedeza*, clover, or alfalfa especially, the effect can be disastrous. In a very short time such a crop is covered with a thick web of filamentous dodder stems making ever more haustorial contacts, and draining nutrients from their hosts. The shading effect is often considerable. In Australia, *C. australis* represents a serious agricultural menace, especially in New South Wales (Walls, 1962).

Dodder is controlled mainly through increased emphasis on the purity of seed sources. The similarity in size and shape of the seeds of dodder to those of the infested crop has been a source of difficulty in the past, but much improvement has been made by using "dodder mills" to separate these weed seeds from the crop seeds. The dodder seeds adhere to felt-covered rollers. Local infestation can be controlled to some extent by using chemical sprays or by manual cutting and burning. The feasibility of control through soil-applied herbicides is being studied actively at present (Dawson, 1967).

The parasitic figworts and broomrapes include many plants which are extremely destructive to planted crops, and others which are potentially dangerous. I have summarized the most important of these in table 4 at the end of this chapter.

Botanists traveling in the Mediterranean region are astounded by the massive development of *Orobanche crenata* on leguminous crops, especially *Vicia faba*. This broomrape profits from the primitive local agricultural methods which usually allow it to disseminate its numerous seeds unmolested. Other species of *Orobanche* of economic import in the Old World are *O. cernua*, *O. minor*, and *O. ramosa*. In North America the last species, introduced long ago from elsewhere, is a troublesome weed in California agriculture (Wilhelm *et al.*, 1958), and of consequence in some other states (Garman, 1903).

Perhaps the most dangerous of all broomrapes are the species of *Aeginetia* which attack sugar cane and sometimes maize or rice, in tropical Asia. The effect on the cane crop is not merely reduction in yield; the parasite also has the ability to convert the host's sucrose to reducing sugars. The efficient vegetative reproduction and the enormous number of seeds produced (40,000 to 70,000 seeds per capsule!) make this parasite a serious threat to tropical agriculture —a threat which is not limited to the Old World.

In the parasitic figworts are several genera which may be of local importance but do not appear to pose a general threat at present. *Rhamphicarpa* is known to parasitize rice, maize, and cowpeas in Madagascar and East Africa. *Rhinanthus* can do much damage to meadows in Europe, and even on cereal grains (especially rye) in some localities. The detrimental effect in the latter case is evident from the Austrian peasant's proverb "Der Klapf frisst das Brot aus dem Ofen heraus" (The Yellow Rattle eats the bread out of the oven) as reported by Heinricher (1898b). Some of the German vernacular names for *Euphrasia*, such as "Milchdieb" and "Wiesenwolf" indicate popular recognition of the damage done to the surrounding vegetation.

The most serious pests among parasitic figworts are *Alectra* and *Striga* (witchweed or Rooibloem). The effects of *Alectra* have scarcely been felt outside the African region. *Striga*, however, was an agricultural challenge of the first order even before its introduction into North America (Shaw *et al.*, 1962). For years *Striga* has caused greater loss than all fungus diseases combined in South African agriculture. The several species attacking a variety of cultivated crops and wild plants are extremely destructive, and especially under primitive conditions of agriculture may drastically reduce the size and quality of yield.

Various methods of controlling these root parasites have been suggested. The variables of hosts, soil types, climatic conditions, and agricultural practices are infinite, however, and control measures of general application have not yet been ascertained.

The most direct method of control, naturally, is the destruction of the parasite wherever it develops. In some areas, so-called catch crops are used which are true hosts; the parasite is pulled out as soon as it appears. If this is done consistently through the years, the concentration of parasitic seeds will decline gradually. Much damage is done, however, even before the *Striga* or *Orobanche* reaches the soil surface.

Biological control by foraging insects has shown no promise in these root parasites, though some damage may occasionally be noted (Williams and Caswell, 1959).

Chemical control has shown promise in both *Striga* and *Orobanche* (Robinson, 1961; King, 1966). Even fertilization with ammonium nitrate affords some control (Shaw *et al.*, 1962). A number of herbicides show promise. That "effective control now and future eradication look favorable" (Shaw *et al.*, 1962) would seem to be an overoptimistic view, however, when we consider how quickly such parasites become established on the surrounding wild vegetation. In the tobacco belt of North America, a special Witchweed Laboratory has been established to combat this menace. In California, soil fumigation with methyl bromide has been effective in control of broomrape (Wilhelm *et al.*, 1958, 1959).

Under a regime of primitive agriculture a simple solution is scarcely possible. In fact, it has been suggested that the *Striga* attack in some areas may

TABLE 4
ROOT PARASITES OF ECONOMIC IMPORTANCE

PARASITES	ECONOMIC HOST AND LOCALITY	REFERENCES
Rhamphicarpa longiflora	maize, cowpeas, rice (Madagascar), sorghum (East Africa)	Bouriquet, 1933; Fuggles-Couchman, 1935
Orobanche spp.	hemp, tobacco, tomato, leguminous crops (various localities)	Micheli, 1723; Garman, 1903; Percival, 1931; Stout, 1938; Marudarjan, 1950; Durbin, 1953; Kadry and Tewfic, 1956
Striga spp.	maize, sorghum, sugar cane, tobacco (paleotropics and parts of eastern North America)	Saunders, 1933; Hedayetullah and Saha, 1942; Wild, 1948; Sharma *et al.*, 1956; Shaw *et al.*, 1962
Aeginetia spp.	sugar cane, rice, maize (India, Pakistan, Ceylon, Southeast Asia)	Kusano, 1908b; McWhorter, 1922; Coert, 1924; Agati and Tan, 1931; Lee and Goseco, 1932; Hedayetullah and Saha, 1942; Espino, 1947; Ling, 1955; Lo, 1955
Alectra spp. and *Melasma* spp.	cowpeas, soybeans, peanuts, and other leguminous crops (Rhodesia, South Africa, etc.), sugar cane (West Indies)	Rattray, 1932; Saunders, 1934; Botha, 1948
Christisonia	sugar cane (Philippines)	Quisumbing, 1940

have long-term beneficial effects on soils (Basinski, 1955). The great losses suffered by native farmers in marginal agricultural lands may be offset by the fact that this makes it impossible for them to farm there indefinitely and thus deplete the soil permanently, leading to the fatal consequences associated with soil erosion.

In broomrape- or witchweed- infested land in South Africa, some higher plants, including crop plants such as *Linum* or *Capsicum* for broomrapes, and cowpeas and soybeans for *Striga*, produce the root exudates required to set off the germination of the parasite's seed, but are immune to attack. By planting such "trap" crops, the infested soil is gradually sanitized. This method is strictly limited in Africa, for the native economy frequently depends upon susceptible crops such as sorghum. A long-range program of breeding sorghum for resistant varieties would seem to be more worthwhile than reliance on trap crops. Some progress in this direction has, indeed, been made (Lo, 1955; Shaw *et al.*, 1962). Trap crops by themselves are of limited effect because too few parasitic seeds are removed with each crop (Nash and Wilhelm, 1960). A combination of a suitable rotating sequence with a good fumigation program will probably be most effective in control.

Parasites may be potential hazards should they succeed in overcoming certain geographical barriers. *Arceuthobium* constitutes such a threat to any country where Pinaceae are a significant source of income. Indeed, even within a country, the movement of dwarf mistletoe may have serious consequences. Pine forests in the South and in eastern North America, for example, might be endangered if dwarf

mistletoes became established within them. That such movements are not impossible has been demonstrated (Offord, 1964). Among other mistletoes, those parasitizing *Eucalyptus* in its country of origin should be watched cautiously, for they constitute a threat to *Eucalyptus* plantations elsewhere, as in California and Chile.

Striga has not by any means occupied its potential territory. The demonstration that temperature by itself does not appear to be the major limiting factor (Robinson, 1960) is cause for a certain amount of anxiety. Its introduction into North America should make us doubly watchful in other countries in the Western Hemisphere. An even more serious potential threat may be seen in *Aeginetia*, especially with regard to sugar cane cultivation in Central America and the Caribbean.

As far as I am aware, only the following parasitic angiosperms have been distributed over long distances by man. The species of *Cuscuta* to which this statement applies may be found in Yuncker's articles (Yuncker, 1932, 1934, 1961; see also Cheeseman, 1925). *Orobanche minor* has reached North America (Garman, 1903; Hitchcock *et al.*, 1959) and New Zealand (Kirk, 1887); *O. picridis* has invaded New Zealand (Kirk, 1887); *O. ramosa* is established in California (Stout, 1938; Wilhelm *et al.*, 1958) and South Africa (Marloth, 1932). *Viscum album*, unbeknownst to California botanists for more than half a century, is securely established in that state (Howell, 1966). *Hydnora* is a doubtful introduction on the island of Réunion (Harms, 1935).

In the figworts, several genera have immigrated to North America. *Odontites rubra* was becoming established in northeastern North America as long ago as 1935 (Pennell, 1935), but I have seen no later reports. *Parentucellia viscosa* and *Bellardia trixago,* originally European, have become a regular part of the weed flora of the Pacific coast of North America (Munz and Keck, 1959). *Striga* in North Carolina and vicinity has been mentioned. On a smaller scale, *Euphrasia arctica* var. *disjuncta* behaves like an introduced weed around Vancouver, Canada; and *Orthocarpus faucibarbatus* seems to have reached Vancouver Island from its more southerly origin (Keck, 1940). Survival is by no means guaranteed for all these genera, but some of the introductions mentioned will almost certainly have to be counted among the natural flora of their new homes.

9

Evolutionary Aspects of Parasitism

THE EVOLUTION OF PARASITISM IN ANGIOSPERMS

If we attempt, in a speculative frame of mind, to construct a unified image of the evolution of parasitism in flowering plants, we must first remove the aberrant genera *Cuscuta* and *Cassytha* from the scene. It seems safe to assume that in their ancestors the roots derived from the radicular pole of the embryo were never involved in parasitism, but have dwindled in importance and, to some extent, changed in function after "captured" nutrients and water began to enter directly into the branches. The earliest haustoria in these two genera almost certainly originated as aggressive adventitious roots on liana-like plants. The twining habit may be presumed to have preceded the parasitic habit, at least in *Cuscuta*. It is unfortunate that no adventitious roots seem to occur on the aerial stems of present-day Convolvulaceae, but this information gap is even more striking in Lauraceae.

It is my thesis that epiphytic parasitism has followed a terrestrial or at least a root-parasitic habit. In Santalalean families I think this is now nearly self-evident. The only defense—although a persuasive one—of the thesis with respect to epiphytic Rafflesiaceae would seem to lie in comparative seed anatomy. The degree of reduction in the seed of Rafflesiaceae tends to be associated with extremely specialized germination requirements. We know such reduction from parasites in which germination depends on chemical stimulants diffusing out from the host organ. Although we must not underestimate the possibilities of chemical exudates and volatile products emitted by the aerial parts of plants, the great variety of organic materials known to be released by the host roots of higher plants, together with the more favorable moisture conditions in the soil, make it more likely that all Rafflesiaceae are also traceable to a root-attacking ancestral form. I admit that this argument would suffer badly if it should be demonstrated conclusively that germination of seeds with reduced, undifferentiated embryos, such as Rafflesiaceae, Balanophoraceae, and Lennoaceae, does not necessarily depend upon root exudates in nature. Actually, two such cases appear in the literature: a casual report of germination of *Pilostyles aethiopica* seeds on moist paper (Soyer-Poskin and Schmitz, 1962); and one of *Cynomorium* seeds in moist soil (Weddell, 1858-1861). Both reports need confirmation. My working hypothesis is

that the seeds of the three above-mentioned families require a chemical stimulant in order to germinate.

But how did these root parasites achieve the parasitic habit to begin with? Here we have no definitive answers. The facultative parasites among Scrophulariaceae do not help us here, for when their haustoria are formed they are of a simple and precise organization which we do not associate with primitive haustoria. One of the unexplored areas of study involves the root anatomy of those autotrophic species which have close affinity to parasitic species. This approach might produce interesting results for both Scrophulariaceae and Olacaceae.

Root grafting has occasionally been suggested as a more possible origin of the parasitic habit. In his recent study of *Orthocarpus*, Atsatt (1966) does not differentiate between root grafting and haustorial attack. Although we are poorly informed on this subject, it is clear that a great deal of root grafting is typical of many tree species, both in temperate and tropical regions (La Rue, 1952; Miller and Woods, 1965; Graham and Bormann, 1966). I am inclined to think that root grafting can be excluded in this regard. I know of only one report of true root grafting in herbaceous plants (Piehl, 1962*b*), and it is obvious that all non-Santalalean parasites, at least, are of herbaceous affinities. Root grafting, furthermore, would seem to be a phenomenon of secondary growth of older roots. It is very difficult to think of primitive haustoria in such areas rather than in the subapical parts of roots where present-day haustoria originate.

A totally different idea may be based on the interesting work of Björkman (1960). It turns out that the mycorrhizal associate of the saprophyte *Monotropa hypopitys* is organically continuous with those of some of the surrounding trees such as the beech, *Fagus silvatica*. By means of trunk injections of C^{14}, passage of organic materials has been demonstrated from those trees, via the mycorrhizal associate, into *Monotropa*, which in Björkman's words, is an "epiparasite" on trees.

Is it not possible that the dependence of a saprophyte might gradually develop into a more precise physiological relationship with the tree? Might not the saprophyte evolve a more specialized absorptive organ which, with greater aggressiveness, could bypass the fungus component to consolidate the growing physiological compatibility between herb and tree? We would then, by elimination of at least the need for the fungus, have a normal parasitic partnership as defined in this book.

For evidence of such a view we might look for traces of early tripartite relationship. I have not been able to discover any indications of this sort. *Melampyrum* seems to show a preference for mycorrhizal roots (Heinricher, 1917*b*), but this is weak evidence indeed. Elsewhere, no signs of mycorrhizae are recorded in parasites, and the fact that some hosts are mycorrhizal is hardly surprising. It would be surprising, rather, if mycorrhizal hosts were avoided by root parasites, most of which are characterized by wide host latitude. The known autotrophic relatives of most parasites—Olacaceae, Scrophulariaceae, Convolvulaceae, and Lauraceae—are not among mycorrhizal families. There are mycorrhizal saprophytes in (or near) Polygalaceae (e.g., *Salomonia*; Penzig, 1901), the family which seems most closely related to Krameriaceae. The origin of parasitism via saprophytism thus seems unlikely.

Another way in which parasitism may have originated is by means of chance establishment of one plant on another. I know of only three reliable reports of this kind.[1] A *Passiflora* has been reported as a parasite on the roots of a *Evonymus* bush (Peé-Laby, 1904). The *Passiflora* organ seems to have been a cauline, tuberous one, and had a haustorium-like body within the host wood. The parasite had scarcely any roots. MacDougal (1911) encountered plants of *Cissus laciniata* and *Opuntia* sp. implanted on a tree of *Yucca,* and one of *O. toumeyi* on *Cercidium microphyllum.* Exploring the idea of an induced type parasitism, he grafted a number of plants onto different species of succulents. The results were very limited, however, and nothing resembling a haustorial union seems to have resulted. A third dramatic instance has been reported by Moran (1966; fig. 1-1). A healthy fruiting plant of a cholla (*Opuntia* sp.) was discovered growing directly out of a trunk of an *Idria* tree in Baja California, Mexico. The cholla extended horizontally from the supporting tree at a point about five meters above soil level. How this "infection" originated is unknown.

We do not know whether the cholla plant had physiological contact with living *Idria* tissues or whether its roots grew in a cavity. It may be significant, as Dr. Moran has pointed out, that the *Idria* branched at this point, an event normally resulting from damage to the leader, and possibly associated with decay. Could such chance occurrences have led to the origin of parasitism as we presently know it? It is noteworthy that the cholla in Dr. Moran's report was bearing fruit. Mutations in such a plant might occur which further adapt the dispersal mechanism, or the germination pattern, or the behavior of the root system to the species of tree inhabited. Moreover, an individual of this sort has moved into

an ecological stratum that is very different from its normal one with respect to many climatic and biotic factors. Such a change would be especially important in certain types of forests, particularly in the multi-level forests of tropical areas. We cannot exclude this type of parasitic origin merely because of its improbability. At the same time, the evidence seems to point to the origin of parasitism as an interaction of the roots of two different species. The fact that in *Cuscuta* and *Cassytha* it was an interaction of the roots of one with the stem of another plant does not affect the hypothetical sequence of events to be discussed next (fig. 9-1).

In some manner which we cannot divine at present, the parasitic habit was established. I conceive this to have happened through a dimorphism of roots, a situation still recognizable in Lennoaceae, possibly in Hydnoraceae, and, if Cannon's (1910) observations are borne out, in *Krameria.* Consider, first, the root system of a normal plant. Through some change in the physiology of the root, or its juxtaposition with the root of an invading or otherwise rare species having certain stimulatory root exudates which hitherto had not been encountered, the growth of the former root is directed to the latter, finally growing into its tissues. At a different level, a similar evolutionary change takes place in the behavior of the radicle of Loranthaceae, Orobanchaceae, and *Striga.* A degree of physiological compatibility and an osmotic gradient would be required to turn this chance event into a primitive parasitic relationship. From here to the type of parasitism described for *Pholisma* is not a great distance; it requires a division of labor between larger nonhaustorial roots and smaller haustorial roots.

As the haustorial root becomes more efficient in this process of specialization, it becomes functional earlier; that is, it becomes shorter. The cortex of the mother root adjacent to the haustorial root also begins to contribute in forming a collar-like mass of parenchyma surrounding the receding haustorial apex. This shift in relative contribution eventually results in a cushion of parenchyma which grows ahead of the haustorial apex. The same situation occasionally materializes in *Cuscuta* haustoria when they do not reach the host surface. (*Cuscuta* and *Cassytha* were excluded from this sequence of thought earlier, but their haustoria may easily have had a similar origin.) When the intrusive haustorial organ appears it comes from the inner tissues of the outgrowth. The farthest forward cells of the early outgrowth differentiate into a glandular layer facilitating adhesion to the host, and we have all the elements of the lateral haustorium as we know it in Santalales, *Cassytha, Cuscuta,* and possibly *Krameria* and some Scrophulariaceae.

We have now reached the level of the terrestrial obligate parasite with only lateral haustoria. How such a habit could have evolved into an epiphytic parasitism I have described in the two Santalalean families (see chaps. 2 and 3). What interests us here is how the changes leading to the haustorial systems of other parasites may be visualized.

---

[1]The behavior of *Pieris phillyreifolia* (Lemon and Voegeli, 1962) represents an unusual ecological condition which cannot be referred to parasitism.

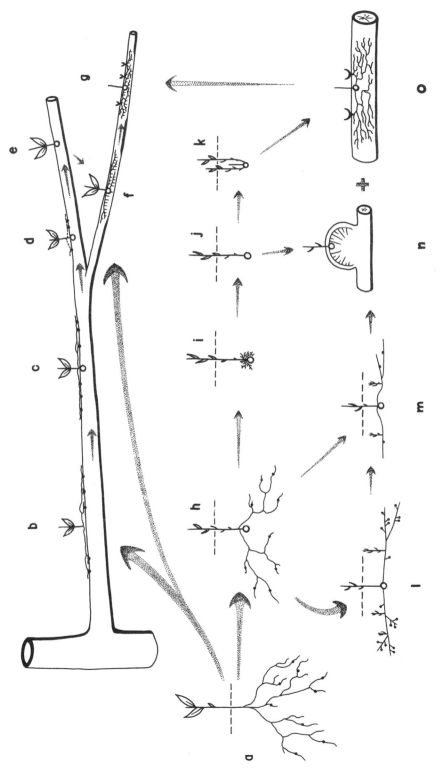

Fig. 9-1. Possible evolutionary derivation of various modes of parasitism in flowering plants, omitting rhizomatous forms, *Cuscuta*, and *Cassytha*: *a*, presumed ancestral condition, terrestrial root parasitism; *b*, transfer to branches of a tree; *c*, evolution of primary haustorium; *d*, gradual disappearance of secondary roots; *e*, primary haustorium as sole haustorial contact; *f*, lateral expansion of endophyte, and dimorphism of latter; *g*, abortion of epicotylar pole of embryo, and transfer of flower-producing function to endophyte; *h*, evolution of primary haustorium under terrestrial conditions; *i*, loss of secondary haustoria; *j*, complete loss of secondary roots; *k*, abortion of epicotylar pole of embryo, with shoots derived from radicular pole; *l*, division of labor of root system into pilot and haustorial roots, and vegetative reproduction from pilot roots; *m*, simplification of root system, reduction in number of haustoria, and increase in complexity; *n*, and *o*, endophytes of *Balanophora* and Rafflesiaceae, respectively, with abortion of epicotylar pole in latter, flowers being produced directly by endophyte.

The first advance from this early stage of terrestrial parasitism is the addition of the primary haustorium to the parasitic equipment, a process involving the direct transformation of the radicular apex into a haustorium. We have seen this happen in Scrophulariaceae (*Striga*), Orobanchaceae, and, in a remarkably parallel fashion, in most epiphytic Santalales. I suspect that Balanophoraceae, Rafflesiaceae, Hydnoraceae, and Lennoaceae also have primary haustoria; I base this speculation on the high degree of embryo reduction in these families.

The effect of this change on the evolution of broomrapes and some figworts has been incalculable. A parasite relying on many small lateral haustoria, most of which are ephemeral, requires a degree of photosynthetic efficiency to carry it through its early development while searching for its first hosts, and perhaps also to bridge occasional periods without hosts. We might even say that the evolution of a primary haustorium makes possible a completely heterotrophic way of life, although this has not been fully achieved in the epiphytic groups. It reduces significantly the independent stage of germination; and, especially where a precise reliance on host-root exudates evolves, large seeds are no longer needed. The stage is thus set for reduction of the embryo. At the same time the primary haustorium is of necessity a permanent one. The reasons for retaining some of the photosynthetic apparatus no longer apply, and the plant is free to advance to a holoparasitic mode of life. The secondary haustoria are progressively more superfluous and eventually disappear.

It might be wondered how *Lathraea* could lose all chlorophyll but not gain a primary haustorium. For one thing, the development of this plant is extremely slow, and it may take a decade to reach the flowering stage. More importantly, many of its haustoria are perennial. Finally, its seeds do require a germination stimulus from the host root. The only way in which *Lathraea* deviates from the pattern described is that it relies on many secondary but perennial haustoria rather than on a single primary haustorium. It is the permanence of haustorial contact which, with the germination requirements described, allows for complete heterotrophism.

It seems equally surprising that none of the epiphytic Santalales, in spite of their evolutionary advances, have quite reached the holoparasitic level. Even the most diminutive *Arceuthobium* species, even *Phrygilanthus aphyllus*, *Viscum minimum*, and *Phacellaria*, have retained a certain amount of photosynthesis. The explanation obviously cannot be sought in the impermanence of haustoria. I would suggest, instead, that the photosynthetic capacity of these plants is not necessarily important in the mature individuals but is simply a carry-over from the seedling stage, where it does play an important role. Since there are no chemical germination requirements here, the seedling's photosynthetic products assist it through the relatively long period of germination. The tremendous importance of photosynthetic activity at this stage is underscored by the evolution, in at least three independent lines of Santalalean

stock, of chlorophyll in endosperm. The chlorophyll present in older plants may thus be no more than a weak reflection of the needs of the seedling.

How a normal radicular apex can change to one capable of becoming a primary haustorium we do not know. Perhaps we should look to the thigmotropic response for clues. The work of Zietz (1954) on *Cuscuta* has demonstrated the relationship between this response and haustorial initiation. The haustorial type of coil in dodders, itself a direct result of a thigmotropic growth response, seems to be a prerequisite to haustorial initiation. The roots of at least some Lennoaceae demonstrate an extreme degree of thigmotropism. This growth response of a subterranean organ is rather unexpected. Actually, it is by no means impossible that the response is to the chemistry of the root rather than to its physical structure. At any rate, the resultant action is felt extremely close to the root apex, which is closely appressed to the host root. Do we not witness here a gradual shift of the response in the direction of the apex? Could not such a shift, in conjunction with the various physiological changes produced in the apex through the effects of touch just below it (Zietz, 1954), set the stage for the drastic change in behavior of the apex? The evolution of the primary haustorium of some parasites may thus result from a complete transfer of thigmotropism to the radicular apex. The growth of the terminal disk on a haustorium is, after all, due to the same growth response.

Loranthaceae having a primary haustorium as well as secondary haustoria show a pronounced thigmotropism in their epicortical roots, whereas comparable roots of *Gaiadendron*, which lacks a primary haustorium, show no such response. Thigmotropism appears to be lacking, however, in the roots of Scrophulariaceae and Orobanchaceae.

Once an efficient primary haustorium is incorporated in the life cycle, the secondary haustoria become nearly superfluous. So it is not surprising that they seem to be in process of disappearing in a number of Orobanchaceae, in *Antidaphne*, and possibly in *Eremolepis* and some species of *Oryctanthus* among the mistletoes. In other mistletoes, especially in *Struthanthus*, the importance of secondary haustoria has instead been increased through the evolution of epicortical roots from stems.

Structural modifications of the radicular apex form a corollary of the latter's changed function; the root cap, in effect, disappears. Some workers have objected that this does not apply to the reduced radicles of broomrapes, but the important fact is that the apices of these small roots which rely on a chemical mechanism of entry are made up of living cells. The dead outer cells of the root cap are an impediment to host entry and have been suppressed.

The similarity in root dimorphism of Hydnoraceae and Lennoaceae is rather puzzling. The latter bear the marks of relatively recent arrivals in the parasitic camp, but the opposite is true of Hydnoraceae. It is unfortunate that we have so little detail about parasitism of the latter family, and cannot be certain that the two situations are truly comparable.

The primary haustorium, if it is to be an efficient and perennial organ, must undergo expansion and differentiation within the tissues of the host. The expansion of the endophyte, in an evolutionary sense, is often balanced by reduction of the aerial portions of the parasite, as is strikingly illustrated by many small mistletoes. A division of labor between longitudinally and radially oriented parts of the endophyte is evident in many of the same plants, but no such dimorphism has evolved on succulent hosts. The cortical strands and sinkers of Viscaceae reflect the internal organization of the woody host, and constitute a response to the latter's cambial activity.

The progressive endophytic fragmentation leads to a further consequence. In the earlier stages of this tendency it is still possible for the epicotyl of the seedling to develop into the aerial plant. At least, the endophytic system is still centralized; it still has its focus on the original court of infection where flowers or inflorescences are formed. This stage is seen in most Viscaceae. In some of these mistletoes the older parts of the endophyte produce additional aerial shoots as a matter of course (e.g., *Phoradendron juniperinum* and *P. californicum*) or when the main stem dies. Finally, the endophytic system takes over the shoot-producing function entirely, as in *Arceuthobium* and probably many other advanced parasites. The entire embryo now concentrates on germination and establishment, and the shoot apex aborts. The resulting diffuse infection is essentially systemic.

Many Balanophoraceae (those with the *Balanophora* type of tuber) seem to have capitalized on the hypersensitive reaction of hosts. Whether the peculiar differentiation of the endophyte in the *Balanophora* tuber was also present in nontuberous ancestral types of parasitism we shall probably never know. The much simpler parasitic organ of somewhat more primitive Langsdorffieae and *Cynomorium* would suggest that tuber and endophytic specialization have evolved apace. In any event, we may look upon the parasitic capsule and strands of vesicular cells associated with *Balanophora* tubers as the nearest approach to cortical strand-sinker organization which can be expected in a globular organ.

EVOLUTION AND THE STRUCTURE OF PARASITIC ANGIOSPERMS

This final section includes items of general interest which are related to the structural evolution of parasites. The topics are concerned mostly with the varieties of evolutionary reduction which the parasitic mode of life has brought about, but the phenomena included are not unique to parasites.

Little serious thought has been directed toward the general effects of parasitism on the structure of plants—we might call these effects the common denominators in the structure of parasites—and it is therefore best to regard the topics outlined as promising avenues of further study. The basic raw materials, the comparative and descriptive data needed to test some of the trends suggested, are lacking.

Since the evolutionary concepts pertaining to all vascular parasites are in an embryonic state, this section must necessarily be fragmentary.

In families which have been associated with Santalales, the flower and allied structures, aside from the haustorial organ itself, have undergone the most striking modifications. The androecium is exceptional in this regard. It has either evolved in a multiplicity of directions (e.g., Viscaceae and Balanophoraceae). or it scarcely differs from the standard angiosperm equipment (e.g., Loranthaceae and Santalaceae). The endothecium seems to have undergone a radical change in three unrelated parasitic groups: in *Arceuthobium* it forms the superficial rather than the subepidermal layer (Cohen, 1968); it is believed to be lacking in Balanophoraceae (Hooker, 1856; Fagerlind, 1945a) and *Pilostyles* (Endriss, 1902; Kummerow, 1962). But since the endothecium is lacking in Ericales (Samuelsson, 1913) and in the saprophytic *Cotylanthera* of Gentianaceae (Oehler, 1927), an evolutionary connection with the parasitic mode of life is not evident.

Some extraordinary modifications in ovule reduction are encountered in the gynoecium of Santalales, Balanophoraceae, and *Prosopanche*. Schellenberg (1932) thinks that ovular reduction is a direct consequence of the parasitic mode of life, and proceeds to draw some very illogical conclusions. It is said, for example, that *Schoepfia* is undoubtedly parasitic, since it lacks normal ovules! Whether or not *Schoepfia* is parasitic scarcely follows from its lack of integuments. Reed (1955) reflects Schellenberg's attitude in Olacaceae. Actually, it is easy to adduce evidence in favor of this notion: the most advanced of Santalales generally have the most highly reduced ovules. In Loranthaceae, Viscaceae, and also Balanophoraceae, integuments have vanished. But what bothers us here is that we cannot visualize any logical connection between the parasitic mode of life in general and the degree of ovular differentiation. Even Reed and Schellenberg do not discuss this problem.

Another point of view is perhaps more meaningful. In my study of *Arceuthobium* (Kuijt, 1960) I have suggested that the great simplicity of its pistillate flower, which is in a nearly meristematic condition during anthesis, may be a prerequisite for the development of the extremely complicated fruit. Perhaps ovular reduction, instead of bearing the vague taint of degeneration, is a reflection of the structural simplicity forming a necessary prelude to the complexity of the fruit of epiphytic parasites. Thus the reduction of ovules would be related to the modes of dispersal and establishment of parasites. The fact that the really striking parallel is between ovular reduction and *epiphytic* parasitism would certainly seem to support this idea. A good example is *Phacellaria* (Stauffer and Hürlimann, 1957), an epiphytic member of Santalaceae, which has an ovarian papilla like that of mistletoes.

The foregoing, of course, does not apply to Balanophoraceae, which are all terrestrial parasites. The real reasons for this evolutionary reduction escape us at present. The only suggestion I can make is based on the earlier discussion of the evolution of parasitism. I pointed out a possible relationship between the germination pattern requiring a chemical stimulant from the host root, on the one hand, and an undifferentiated embryo, on the other hand. In Balanophoraceae it would not be surprising to find a similarly complex germination mechanism which, being a chance affair as in *Orobanche*, requires the production of great numbers of seeds which in consequence must be very small. We might say that the differentiation of the embryo has been held back, in its evolutionary process, in an effort to reduce the size of the seed, and this, in turn, has been followed or paralleled by decreases in the size and complexity of the flower as a whole.

We must be careful not to extend such an explanation beyond the limit of facts. A similar increase in the number of seeds and decrease in their size and complexity in *Hydnora* and Rafflesiaceae have not been achieved at the expense of the integuments, which are still of normal appearance even .though only one is formed. In *Agonandra* (Olacaceae) the gynoecium has an ovarian papilla as known from *Phacellaria* and mistletoes, but there is no evidence in regard to the parasitism of this tree. In at least one completely autotrophic genus, *Houstonia* of Rubiaceae, integuments are also lacking (Lloyd, 1902).

Similar situations obtain in saprophytic Gentianaceae (Oehler, 1927). In *Leiphaimos* and *Cotylanthera* there are no integuments at all; in *Voyria* and *Voyriella* only one remains. The structure of the seed in *Voyria coerulea* (fig. 9-2a) is nearly indistinguishable from that of *Pilostyles thurberi* (fig. 5-8c). These examples teach us not to look exclusively in the phenomenon of parasitism for explanations of reduction. It is still possible, of course, that extremely reduced seeds and embryos are related to very specialized germination requirements.

The reduction of reproductive (and vegetative) features to a minimum may cause difficulty for taxonomists, for it tends to limit the number of useful criteria for classification. The fewer features that are present, the easier it is to make errors in assignment. The most spectacular of such errors is the erection of a separate genus and species, *Sarcopus aberrans* Gagn., and even a new family of gymnosperms, Sarcopodaceae, on the basis of what turned out to be only a known member of Santalaceae, *Exocarpos longifolius* (Gagnepain and Boureau, 1947; Stauffer, 1959). This case was carried into the realm of the ludicrous when Lam (1948) elevated the group to the ordinal level by the creation of Sarcopodales. Let us hope that this is the only instance in the plant kingdom where an order must be cited in synonymy of a species! It must be admitted, however, that in the study of individual features in parasitic angiosperms, the advanced sometimes seems to descend "again" to the level of the primitive, and the two nearly indistinguishable features can be understood only when we widen our scope to encompass the entire plant and its relatives. Another in-

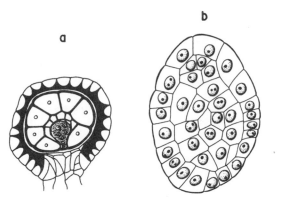

Fig. 9-2. *a, Voyria coerulea*, a saprophytic member of Gentianaceae, longisection of mature seed (embryo shaded; after Oehler, 1927; x 115); *b*, *Orobanche cernua*, longisection through mature embryo (after Tiagi, 1951*b*; x 412).

stance of this may be seen in the skeletal structure of leaves.

It is not surprising that the leaf, being the major photosynthetic organ, has frequently diminished in size in parasites. We see little of this diminution in relatively recent arrivals in the parasitic camp, but even there it is not altogether lacking. In Scrophulariaceae the leaves range from very finely dissected, as in *Rhamphicarpa* and various *Gerardia* and *Seymeria* species, to entire leaves. *Gerardia aphylla*, however, has attained the squamate condition, and the same is true of some species in the genera *Striga* and *Harveya*. Totally squamate are the most advanced of parasitic figworts, *Hyobanche* and *Lathraea*. All Orobanchaceae have their leaves reduced to scales, and so do *Cassytha*, *Cuscuta*, Lennoaceae, Balanophoraceae, and Rafflesiaceae. In several genera of the latter two families, plants have so decreased in size and complexity that only a few scale leaves remain or none are recognizable (*Ombrophytum*, *Lathrophytum*, and *Chlamydophytum*). In *Hydnora* and *Prosopanche* all accessory leafy organs, aside from the single whorl of perianth members, have vanished.

In Santalales the squamate habit has evolved in a number of independent genera, (figs. 3-2,c; 3-7,a; 3-25,a; 9-3). It is a generic feature in *Korthalsella*, *Arceuthobium*, *Ixocactus*, and no less than six genera of Santalaceae. In a number of other genera, squamate species are common, or the types of heterophylly may be interpreted as leading up to the "leafless" condition (see chaps. 2 and 3). With the exception of some genera like *Exocarpos*, whose squamate habit is an obvious adaptation to arid climates, the "leafless" condition is nearly restricted to epiphytic parasites in this order. The epiphytic habitat is often a xerophytic habitat also, as structural adaptations of various sorts demonstrate, and we may not have to resort to parasitism per se to explain this phenomenon in the genera mentioned.

I have gone to some trouble to illustrate the vasculature of scale leaves of parasites, as information of this sort does not seem to be available. The material has been selected in a rather arbitrary fashion from several groups and is by no means comprehensive.

The main point I wish to make is that we should look beyond the factor of size. It is true that the smaller scale leaves have a much reduced vasculature; but considerable reduction in vascular complexity may occur *before* the size of leaves diminishes.

In the small scale leaves of *Cuscuta* and *Cassytha* only a single strand of xylem remains, although Mirande (1900) reports minute lateral veinlets consisting of phloem only. The same is possibly true of squamate mistletoes, as demonstrated by *Arceuthobium* and *Ixocactus*. In the scalelike bracts of inflorescences of *Dendrophthora* and *Phoradendron* the single median vein is sometimes joined by two small lateral veins (Kuijt, 1959). About the scale leaves of *Phacellaria* and *Myzodendron* we know nothing.

The vasculature of *Pholisma* leaves is more elaborate, although the smallest ones near the base of the stem may have only a slender median vein giving off two small laterals (fig. 9-4a). Higher up, where leaves are larger, there is much variation, although a single initial strand supplies the whole system. Most of the branching takes place at the point of insertion of the leaf. Some leaves show a few anastomoses; others, none at all. The median vein always retains its dominance.

In Orobanchaceae I have only the leaves of *Boschniakia* before me, taken from the flower-bearing part of the stem (fig. 9-4,b). Each bract is supplied by two independent traces which undergo a series of dichotomous ramifications until a more sympodial tendency becomes established in the upper part of the bract. Fusions are sporadic; most of the vasculature is quite open. The first branch given off by one of the two original traces runs directly into the leaf apex and bears only a few small branches near its tip. It would not surprise me if more intensive study should reveal that this median vein is sometimes supplied by branches from both the original traces; this is commonly the origin of the median vein in flowering plants and may be presumed to be the ancestral condition of the *Boschniakia* leaf.

The size of most Rafflesiaceae makes an analysis of this sort very difficult. I cannot supply data for *Pilostyles*, where the vasculature must be rather simple. A few floral fragments of *Rafflesia patma* showed a surprisingly sparse system of mostly uniseriate xylem strands, separated from one another by an average distance of 2 to 3 mm. How such a massive, fleshy organ as the *Rafflesia* corolla can develop on the basis of so feeble a vasculature supply is a miracle. I also present (fig. 9-5) the skeletal system of petals and leaf scales of *Cytinus hypocistis* collected near Caldas de Monchique, Portugal, in 1959. Aside from a greater proliferation of veins in

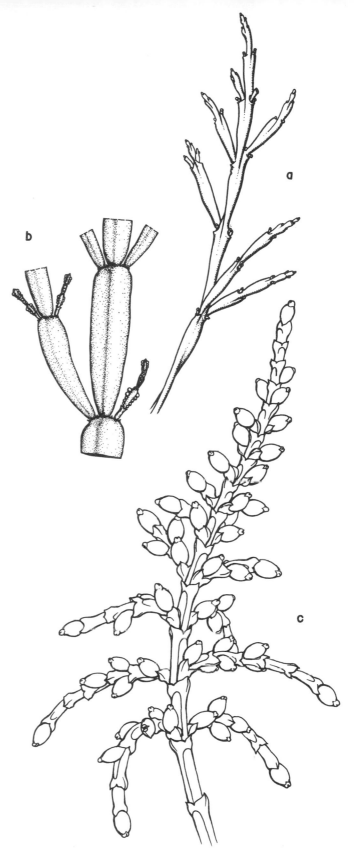

Fig. 9-3. Squamate mistletoes: *a, Ixocactus hutchisonii* (after Kuijt, 1967a; × 1); *b, Dendrophthora opuntioides* (after Kuijt, 1961c; × 1); *c, Arceuthobium bicarinatum* (Ekman No. 12024, Hispaniola, S; × 2).

214

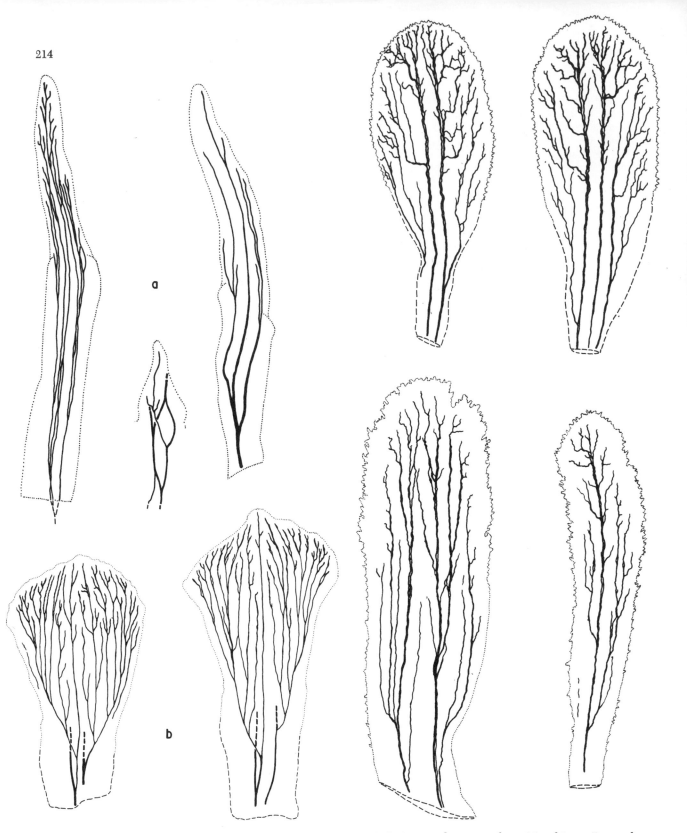

Fig. 9-4. a, *Pholisma depressum*, vasculature of leaves, two near middle portion of stem (left and right), one near its base (middle) (material from Morro Bay, California; × 4); b, *Boschniakia hookeri*, vasculature of two floral bracts (material from Horseshoe Bay, B.C.; × 4).

Fig. 9-5. *Cytinus hypocistis* from Monchique, Portugal, foliar organs cleared to show vascular skeleton; upper two are petals, lower two, leaves (× 8).

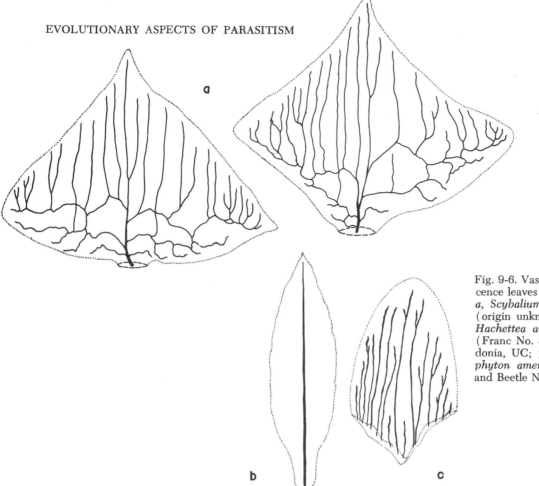

Fig. 9-6. Vasculature of inflorescence leaves of three parasites; a, *Scybalium jamaicense* (origin unknown; × 5); b, *Hachettea austro-caledonica* (Franc No. 3027, New Caledonia, UC; × 2); c, *Bdallophyton americanum* (Morrison and Beetle No. 8762, UC; × 5).

the upper half of the petals, it is difficult to find much difference between the two kinds of organs, especially because of the great variability in the leaves. The frequently double nature of veins is peculiar. The petals have a fairly distinct midvein, but this is true of only one of the vegetative leaves. The other might be regarded as a fusion of two leaves, but the diverse skeleton in each of the other three illustrated foliar organs makes this supposition unnecessary. To be noted also is the curiously twisting course of nearly all the smaller veins.

In the fleshy, deciduous scale leaves of *Scybalium jamaicense* (fig. 9-6,a) a single strand of xylem leads into the base of each leaf and produces laterals left and right in alternating fashion. One or two of the laterals, after having been joined by one or more other branches, reach sideways into the corners of the deltoid leaf. The largest lateral, on its way, issues a series of remarkably parallel veins in the length-direction of the leaf, nearly to the margin. An occasional longitudinal vein also springs from a lower lateral vein and seems to bypass the upper one. Anastomoses in the parallel system of veins must be rare, as only one is visible in the two leaves illustrated.

The vascular system of *Hachettea* and *Balanophora,* belonging to the same family, is quite dif-ferent. In *Balanophora harlandi* from Hong Kong I have been able to find no more than three or four straight, extremely slender bundles, apparently uniseriate, reaching into the upper portions of the leaf scale without branching. In *Hachettea* (fig. 9-6,b) a single straight median bundle continues into the apex. The same has been reported for the leaves of the genus *Mystropetalon* (Harvey-Gibson, 1913).

My information on *Bdallophyton* is very incomplete. A number of bundles, showing very few branches, have a longitudinal course in these fleshy leaves, with no fusions taking place (fig. 9-6,c). I have not been able to ascertain the manner in which these traces unite in the basal part of the leaf, but it is possible that the organization is similar to that in *Scybalium*.

The expanded foliage of most Santalalean plants shows a structural complexity and variability in detail which have scarcely been tapped. The early work of von Ettingshausen (1872) outlined the major venation patterns of mistletoes but gave scarcely a hint of the diversity of sclerenchyma. Some of the anatomical leaf peculiarities in mistletoes may be seen in *Oryctanthus* (Kuijt, 1961c).

The only mistletoe genus (aside from squamate types) which shows unmistakable evidence of a reduction in vascular complexity is *Lepidoceras* from

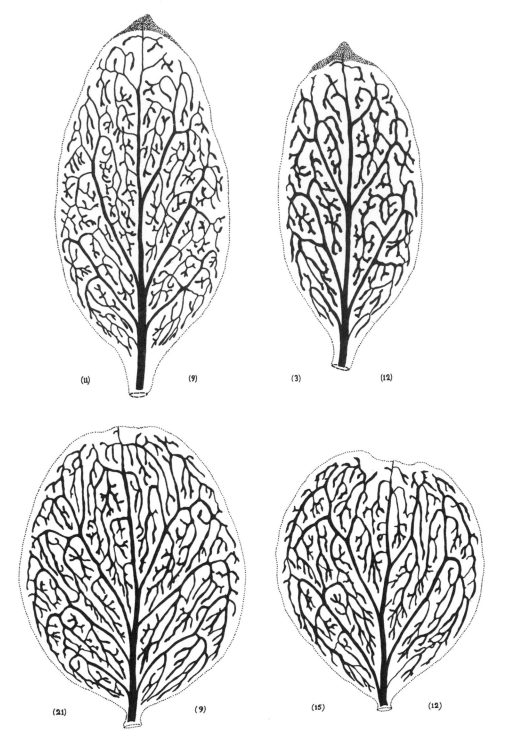

Fig. 9-7. *Lepidoceras* spp., leaves, showing degree of
fusion of vascular skeleton: above, *L. squamifer*
(Werdermann No. 315, Chile, U; × 9.5); below,
L. kingii (Andreas No. 718, Chile, U; × 9.5) (bracketed
numbers represent number of fusions on each side
of lamina).

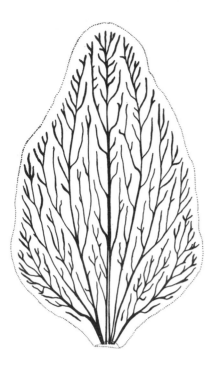

Fig. 9-8. *Myzodendron brachystachyum,* leaves (bracts) of inflorescence, showing vascular structure (one fusion indicated by arrow) (Andreas No. 259, Chile, U; × 7.3).

Fig. 9-9. *Myzodendron brachystachyum,* foliage leaves of innovations, showing vasculature (each arrow indicates one vein fusion) (Andreas No. 259, Chile, U; × 7.3).

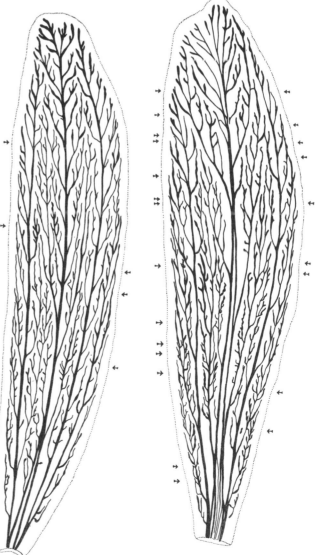

South America (fig. 9-7). The leaf has a peculiar developmental history (Engler and Krause, 1935); the situation in this genus may represent a halfway station to a squamate condition. The veins of the rather small leaves are reinforced with fibers and form an interesting pattern, especially in *L. kingii,* as the smaller veins recurve toward the petiole. The degree to which the ultimate veins have failed to anastomose is realized when one counts the anastomoses on the left and right sides of each leaf, as indicated in brackets below the drawings. Entire sectors of the blade may be found without a single fusion. We can thus assume that one of the first steps in reduction of leaf vasculature is the "dropping out" of connecting links between contiguous major branches.

This fact is strikingly illustrated in the leafy species of *Myzodendron* also (fig. 9-8; 9-9). These plants have two types of leaves (see chap. 3): the more elongated leaves of vegetative innovations and the ovate bracts of the inflorescences. Each type is characterized by a highly complex vasculature. Here, also, the number of vein fusions is disproportionately low. I have indicated each fusion by means of a small arrow, ignoring the midvein. The latter is constituted from equal branches of what appear to be two leaf traces; the level of union is variable. The appearance of the vasculature is dominated by the midvein. The two foliage leaves show that we must expect wide variation in fusion from leaf to leaf. Even in the leaf having twenty-three fusions, however, the degree of openness is remarkable. In the bractlike leaves, fusions are virtually absent, and the beautiful dendritic pattern is reminiscent of some ferns. We might well ask to what group a paleobotanist would assign this leaf if it were found in the fossil form. With respect to the degree of reticulation, this highly advanced plant seems to approach what is often regarded as a primitive condition. It may, of course, be held that the irregularity in the leaves distinguishes them from the primitive situation. But is there any reason to believe that, in the course of evolution, such a derived, open vasculature could not reach a level of stability and regularity? Similar questions had to be faced in the studies of *Kingdonia* and *Circaeaster,* where a decision as to the primitive or derivative nature of the remarkable open venation patterns is not so clear-cut (Foster 1959, 1963).

The phenomena we are discussing are not necessarily unique to parasites, but may be shared by other plants undergoing reduction, for whatever evolutionary reason. Figure 9-10 represents the vascular framework of leaves of *Allotropa virgata,* a saprophyte belonging to Ericaceae. In these rather large leaves fusions are totally lacking. Considering the differences in leaf shape, there is a marked similarity to *Boschniakia* leaves.

The reduction of the foliage leaf to a scale leaf does not inevitably signify the end of photosynthetic capacity. The internodes of the stem and even inflorescence have sometimes taken over this function, and have been structurally altered in accordance. In the flattened, photosynthetic internodes of *Ixocactus* we can piece together the development of this accessory vasculation (fig. 9-11, *c, d;* 9-12). First to be differentiated is the solitary leaf trace. Some slender and rather convoluted xylem strands then differentiate upward from the previous internode, one growing out to join the leaf trace just below its tip, which is somewhat club-shaped. At the same time the internode elongates, and soon, from the shoulder of the second leaf trace, small tracheal strands begin to differentiate upward, into the following internode. It is probably these small traces which eventually produce the leaf traces two internodes higher. From approximately the same source two other small strands go forth to supply the axillary region where flowers or other laterals will develop. The primary leaf traces rarely give off a small vascular branch, but usually are remarkably separate from other vascular contact along their length. Somewhat later the leaf traces are matched by a straight fiber bundle each just to the outside and separated from it. An essentially leaflike type of vasculature (fig. 9-12) is progressively elaborated in the "blade" of the internode. Largely a product of the slender tracheal strands "entering" earlier from below, a complex and irregular network is fashioned in the region between leaf traces and percurrent veins, including many vein endings with storage tracheids. The development of axillary organs brings about a great strengthening of veins leading in that direction.

It would be interesting to compare this situation to that in the species of *Korthalsella* and *Viscum* with flattened stems, but no information is available at present. We are sure to find marked differences from *Ixocactus* at least in species of *Korthalsella* where all internodes are arranged in the same plane.

Comparisons can be made to stems of three members of Phoradendreae, but they are very incomplete comparisons. In *Dendrophthora squamigera (D. biserrula)* "extrastelar" ramifications in inflorescence internodes presumably supply the flowers (Kuijt, 1959). An extremely intricate vascular reticulum is present in the inflorescence internodes of *D. flagelliformis* (fig. 9-11,*a*). The percurrent bundles (drawn as one but probably double) are only occasionally involved in the elaborate, leaflike vascular reticulum which has formed to their left and right. Particularly impressive are the aggregations of vein endings around the floral cups. The basal part of this species at least at times bears leaves; in neither of these species of *Dendrophthora* do we know whether similar modifications are present in vegetative internodes. In the vegetative internodes of *Phoradendron californicum* all but two prominent percurrent bundles give off numerous branching traces which terminate in the cortex in club-shaped vein endings reminiscent of those in the leaves of many mistletoes (fig. 9-11,*b*). All these instances of the evolution of novel vascular equipment surely reflect the

Fig. 9-10. *Allotropa virgata,* vasculature of two leaves, from middle stem region of saprophyte (Gibson, B. C., UBC; × 4).

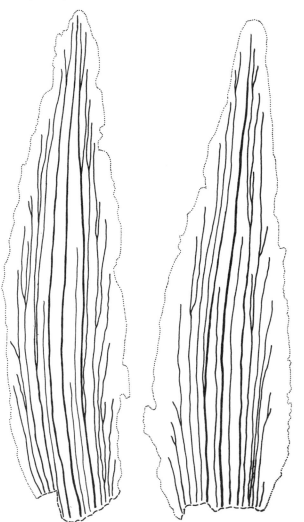

fact that the stem has assumed the main share of photosynthesis. The stem vasculature of many genera of Chenopodiaceae has evolved in similar directions; although the details differ (Fahn and Arzee, 1959), it again denotes photosynthetic activity in the stem.

A separate consequence of changes in photosynthetic activity has often been mentioned: reduction in stomata. Unfortunately, the literature is somewhat unreliable here, and there is danger of repeating some of these errors. The facts as I have gathered them from past work are as follows.

In Olacaceae no really unusual stomatal features are known (Reed, 1955), and perhaps the same can be said of other Santalales. In mistletoes the cauline stomata are said to be aligned at right angles to the stem, but no careful survey exists. In some species of *Myzodendron* the stomata are elevated on curious epidermal mounds for which no ecological or physiological explanation has yet been given. In no Santalales has a decrease in the number of stomata been reported as a parallel to advanced parasitic status. In fact, even the radicles in some epiphytic forms bear stomata (Kuijt, 1960).

In Scrophulariaceae the only irregularities known to me in this respect are in the peculiar leaves of *Lathraea* (see chap. 4). All the investigated species of Orobanchaceae are reported as having stomata on the leaves (Metcalfe and Chalk, 1950), but *Conopholis* may not have any on either surface of its leaves (Wilson, 1898). In *Aeginetia*, *Christisonia*, and *Orobanche uniflora* one leaf surface or the other may be without stomata (Smith, 1901; Juliano, 1935).

Cuscuta is variously said to be without stomata

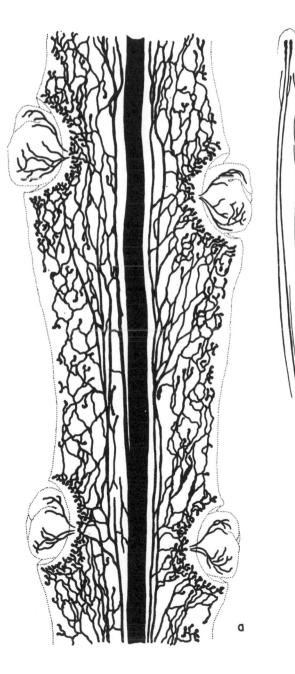

Fig. 9-11. Squamate mistletoes, skeletal elaborations in stems: a, *Dendrophthora flagelliformis*, section of fertile internode (after Kuijt, 1959; x 12); b, *Phoradendron californicum*, thick longisection of young vegetative internode (sclereids in pith; material from California; x 20); c and d, *Ixocactus hutchisonii*, two early stages of vascularization (Hutchison and Idrobo No. 3008, Colombia, UC; x 20). (For mature structure see fig. 9-12).

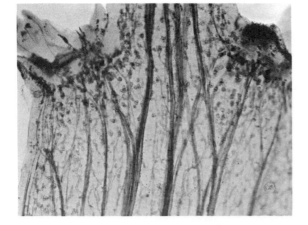

Fig. 9-12. *Ixocactus hutchisonii*, cleared nodal region of stem, with entire flower visible at left (Hutchison and Idrobo No. 3008, Colombia; x 31).

or to have stomata which act as hydathodes (Renner, 1934). A recent study leaves no doubt as to their existence (Pant and Banerji, 1965); in fact, stomata may be found even on the root (Haccius and Troll, 1961). *Cassytha*, similarly, has normally abundant stomata in series along the stem (Schmidt, 1902; fig. 6-8,*c*).

Even in his monograph of Balanophoraceae, Hooker (1856) spoke of the absence of stomata in that family, an opinion repeated by Metcalfe and Chalk (1950), where it is said, however, that stomata on *Cynomorium* have been recorded. Of Rafflesiaceae we know very little. *Pilostyles ulei* has stomata on its outer surfaces (Endriss, 1902). *Cytinus hypocistis* is credited with functionless stomata (Metcalfe and Chalk, 1950). Similarly, in *Rafflesia* and *Rhizanthes* reduced stomata are reported (Cammerloher, 1920). No stomata have, to my knowledge, been described from Hydnoraceae.

In summary we can only say that no clear picture emerges. There remains an impression that among the most advanced holoparasites stomata tend to become fewer, possibly vestigial, or to disappear. Even if this impression is valid, we cannot easily explain the matter. Even holoparasites require from their host great amounts of water, and this water-uptake in all probability is achieved in the normal way, with the assistance of transpiration through stomata. A thorough study of this topic cannot afford to ignore the parallel situations in saprophytes, where stomata often are said to be lacking (Esau, 1953).

When considering the importance of roots in the scheme of parasitic plants we may focus on two notions. It is my opinion that haustoria represent modified roots. If this be so, there must have been, at one time, structures intermediate between these two categories: actually, there is no reason to think that there may not be transitional stages in contemporary parasites. Second, if some roots of a plant are in process of becoming progressively more haustorial,

they will be in active competition with the remaining nonhaustorial roots, and this may bring about structural change in the latter. The more efficient haustoria are, the more they will usurp the dual function of absorption and anchorage from normal roots. The various reports on the reduction of root caps and root hairs as summarized below are in accordance with such trends.

Neither *Cuscuta* nor *Cassytha* has a recognizable root cap. Although McLuckie (1924) speaks of a partly developed root cap in *Cassytha*, the careful illustrations of Mirande allow us to call at least the lateral roots tunicate (Mirande, 1905*a*; fig. 6-8,*f*). The roots of Orobanchaceae also lack caps (Koch, 1887*b*; Smith, 1901; Cooke and Schively, 1904). Parasitic Scrophulariaceae have root caps, but I have not encountered literature on their structure. In *Pholisma* the pilot roots are provided with a distinct cap; haustorial roots have not been studied, however (Kuijt, 1967*b*). Of all Balanophoraceae a root cap seems to be present only in *Thonningia* (Mangenot, 1947). The situation in Hydnoraceae is not altogether clear, but a caplike structure may be present in *Prosopanche* (Schimper, 1880).

In the terrestrial plants of Santalalean families I have discovered no anatomical work on the root apex, but in *Gaiadendron* a well-formed root cap is known to exist (Kuijt, 1965*a*). It is perhaps safe to extrapolate this to other terrestrial Santalales.

In the root apices of most epiphytic Santalalean parasites there is a significant difference between the radicle and the epicortical roots. The radicle lacks any vestige of root cap, whereas epicortical roots have a small but distinguishable root cap (Kuijt, 1965*b*). This root dimorphism is meaningful from a functional and an evolutionary point of view: the apical meristem of the primary root transforms itself directly into a haustorium; that of epicortical roots does not. The haustoria of epicortical roots are lateral organs (see chap. 7). The cells of a young haustorium first contacting the host may have a variety of functions, such as the adoption of the shape of the host contours, the secretion of a cementing substance, and possibly, as in *Cuscuta*, some initial digestive action. The outer cells of a root cap, however, are moribund and cannot fulfill any of these functions. The nonliving or moribund cells of the root cap are therefore no more than an impediment to the terminal haustoria; the cap, consequently, has diminished in size or disappeared.

Some nonparasitic plants that are highly specialized in other directions may also have lost their root caps (e.g., some Podostemaceae: Cario, 1881; *Listera cordata*, *Neottia nidus-avis*, *Asplenium*, and other ferns: Goebel, 1932). Again, this type of reduction is not confined to parasitic angiosperms.

In root hairs also, there are variations, as might be expected of parasites which have arrived at their mode of life in different ways, and have reached varying degrees of dependence upon their hosts. To my knowledge, root hairs have not been reported in *Krameria*, Lennoaceae, or Orobanchaceae—in fact, quite the reverse (Koch, 1887b; Smith, 1901; Cannon, 1910; Kuijt, 1967b). The unusual tendril-like epidermal hairs in the *Aeginetia* seedling could be thought of as an exception. In Santalales, root hairs have been mentioned for *Buckleya* (Kusano, 1902); and a few were seen by Barber (1906, 1907b) on roots and haustoria of *Santalum* and *Olax,* and by Piehl (1965c) on *Comandra. Exocarpos bidwillii* has numerous root hairs (Fineran, 1963c), but their curiously straight form suggests that they may not be normally functional. Virtually none exists on the roots of *Leptomeria spinosa* and *Exocarpos sparteus* (Herbert, 1925). In the primitive mistletoes the early reports of root hairs in *Atkinsonia* (Blakely, 1922-1928) have been emphatically denied by Menzies and McKee (1959). The closely related *Gaiadendron* also lacks root hairs, as do, of course, all epicortical roots. Even in the Costa Rican *Heisteria longipes,* an autotrophic member of Olacaceae, there is no sign of root hairs.

In *Cuscuta* and *Cassytha* it seems to be a matter of interpretation whether the somewhat papillate epidermal cells of the former (Haccius and Troll, 1961; Truscott, 1966) or the slightly longer hairs of the latter (Mirande, 1905a) should be called root hairs. In neither case can we expect much physiological significance; in each we recognize a reduced condition.

For Scrophulariaceae, observations have varied greatly with regard to root hairs and their significance as indicators of the degree of independence of the plant. No root hairs were seen on *Striga lutea* (Stephens, 1912) or *Castilleja affinis* (Heckard, 1962). In several instances, only a few hairs are reported near root tips or haustoria: *Euphrasia officinalis* (Koch, 1891); *E. minima* (Heinricher, 1898b), *Melampyrum arvense* (Heinricher, 1908a), *M. lineare* (Piehl, 1962a), *Pedicularis canadensis* (Piehl, 1963), and several species of *Castilleja* (Heckard, 1962). Root hairs are abundant in some species of *Euphrasia* and *Odontites* (Härtel, 1959), in *Gerardia* (Boeshore, 1920), and in *Castilleja franciscana* (Heckard, 1962).

Does any recognizable pattern emerge? More particularly, do these observations point to a relationship between abundance of root hairs, on the one hand, and degree of parasitism, on the other hand? This was the conclusion drawn from a comparison between species of *Euphrasia*: some were able to set fruit without the benefit of host contacts and had abundant root hairs (e.g., *E. minima*); others were far more dependent on hosts and did not have abundant root hairs (Heinricher, 1898a, 1898b). From such observations Heinricher reached the conclusion that root-hair reduction had followed the formation of haustoria in the course of evolution and not vice versa. It is doubtful whether this principle has general application, however, as shown by the Costa Rican *Heisteria longipes* mentioned earlier.

The wide variety with regard to root-hair abundance should, in fact, give us pause. Heckard's work (1962) is explicit here: in some species of *Castilleja*

great variation exists from plant to plant. What is more, *C. franciscana,* which seems to lack root hairs, may nevertheless be raised to maturity without a host. The possibility that host exudates have morphogenetic effects including the formation of root hairs on the parasitic roots has been demonstrated (Williams, 1961b). While Heinricher's notion would seem to be a logical extension of the "water parasitism" of green parasites, the evidence in support of it is very weak at present.

On the topic of reduction in stelar tissues of parasites, ample information is available, but, as its various manifestations have never before been assembled, I may have overlooked important items. Some of even the most recent articles on the subject leave much to be desired as to precision and reliability, and I, also, may be guilty of perpetuating statements based on misinterpretations which I have been unable to check. Admittedly, the study of phloem is difficult at best, and is certainly so where evolutionary reduction is concerned.

The endodermis either has received little attention in these plants or I have missed its appearance in the literature. In roots of *Pholisma* (Kuijt, 1967b) and *Epifagus* (Cooke and Schively, 1904) the endodermis is definitely absent and the same might be suspected in other Orobanchaccac and in the stems of *Cuscuta* and *Cassytha.* The endodermis has been reported in the roots of *Gerardia* (Boeshore, 1920), a fact which may perhaps be extended to cover Rhinanthoideae in general. In *Exocarpos* the endodermis is present but the Casparian strips are rather indistinct (Fineran, 1963c). In both *Gaiadendron* (Kuijt, 1965a) and *Atkinsonia* (Menzies and McKee, 1959) it is also recognizable, but in the epicortical roots of mistletoes there is no sign of it.

From the mass of information on the haustoria of parasitic angiosperms (chap. 7) we can draw the general conclusion that *phloem does not occur in haustoria.* As with any generality, this one does not quite cover all the facts. For example, phloem has been reported in *Cuscuta* haustoria (Truscott, 1958) and older self-parasitic ones in *Olax* (Barber, 1907b), but also in some Viscaceae (*Phoradendron*: Calvin, 1966), while it has been denied in others (*Viscum album*: Peirce, 1893). In *Pedicularis* haustoria Sprague (1962a) speaks of a rather large amount of phloem where Maybrook (1917) had denied its presence. Even in *Cuscuta* and *Phoradendron,* however, there is no phloem continuity between host and parasite: a "phloem gap" always exists (Kusano, 1902; Reiche, 1904; Engler and Krause, 1908; Stephens, 1912; Schumacher and Halbsguth, 1938; Fineran, 1963f; Kuijt, 1965a, 1965b, 1967b; Thurman, 1965; Okonkwo, 1967). I hope I do no injustice to Peirce (1893), who seems to have observed phloem in haustoria where subsequent workers could not, as in *Cuscuta, Lathraea,* Balanophoraceae, and Rafflesiaceae.

The literature gives one the impression that a progressive reduction of phloem occurs, starting from the haustoria, and that this process has reached different points in the various groups of parasites. The numerous parasites in which normally appearing sieve tubes are known from the shoot probably include virtually all Santalales, and certainly *Cuscuta* (Mirande, 1898) and *Cassytha* (Boewig, 1904; Mirande, 1905*a*). In Orobanchaceae the situation is not altogether clear. There are reports of sieve tubes in stems of *Conopholis* (Wilson, 1898), *Orobanche* (Kadry and Tewfic, 1956), and *Epifagus* (Cooke and Schively, 1904). Yet we hear of the absence of sieve tubes in the stems of *O. uniflora* (Smith, 1901), and in the above-mentioned work on *Epifagus* the authors are not certain about the presence of sieve plates in longitudinal sections: "It may be that degeneration has caused the loss of these structures, while the cells that possessed them remain." Statements which are similarly suspicious are made with respect to phloem in the tuber and root. We become uncomfortably aware of the fact that many students, when referring to phloem, mean little more than the cells occupying the phloic region. This also applies to some extent to the two reports I have been able to find on phloem in Rafflesiaceae: in the flowers of *Rafflesia patma* and *Pilostyles ulei* "phloem" is reported to consist of narrow elongated elements which have nuclei (Endriss, 1902; Hunziker, 1926). In the highly advanced mistletoe genus *Arceuthobium* sieve tubes are totally lacking (Kuijt, 1955). In *Balanophora*, similarly, there is nothing resembling sieve tubes (Fagerlind, 1948*b*); in *Dactylanthus* the situation is not clear (Moore, 1940). In *Pholisma* stems true sieve plates are present (Kuijt, 1967*b*), but they seem to be unusually scarce, which explains Copeland's (1935) failure to find sieve tubes.

In the roots of parasites the situation is equally ambivalent. In some of the most highly advanced parasites, like *Prosopanche* (Schimper, 1880) and *Helosis* (Umiker, 1920), sieve tubes are present. The statements with regard to phloem in roots of parasitic Scrophulariaceae and Orobanchaceae are confusing. Boeshore (1920), when speaking of the root of *Orobanche uniflora*, mentions large amounts of phloem, and in the root of *Gerardia* the phloem consists "of the usual elements." Smith (1901), on the contrary, says that there are no sieve tubes in the roots of *O. uniflora*, or, indeed, in the stem; yet this author discusses the phloem continuity between host and parasite! In recent work on the root of Rhinanthoideae (Rogers and Nelson, 1962) sieve tubes were traced through the roots of *Striga* to the base of the haustorium; this contradicts earlier workers who reported this cell type to be absent in the root of *Striga* (Stephens, 1912; Uttaman, 1950), and *Pedicularis* (Maybrook, 1917). In Santalales, *Buckleya* roots have sieve tubes (Kusano, 1902); in *Exocarpos* both sieve tubes and companion cells are said to be present, even though sieve plates were not visible (Fineran, 1963*f*); in the epicortical roots of *Struthanthus* even a persistent search could not reveal sieve plates, and the conclusion had to be drawn

that sieve tubes are either much reduced or not formed at all (Heil, 1926).

This enumeration shows the great need for a careful, comprehensive study of the nature and occurrence of phloem in the organs of angiospermic parasites. We cannot take much of the literature for granted in this respect, for it is permeated with carelessness. It seems certain that the focus of phloem reduction lies in the haustorium. While it appears illogical that even in holoparasites where phloem is probably formed in the stem there is no phloem bridge to the host, it is possible that the nature of food transfer from the host to the parasite is incompatible with the differentiation and/or food-conduction mechanism between sieve tube members. Perhaps it is this apparent paradox between the existence of a phloem gap and the dependence on organic materials from the host which has colored the vision of students like Pierce (1893) and others after him who felt the pressure of this paradox and thought to see sieve tubes where none may have existed.

The statement of Fineran (1963*f*) to the effect that the absence of phloem in a haustorium indicates "water parasitism" thus has little meaning, for there is no evidence of phloem in the haustoria of the most clearly holoparasitic plants such as Rafflesiaceae, Balanophoraceae, Lennoaceae, and Orobanchaceae.

Beyond the parasites there are probably very few angiosperms that lack sieve tubes. I know of two such reports in Chenopodiaceae, *Salicornia australis* (Cooke, 1912) and *Anabasis aretioides* (Hauri, 1912), and one in the greatly reduced Podostemaceae, *Mourera aspera* (Steude, 1935). The phloem in climbing roots of the ivy, *Hedera,* is said to be much reduced and often disorganized (Bruhn, 1910). I do not know whether these observations have been subsequently confirmed.

Contemporary thought in plant anatomy would lead one to postulate that in the haustorium, surely the most advanced organ in parasitic plants, the xylem might consist mostly of vessel members, not of tracheids. This notion, I think, turns out to be correct in general, with some apparent exceptions, though here again we have to deal with the shadows of past misinterpretations and vague descriptions.

The "phloeotracheids" reported from some Santalalean haustoria are said to be, both in structure and function, a combination of sieve elements and tracheal elements. The mind boggles at this notion. I do not mean to imply that misinterpretations are necessarily involved here, but vagueness certainly attaches to the treatment of this cell type in the literature. No adequate illustrations exist (Kuijt, 1965*b*).

I have reviewed the information available on the xylem of Santalalean haustoria (Kuijt, 1965*b*). On the basis of that résumé it can be safely stated, notwithstanding early references to the contrary, that lignified xylem in these organs consists exclusively of vessel members with simple perforation plates (see also Calvin, 1966). Both the position of the aperture and the shape and wall contours of the vessel members vary within wide limits, except that the vessel members are approximately isodiametric.

Two separate studies speak of tracheids in the haustorium of Orobanchaceae, in *Epifagus* (Cooke and Schively, 1904) and *Orobanche* (Kadry and Tewfic, 1956). In the latter it is unambiguously said that only tracheids occur. I have observed, however, that the lignified xylem of the *Boschniakia* tuber, in contrast, consists of nothing but vessel members with simple perforation plates, a situation that seems irreconcilable with the above-mentioned statements. In Balanophoraceae there is mention of short spiral or reticulated tracheids in the tuber of *Balanophora* (Strigl, 1908), which is again diametrically opposite to Mangenot (1947), who found only reticulated vessel members in *Thonningia*.

About the haustoria of Scrophulariaceae there is controversy also. Haustorial tracheids are cited in *Striga* (Stephens, 1912; Rogers and Nelson, 1962) and probably in other work which I have missed. The occurrence of vessels is known from *Dasistoma* and *Pedicularis* (Piehl, 1962a, 1962b, 1963) and *Melampyrum* (Cantlon *et al.*, 1963). In another paper on *Pedicularis* we hear of both tracheids and vessel members in haustoria (Sprague, 1962a). Thurman (1965) has demonstrated that all haustorial xylem in *Orthocarpus* is made up of vessel members only.

For the remaining organs of parasites we must work with similarly inadequate material. In the stems of *Cuscuta* and *Myzodendron* (Metcalfe and Chalk, 1950), *Cassytha, Melampyrum lineare, Pedicularis canadensis, Castilleja coccinea, Amyema scandens* (Bierhorst and Zamora, 1965), *Phoradendron* (Calvin, 1966), *Arceuthobium* (Kuijt, 1960), *Thonningia* (Mangenot, 1947), and *Pholisma* (Kuijt, 1967b) the xylem appears to exist only in the shape of vessel members. Evidence for tracheids comes from the earliest xylem in *Castilleja coccinea* (Bierhorst and Zamora, 1965), from *Pilostyles ulei* (Metcalfe and Chalk, 1950), *Rafflesia patma* (Hunziker, 1926), and *Orobanche* (Kadry and Tewfic, 1956). In *Santalum* both perforate and imperforate elements are known (Metcalfe and Chalk, 1950). Perforation plates, where referred to, are simple except for the scalariform plates in *Heisteria* (Sleumer, 1935), perhaps unique in the family in this regard.

When reporting on haustorial phloem, it is easy to fall into the same sort of trap which has caught others, and to get a distorted view of reality under the influence of contemporary evolutionary notions about xylem anatomy. The idea that vessel members are evolutionary derivatives of tracheids and therefore constitute one criterion of advancement meets few objections nowadays. It must be admitted that such ideas are generally confirmed in haustoria of parasites and to some extent in the rest of the plant. I would predict that some of the statements cited above which seem to clash in this respect will turn out to be unfounded or otherwise erroneous. In some

other highly advanced plants, Cactaceae, a trend has been demonstrated from large to very small perforations or none at all (Carlquist, 1961)—a return, in a secondary fashion, to a small tracheid-like element. It is not impossible that a similar trend has created confusion of interpretation in some parasites also. In *Arceuthobium*, indeed, the vessel members of the endophyte have nearly reached that point (Kuijt, 1960).

The topic of evolutionary convergence seems, at first glance, to be out of context here, but I would point out the great importance of reduction in the achievement of present-day convergence. In some of the cases cited it is as if "peripheral" modifications otherwise used to characterize large taxa have been stripped away to leave only the essential elements. Viewing it in another way, we may say that the end point in ontogenetic development has come progressively earlier. Convergence is most apparent at this lower level of structural complexity. In no case do I maintain that convergence is due to reduction alone; this view would be too simplistic. At any rate, the line between reduction and elaboration is rarely marked clearly enough for such a decision. Only the briefest comments suffice here.

1. *Reduction in size of foliage leaves.*—The trend finds its extreme expression in *Prosopanche, Hydnora, Lathrophytum, Ombrophytum,* and *Chlamydophytum,* all of which are perhaps totally leafless.

2. *Evolution of the inferior ovary from a superior one.*—This is a well-recognized trend in angiosperms generally. We see evidence for it in various Santalales (Santalaceae, *Myzodendron, Antidaphne,* Olacaceae) and Rafflesiaceae.

3. *Reduction of ovule and placenta.*—Long dominant in systematic thought in Santalalean families, these trends must be considered with caution (see earlier in this chapter and in chap. 3).

4. *Reduction of the seed.*—The striking similarity of the seeds of *Balanophora, Pilostyles,* and *Orobanche* is largely the result of the reduction of embryo and endosperm. Compare the seed of the saprophyte *Voyria* (fig. 9-2,*a*).

5. *Pollen exine.*—Sculptural features have been stripped away, for the most part, in Lennoaceae, Balanophoraceae, Rafflesiaceae, and Scrophulariaceae. Even in other families, species with exine elaboration are very rare.

6. *Isophasic behavior.*—The isophasic growth pattern of the endophyte, and the correlated shoot-emergence pattern of *Arceuthobium* (in some host-parasite combinations), *Pilostyles haussknechtii,* and possibly other members of Rafflesiaceae is one of the most remarkable convergences in the plant kingdom.

7. *Germination in* Cuscuta *and* Cassytha.—Not only the adult behavior and general appearance of these two genera are incredibly similar, but the patterns of germination, with many of the consequent modifications, also coincide.

8. *The haustoria in Santalales and* Cassytha.—These follow much the same ontogeny, although some individual details differ.

9. *Root dimorphism.*—The subdivision of the root system into two types of roots, each with a separate function, I interpret to be nearly the same in Hydnoraceae and *Pholisma.*

10. *Addition of extra-ovular tissues to the seed.*—This

phenomenon occurs in Lennoaceae, Loranthaceae, Viscaceae, *Prosopanche,* probably in *Sarcophyte* and *Chlamydophytum,* and possibly in epiphytic Santalaceae.

11. *Evolution of the primary haustorium.*—The addition of the primary haustorium, sometimes followed by the eventual loss of secondary haustoria, is seen in Loranthaceae, *Antidaphne, Eremolepis, Phacellaria,* all Orobanchaceae, and a few Scrophulariaceae.

12. *Endogenous flowers.*—This remarkable tendency is found in many Rafflesiaceae, in Hydnoraceae, and sometimes in *Phacellaria* and *Cuscuta.* Except in the latter two it may be viewed as an extension of the endogenous emergence of inflorescences in some other Rafflesiaceae, in Balanophoraceae, Orobanchaceae, and Lennoaceae—a phenomenon indicative of the rootlike nature of the mother organs.

Literature Cited

Abbiatti, D. 1946. Las Lorantáceas Argentinas. Rev. Mus. La Plata, N.S., Sec. Bot. 7:1-110.

Agarwal, S. 1961. The embryology of *Strombosia* Blume. Phytomorph. 11:269-272.

————. 1962. Embryology of *Quinchamalium chilense* Lam. Proc. Symp. Plant Embryol., 1960 (New Delhi), pp. 162-169. (See also Johri and Agarwal, Phytomorph. 15:360-372. 1966.)

————. 1963. Morphological and embryological studies in the family Olacaceae. I. *Olax* L. Phytomorph. 13:185-196.

Agati, J. A., and J. P. Tan. 1931. Controlling the *Aeginetia indica* in cane fields. Sug. News 12:852.

Ali, S.A. 1931. The role of the sunbirds and the flowerpeckers in the propagation and distribution of the tree-parasite, *Loranthus longiflorus* Desr., in the Konkan. Jour. Bombay Nat. Hist. Soc. 35:144.

Allen, G. M. 1962. Bats. New York: Dover Publ., Inc.

Anonymous. 1924. *Melasma:* a root parasite on peanuts. Jour. Dept. Agr. S. Afr. Transv., 3 pp.

————. 1955. An evaluation of forest insect and disease research needs in Oregon and Washington. Part II. Diseases. (Not seen; cited in Shea, For. Sci. 8:298-302. 1962.)

————. 1962. Documented chromosome numbers in plants. Madroño 16:267.

————. 1964. Trees and shrubs of the Witwatersrand. Johannesburg: Witwatersrand Univ. Press.

Anselmino, E. 1933. Die Stammpflanzen von Muirapuama. Inaug. Diss., Berlin. (Not seen.)

————. 1934. Geschichtliche Uebersicht der Stellung der Olacaceen bei den verschiedenen Systematikern. Feddes Rept. 33:285-297.

Arekal, G. D. 1963. Embryological studies in Canadian representatives of the tribe Rhinantheae, Scrophulariaceae. Can. Jour. Bot. 41:267-302.

Asplund, E. 1928. Eine neue Balanophoraceen-Gattung aus Bolivien. Sv. Bot. Tidskr. 22:261-277.

Atsatt, P. R. 1965. Angiosperm parasite and host: coördinated dispersal. Science 149:1389-1390.

————. 1966. The population biology of closely related species of *Orthocarpus*. Ph. D. Thesis, Univ. Calif., Los Angeles.

Baccarini, A. 1967. Dodder's autotrophy in parasitizing state. Zeits. Pfl. Physiol. 57:201-202.

Babington, C. C. 1844. On some species of *Cuscuta*. Ann. and Mag. Nat. Hist. 13:249.

Bajaj, Y. P. S. 1966. Behavior of embryo segments of *Dendrophthoe falcata* (L.f.) Ettings, in vitro. Can. Jour. Bot. 44:1127-1131.

Baker, E. G. 1888. On a new species of *Cytinus* from Madagascar, constituting a new section of that genus. Jour. Linn. Soc., Bot. 24:465-469.

Baldev, B. 1959. *In vitro* responses of growth and development in *Cuscuta reflexa* Roxb. Phytomorph. 9:316-319.

Baldwin, J. T., and B. M. Speese. 1957. *Phoradendron flavescens:* chromosomes, seedlings, and hosts. Amer. Jour. Bot. 44:136-140.

Balle, S. 1954. Sur quelques Loranthoidées d'Afrique. Bull. Séanc. Acad. Roy. Sci. Col., Brussels, 25:1619-1635.

————. 1956. A propos de la morphologie des "Loranthus" d'Afrique. Webbia 11:541-585.

————. 1960. Contribution à l'étude des *Viscum* de Madagascar. Lejeunia, Mém. 11:1-151.

————. 1964. Loranthacées de Madagascar. Adansonia 4:105-141.

Balle, S., and N. Hallé. 1961. Les Loranthacées de la Côte d'Ivoire. Adansonia 1:208-265.

Banerji, I. 1961. The endosperm in Scrophulariaceae. Jour. Ind. Bot. Soc. 40:1-11.

Barber, C. A. 1906. Studies in root-parasitism. The haustorium of *Santalum album*. 1. Early stages, up to penetration. Mem. Dept. Agr. India, Bot. Ser. 1(1):1-30.

————. 1907a. *Ibid*. 2. The structure of the mature haustorium and the inter-relations between host and parasite. *Idem, 1* (1, Part II):1-58.

————. 1907b. Studies in root-parasitism. The haustorium of *Olax scandens. Idem,* 2(4): 1-47.

————. 1907c. Parasitic trees in southern India. Proc. Cambridge Phil. Soc. 14:246-256.

————. 1908. Studies in root parasitism. 4. The haustorium of *Cansjera rheedii*. Mem. Dept. Agr. India, Bot. Ser. 2:1-36.

Barlow, B. A. 1962. Studies in Australian Loranthaceae. I. Nomenclature and new additions. Proc. Linn. Soc. N. S. W. 87:51-61.

————. 1963a. *Ibid*. III. A revision of the genus *Lysiana* Tiegh. *Idem.* 88:137-150.

————. 1963b. *Ibid*. IV. Chromosome numbers and their relationships. *Idem,* 88:151-161.

————. 1964. Classification of the Loranthaceae and Viscaceae. *Idem,* 89:268-272.

————. 1966. A revision of the Loranthaceae of Australia and New Zealand. Aust. Jour. Bot. 14:421-499.

Barnes, E. 1941. A note on the root-parasitism of *Centranthera humifusa* Wall. Jour. Bombay Nat. Hist. Soc. 42:668-669.

Bartling. F.T. 1830. Ordines naturales plantarum. Göttingen.

Basinski, J. J. 1955. Witchweed and soil fertility. Nature (London) 175:431.

Beccari, O. 1869. Illustrazione di nuove specie di piante Bornensi: *Balanophora reflexa* e *Brugmansia Lowi*. Nuovo Giorn. Bot. Ital. 1:65-91. (Not seen.)

Beck von Mannagetta, G. 1890. Monographie der Gattung *Orobanche*. Bibl. Bot. 19:1-275.

————. 1930. Orobanchaceae, *in* A. Engler, Das Pflanzenreich, 96(IV.261):1-348.

Bedi, R. 1967. *Alectra parasitica* var. *chitrakutensis*. Econ. Bot. 21:276-283.

Bennett, C. W. 1944. Studies of dodder transmission of plant viruses. Phytopath. 34:905-932.

Bentham, G., and J. D. Hooker. 1862-1883. Genera plantarum. London.

Berg, R. Y. 1954. Development and dispersal of the seed of *Pedicularis silvatica*. Nytt Mag. Bot. 2:1-60.

Bergdolt, E. 1927. Ueber die Saugkräfte einiger Parasiten. Ber. Deuts. Bot. Ges. 45:293-301.

Bertossi, F., A. Baccarini, and N. Bagni. 1964. Rapporto tra le clorofille *a* e *b* in *Cuscuta australis*. Giorn. Bot. Ital. 71:517-521.

Bhatnagar, S. P. 1965. Studies in angiospermic parasites. No. 2. *Santalum album*, the Sandalwood tree. Bull. Nat. Bot. Gardens 112:1-90.

Bierhorst, D. W., and P. M. Zamora. 1965. Primary xylem elements and element associations of angiosperms. Amer. Jour. Bot. 52:657-710.

Björkman, E. 1960. *Monotropa hypopitys* L., an epiparasite on tree roots. Physiol. Plant. 13:308-327.

Black, J. M. 1948. Flora of South Australia. 2d ed. Adelaide: Government Printer.

Blake, S. F. 1926. *Lennoa caerulea* in Colombia. Proc. Biol. Assoc. Wash. 39:146.

Blakely, W. F. 1922-1928. The Loranthaceae of Australia. Parts 1 - 7. Proc. Linn. Soc. N. S. W. 47-53.

Blume, C. L. 1823. Catalogus van eenige der merkwaardigste zo in- als uitheemsche gewassen te vinden in 's Land's Plantentuin te Buitenzorg.

————. 1827. Enumeratio plantarum Javae et inss. adjucentium. *Balanophora:* Fasc. I, p. 36. (Not seen)

————. 1830. Flora Javae. Loranthaceae: 3 (Not seen)

Boerhave Beekman, W. 1949-1955. Hout in alle tijden. Deventer: Kluwer.

Bonneville, M. A., and B. R. Voeller. 1963. A new cytoplasmic component of plant cells. Jour. Cell Biol. 18:703-708.

Boeshore, I. 1920. The morphological continuity of Scrophulariaceae and Orobanchaceae. Contr. Bot. Lab. Univ. Penn. 5:139-177.

Boewig, H. 1904. The histology and development of *Cassytha filiformis*, L. Contr. Bot. Lab. Univ. Penn. 2:399-416.

Bonnier, G. 1891. Sur l'assimilation des plantes parasites à chlorophylle. C. R. Acad. Sci. 113:1074.

————. 1893. Recherches physiologiques sur les plantes vertes parasites. Bull. Sci. France Belg. 25:77-92.

Boodle, L. A. 1913. The root and haustorium of *Buttonia natalensis*. Kew Bull. 6:240-242.

Bornmüller, J. 1909. Plantae Straussianae. Beih. Bot. Centralbl. 24:91.

————. 1913. Collectiones Straussianae novae. Beih. Bot. Centralbl. 28:458-535.

Botha, P. J. 1948. The parasitism of *Alectra vogelii* Benth., with special reference to the germination of its seeds. Jour. S. Afr. Bot. 14:63-80.

————. 1950a. The germination of the seeds of angiospermous root-parasites. Part I. The nature of the changes occurring during pre-exposure of the seed of *Alectra vogelii* Benth. *Idem*, 16:23-28.

————. 1950b. *Ibid*. Part II. The effect of time of pre-exposure, temperature of pre-exposure and concentration of the host factor on the germination of the seeds of *Alectra vogelii* Benth. *Idem*, 16:29-38.

————. 1951a. *Ibid*. Part III. The effect of time of exposure to the host factor on the germination of the seed of *Alectra vogelii* Benth. *Idem*, 17:49-58.

————. 1951b. *Ibid*. Part IV. The properties and physiological significance of the host factor necessary for the germination of the seeds of *Alectra vogelii* Benth. *Idem*, 17:59-72.

Bouriquet, G. 1933. Une Scrophulariacée parasite du riz à Madagascar. Rev. Path. Vég. Ent. Agr. 20:149-151.

Bowman, J. E. 1833. On the parasitical connection of *Lathraea squamaria* and the peculiar structure of its subterranean leaves. Trans. Linn. Soc. London 16:399-420.

Britton, N. L. 1930. Krameriaceae. N. Amer. Flora 23:195-200.

Brongniart, A. 1824. Observations sur les genres *Cytinus* et *Nepenthes*. Ann. Sc. Nat. 1:29-52.

Brown, R. 1822. An account of a new genus of plants named *Rafflesia*. Trans. Linn. Soc. London 13:201-234.

————. 1834. Description of the female flower and fruit of *Rafflesia arnoldi*, with remarks on its affinities; and an illustration of the structure of *Hydnora africana. Idem*, 19:221-247.

Brown, R., A. W. Johnson, E. Robinson, and A. R. Todd. 1949. The stimulant involved in the germination of *Striga hermontica*. Proc. Roy. Soc. 136B:1-12.

Bruch, C. 1923. Coleopteros fertilizadores de *Prosopanche burmeisteri* De Bary. Physis 7:82-88.

Bruhn, W. 1910. Beiträge zur experimentellen Morphologie, zur Biologie und Anatomie der Luftwurzeln. Flora, N.F., 1:98-166.

Bünning, E. 1947. In den Wäldern Nordsumatras. Bonn: Dümmler.

Bünning, E., and R. Kautt. 1956. Ueber den Chemotropismus der Keimlinge von *Cuscuta europaea*. Biol. Zentralbl. 75:356-359.

Burkart, A. 1963. Nota sobre *Prosopanche bonacinae* Speg. (Hydnoraceae) su área y parasitismo sobre algodón. Darwiniana 12:633-638.

Butcher, D. N., and H. E. Street. 1964. Excised root culture. Bot. Rev. 30:513-586.

Calvin, C. L. 1966. On the anatomy of the mistletoe (*Phoradendron flavescens*). Ph. D. Thesis, Univ. California, Davis.

Cammerloher, H. 1920. Der Spaltöffnungsapparat von *Brugmansia* und *Rafflesia*. Oester. Bot. Zeit. 69:153-164.

————. 1921. Blütenbiologische Beobachtungen an *Loranthus europaeus* Jac. Ber. Deuts. Bot. Ges. 39:64-70.

Cannon, W. A. 1909. The parasitism of *Orthocarpus purpurascens* Benth. Plant World 12:259-261.

————. 1910. The root habits and parasitism of *Krameria canescens* Gray. Carn. Inst. Wash. Publ. 129:5-24.

Cantlon, J. E., E. J. C. Curtis, and W. M. Malcolm. 1963. Studies of *Melampyrum lineare*. Ecology 44:466-474.

Cario, R. 1881. Anatomische Untersuchung von *Tristicha hypnoides* Spreng. Bot. Zeit. 39:25-33, 41-48, 57-64, 73-82.

Carlquist, S. 1953. Documented chromosome numbers in plants. Madroño 12:31.

————. 1961. Comparative plant anatomy. New York: Holt, Rinehart and Winston.

Cartellieri, E. 1926. Das Absorptionssystem der Rafflesiacee *Brugmansia*. Bot. Arch. 14:284-311.

————. 1928. Das Haustorium von *Cassytha pubescens* R. Br. Planta 6:162-182. Springer-Verlag.

Cassera, J. D. 1935. Origin and development of the female gametophyte, endosperm and embryo in *Orobanche uniflora*. Bull. Torrey Bot. Club 62:455-466.

Caullery, M. 1952. Parasitism and symbiosis. London: Sidgewick & Jackson Ltd.

Cavaco, A. 1954. Sur le genre *Phanerodiscus* gen. nov.

(Olacacées) de Madagascar. Not. Syst. Paris 15:10-14.

Chabrolin, C. 1939. Contribution à l'étude de la germination des graines de l'Orobanche de la Fève. Ann. Serv. Bot. Tunis 14-15:92. (Not seen)

Cheeseman, T. F. 1925. Manual of the New Zealand Flora. Wellington.

Chesnut, V. K. 1902. Plants used by the Indians of Mendocino County, California. Contr. U. S. Nat. Herb. 7:295-408.

Chodat, R. 1890. Sur la famille de Krameriacées. Arch. Sci. Phys. Nat. 24:495-499.

————. 1915. Les espèces du genre Prosopanche. Bull. Soc. Bot. Genève, Sér. 2, 7:65-66.

Choudhuri, J. C. B. 1963. Sandalwood tree (Santalum album L.) and its diseases. Ind. For. 89:456-462.

Christmann, C. 1960. Le parasitisme chez les plantes. Paris: Leclerc & Cie.

Claus, E. P. 1955. A pharmacognostical study of Pyrularia pubera Michx. Amer. Pharm. Assoc. Jour. 44:39-42.

Cocucci, A. E. 1965. Estudios en el género Prosopanche (Hydnoraceae). I. Revisión taxonómica. Kurtziana 2:53-73.

Coert, J. H. 1924. Aeginetia species: a root parasite of sugar cane. Meded. Proefst. Java-Suikerind. 13:437-447.

Cohen, L. I. 1968. Development of the staminate flower in the dwarf mistletoe, Arceuthobium. Amer. Jour. Bot. 55:187-193.

Colbatch, Sir John. 1719. A dissertation concerning Mistletoe; a most wonderful specifick remedy for the cure of convulsive distempers. Calculated for the benefit of the poor as well as the rich and heartily recommended for the good of mankind. 8 + 30 pp. London. (Not seen).

Coleman, E. 1934. Notes on Exocarpus. Vict. Natural. 51:132-139.

Conzatti, C., and L. Smith. 1910. Flora synoptica Mexicana. Mexico.

Cooke, E., and A. F. Schively. 1904. Observations on the structure and development of Epiphegus virginiana. Contr. Bot. Lab. Univ. Penn. 2:352-398.

Cooke, F. W. 1912. Observations on Salicornia australis. Trans. N. Z. Inst. 44:349-362.

Copeland, H. F. 1935. The structure of the flower of Pholisma arenarium. Amer. Jour. Bot. 22:366-383.

Corner, E. J. H. 1958. Transference of function. Jour. Linn. Soc. London (Bot.) 56:33-40.

Coutinho, L. de A. 1957. Observacoes carioloicas en Viscum cruciatum Sieber. Algunas aspectos da estrutura do centromero. Genet. Iber. 9:117-132.

Covas, G., and B. Schnack. 1946. Número de cromosomas en Antofitas de la región de Cuyo. Rev. Arg. Agron. 13:153-166.

Cowles, R. B. 1936. The relation of birds to seed dispersal of the desert mistletoe. Madroño 3:352-356.

Cronquist, A. 1965. The status of the general system of classification of flowering plants. Ann. Miss. Bot. Gard. 52:281-303.

Crosby-Browne, A. J. 1950. The root parasitism of Euphrasia salisburgensis Funck. Watsonia 1:354-355.

Crouch, J. E. 1943. Distribution and habitat relationships of the Phainopepla. Auk 60:319-333.

Cunningham, A. M. 1898. Morphological characters of the scales of Cuscuta. Proc. Ind. Acad. Sci., pp. 212-213.

Curtis, E. J. C., and J. E. Cantlon. 1965. Studies of the germination process in Melampyrum lineare. Amer. Jour. Bot. 52:552-555.

Curtis, E. J. C., and J. E. Cantlon. 1968. Seed dormancy and germination in Melampyrum lineare. Amer. Jour. Bot. 55:26-32.

Dalziel, J. M. 1937. The useful plants of West Tropical Africa. London.

Danser, B. H. 1929. On the taxonomy and the nomenclature of the Loranthaceae of Asia and Australia. Bull. Jard. Bot. Buitenz., Sér. 3, 10:291-374.

————. 1931a. On the harmony and nomenclature of the Loranthaceae of Asia and Australia. Idem, Sér. 3, 10:291.

————. 1931b. The Loranthaceae of the Netherlands Indies. Idem, Sér. 3, 11:233-519.

————. 1933a. A new system for the genera of Loranthaceae-Loranthoideae, with a nomenclator for the Old World species of this subfamily. Verh. K. Akad. Wetens. Amsterdam, Afd. Natuurk., Sect. 2, 29(6):1-128.

————. 1933b. Vernacular names of Loranthaceae in the Malay Peninsula and the Netherlands Indies. Bull. Jard. Bot. Buitenz., Sér. 3, 13:487-496.

————. 1934a. Miscellaneous notes on Loranthaceae, 1-6. Rec. Trav. Bot. Néerl. 31:223-236.

————. 1934b. The Loranthaceae of the Oxford University Expedition of Sarawak in 1932. Idem, 31:237-247.

————. 1934c. Miscellaneous notes on Loranthaceae, 7-8. Idem, 31:751-760.

————. 1935. A revision of the Philippine Loranthaceae. Philip. Jour. Sci. 58:1-149.

————. 1936a. Miscellaneous notes on Loranthaceae, 9-15. Blumea 2:34-59.

————. 1936b. New Papuan Loranthaceae. Brittonia 2:131-134.

————. 1936c. The Loranthaceae Loranthoideae of the tropical archipelagoes east of the Philippines, New Guinea, and Australia. Bull. Jard. Bot. Buitenz., Sér. 3, 14:73-98.

————. 1937. A revision of the genus Korthalsella. Idem, Sér. 3, 14:115-159.

————. 1938a. The Loranthaceae of French Indo-China and Siam. Idem, Sér. 3, 16¹:1-63.

————. 1938b. Miscellaneous notes on Loranthaceae, 16-18. Blumea 3:34-59.

————. 1939. A revision of the genus Phacellaria (Santalaceae). Idem, 3:212-235.

————. 1940a. Miscellaneous notes on Loranthaceae, 19-24. Idem, 3:389-404.

————. 1940b. A supplement to the revision of the genus Korthalsella (Lor.). Bull. Jard. Bot. Buitenz., Sér. 3, 16³:329-342.

————. 1940c. On some genera of Santalaceae Osyrideae from the Malay Archipelago, mainly from New Guinea. Nova Guinea, N.S., 4:133-150.

————. 1941a. The British-Indian species of Viscum revised and compared with those of South-Eastern Asia, Malaysia, and Australia. Blumea 4:261-319.

————. 1941b. Miscellaneous notes on Loranthaceae, 25. Idem, 4:259-260.

————. 1955. Supplementary notes on the Santalaceous genera Dendromyza and Cladomyza (with pictures of these genera and of Hylomyza). Nova Guinea, N.S. 6:261-277.

Darwin, E. 1825. The Botanic Garden, etc. London: Jones & Co.

Dastur, R. H. 1921-1922. Notes on the development of the ovule, embryosac, and embryo of Hydnora africana Thunb. Trans. Roy. Soc. S. Afr. 10:27-31.

Datta, R. M. 1951. Occurrence of a hermaphrodite flower

in *Arceuthobium minutissimum* Hook. f., the smallest known dicotyledonous plant. Nature (London) *167*: 203-204.

Dawson, J. H. 1965. Prolonged emergence of field dodder. Weeds *13*:373-374.

————. 1967. Soil-applied herbicides for dodder control; initial greenhouse evaluation. Wash. Agr. Expt. Sta. Bull. *691*:1-7.

Dean, H. L. 1937*a*. An addition to bibliographies of the genus *Cuscuta*. Univ. Iowa Stud. Nat. Hist. *17*:191-197.

————. 1937*b*. Gall formation in host plants following haustorial invasion by *Cuscuta*. Amer. Jour. Bot. *24*:167-173.

————. 1942. Total length of stem developed from a single seedling of *Cuscuta*. Proc. Iowa Acad. Sci. *49*:127-128.

————. 1954. Dodder overwintering as haustorial tissues within *Cuscuta*-induced galls. *Idem, 61*:99-106.

De Bary, A. 1868. *Prosopanche burmeisteri*, eine neue Hydnoree aus Süd-Amerika. Abhandl. Naturf. Ges. Halle *10*:243-272. (Not seen)

Decaisne, M. J. 1847. Sur le parasitisme des Rhinanthacées. Ann. Sci. Nat., Sér. 3, *8*:5-9.

De Candolle, A. P. 1801. Mémoire sur la végétation du guy. Bull. Sci. Soc. Philom. *2*:162-163 (trans. Phil. Mag. *9*:176-177. 1801).

————. 1813. Théorie élémentaire de la botanique. Paris.

De Laubenfels, D. J. 1959. Parasitic conifer found in New Caledonia. Science *130*:97.

Denffer, D. von. 1947. *Cuscuta gronovii* Willd als Endoparasit. Nachr. Akad. Wiss. Göttingen, Math.-Phys. Kl., Biol.-Phys.-Chem. Abt., *1947*:21-23.

————. 1948. Ueber die Bedeutung des Bliihtermins der Wirtspflanzen von *Cuscuta gronovii* Willd. für die Blütenbildung des Schmarotzers. Biol. Zentralbl. *67*:175-189.

Desselberger, H. 1931. Der Verdauungskanal der Dicaeiden nach Gestalt und Funktion. Jour. f. Ornithol. *79*:353-370.

Diem, J. 1950. Las plantas huespedes de la loranthacea *Phrygilanthus tetrandrus* (Ruiz & Pavon) Eichl. Bol. Soc. Arg. Bot. *3*:177-179.

Dieterici, F. 1861. Die Naturanschauung und Naturphilosophie der Araber im 10. Jahrhundert. Berlin (not seen).

Dixit, S. N. 1962. Rank of the subfamilies Loranthoideae and Viscoideae. Bull. Bot. Survey India *4*:49-55.

Docters van Leeuwen, W. M. 1954. On the biology of some Javanese Loranthaceae and the role birds play in their life-history. Beaufortia, Misc. Publ. *4*:105-207.

Dorner, J. von. 1867. Die Cuscuten der Ungarischen Flora. Linnaea *35*:125-151.

Dressler, R. L., and Job Kuijt. 1968. A second species of *Ammobroma* (Lennoaceae) in Sinaloa, Mexico. Madroño *19*:179-182.

Drugg, W. S. 1962. Pollen morphology of the Lennoaceae. Amer. Jour. Bot. *49*:1027-1032.

Durbin, R. D. 1953. Hosts of the branched broomrape and its occurrence in California. Plant Dis. Rep. *37*:136-137.

Du Rietz, G. E. 1931*a*. The long-tubed New Zealand species of *Euphrasia* (= *Siphonidium* Armstr.). Sv. Bot. Tidskr. *25*:108-125.

————. 1931*b*. Two new species of *Euphrasia* from the Philippines and their phytogeographical significance. *Idem, 25*:500-542.

Eichler, A. W. 1868*a*. Loranthaceae, *in* Martius, Flora Brasil. 5(2).

————. 1868*b*. *Lathrophytum*, ein neues Balanophoreengeschlecht aus Brasilien. Bot. Zeit. *26*:513-520, 529-537, 545-551.

————. 1878. Blüthendiagramme. Leipzig.

Endlicher, S. L. 1836-1850. Genera plantarum. Vienna.

Endriss, W. 1902. Monographie von *Pilostyles ingae* (Karst.) (*Pilostyles ulei* Solms-Laubach) Flora (Erganz. Bd.) *91*:209-236.

Engler, A. 1889. Olacaceae, *in* Engler & Prantl, Die Nat. Pfl. Fam. *3*(1):231-242.

————. 1894. Loranthaceae, *in* Engler & Prantl, *idem*, *3*(1):156-198.

Engler, A., and K. Krause. 1908. Ueber die Lebensweise von *Viscum minimum* Harvey. Ber. Deuts. Bot. Ges. *26a*:524-530.

————. 1935. Loranthaceae, *in* Engler & Prantl, Die Nat. Pfl. Fam., 2d ed., *16b*:98-203.

Engler, A., and G. Volkens. 1897. Ueber das wohlriechende ostafrikanische Sandelholz (*Osyris tenuifolia* Engl.). Notizbl. K. Bot. Gart. Mus. Berlin *1*:269-275.

Erbrich, P. 1965. Ueber Endopolyploidie und Kernstrukturen in Endospermhaustorien. Oesterr. Bot. Zeits. *112*:197-262.

Erdtman, G. 1944. The systematic position of the genus *Diclidanthera*. Bot. Notiser *1944*:80-84.

————. 1952. Pollen morphology and plant taxonomy. Angiosperms. Stockholm: Almquist & Wiksell.

Ernst, A. 1913. Embryobildung bei *Balanophora*. Flora *106*:129-159.

Ernst, A., and E. Schmid. 1913. Ueber Blüte und Frucht von *Rafflesia*. Ann. Jard. Bot. Buitenz., Sér. 2, *12*:1-58.

Esau, K. 1953. Plant anatomy. New York: John Wiley & Sons.

————. 1960. Anatomy of seed plants. New York: John Wiley & Sons.

Espino, R. B. 1947. Eleven years study on "bungang tubo": a résumé. Philip. Agr. *31*:151-153.

Ettingshausen, C. von. 1872. Ueber die Blattskelete der Loranthaceen. Denkschr. Akad. Wiss. Wien, Math.-Nat. Kl., *32*:51-84.

Ewart, A. J. 1930. Flora of Victoria. Melbourne: Government Printer.

Fagerlind, F. 1938*a*. *Ditepalanthus*, eine neue Balanophoraceen-Gattung aus Madagaskar. Arkiv Bot. *29A*:1-15.

————. 1938*b*. Bau und Entwicklung der floralen Organe von *Helosis cayennensis*. Sv. Bot. Tidskr. *32*:139-159.

————. 1940. Beobachtungen über die Kletterorgane bei *Olax*. *Idem, 34*:26-34.

————. 1945*a*. Blüte und Blütenstand der Gattung *Balanophora*. Bot. Notiser *1945*:330-350.

————. 1945*b*. Bildung und Entwicklung des Embryosacks bei sexuellen und agamospermischen *Balanophora*-Arten. Sv. Bot. Tidskr. *39*:65-82.

————. 1945*c*. Bau der floralen Organe der Gattung *Langsdorffia*. *Idem, 39*:197-210.

————. 1948*a*. Beiträge zur Kenntnis der Gynäceummorphologie und Phylogenie der Santalales-Familien. *Idem, 42*:195-229.

————. 1948*b*. Bau und Entwicklung der vegetativen Organe von *Balanophora*. K. Sv. Vetensk. Akad. Handl. *25*(3):1-72.

————. 1959. Development and structure of the flower and gametophytes in the genus *Exocarpus*. Sv. Bot. Tidskr. *53*:257-282.

Fahn, A., and T. Arzee. 1959. Vascularization of articulated Chenopodiaceae and the nature of their fleshy cortex. Amer. Jour. Bot. *46*:330-338.

Fawcett, W., and A. B. Rendle. 1914. Flora of Jamaica.

London: Longmans, Green & Co.

Fedortschuk, W. 1931. Embryologische Untersuchung von *Cuscuta monogyna* Vahl und *Cuscuta epithymum* L. Planta *14*:94-111.

Ferrarini, E. 1950. Il parassitismo di *Osyris alba*. Nuovo Giorn. Bot. Ital. *57*:351-381.

Fineran, B. A. 1963a. Root parasitism in Santalaceae. Nature *197*:95.

————. 1963b. Parasitism in *Exocarpus bidwillii* Hook. f. Trans. Roy. Soc. N. Z., Bot. *2*:109-119.

————. 1963c. Studies on the root parasitism of *Exocarpus bidwillii* Hook. f. I. Ecology and root structure of the parasite. Phytomorph. *12*:339-355.

————. 1963d. *Ibid*. II. External morphology, distribution and arrangement of haustoria. *Idem, 13*:30-41.

————. 1963e. *Ibid*. III. Primary structure of the haustorium. *Idem, 13*:42-54.

————. 1963f. *Ibid*. IV. Structure of the mature haustorium. *Idem, 13*:249-267.

————. 1965a. *Ibid*. V. Early development of the haustorium. *Idem, 15*:10-25.

————. 1965b. *Ibid*. VI. Haustorial attachment to non-living objects and the phenomenon of self-parasitism. *Idem, 15*:387-399.

Follmann, G. 1963. Ueber eine gelbe Form von *Phrygilanthus aphyllus* (Miers) Eichl. Ber. Deuts. Bot. Ges. *76*: 344-348.

Follmann, G., and M. Mahn. 1964. Las plantas huespedes de *Phrygilanthus aphyllus* (Miers) Eichl. Bol. Univ. Chile, Ci., 7:39-41.

Foster, A. S. 1959. The morphological and taxonomic significance of dichotomous venation in *Kingdonia uniflora* Balfour F. et W. W. Smith. Notes Roy. Bot. Gard. Edinb. *23*:1-12.

————. 1963. The morphology and relationships of *Circaeaster*. Jour. Arn. Arb. *44*:299-327.

Fournier, E. 1868. Sur le genre *Lennoa*. Bull. Soc. Bot. France *15*:163-164.

Fratianne, D. G. 1965. The interrelationship between the flowering of dodder and the flowering of some long and short day plants. Amer. Jour. Bot. *52*:556-562.

Frazer, J. G. 1900. The Golden Bough: a study in magic and religion. New York and London: Macmillan.

Freeland, R. O. 1943. The American mistletoe with respect to chlorophyll and photosynthesis. Plant Physiol. *18*:299-302.

Fritsché, E., M. Bouillenne-Walrand, and R. Bouillenne. 1958. Quelques observations sur la biologie de *Cuscuta europaea* L. Acad. Roy. Belg., Bull. Classe Sci., Sér. 5, *44*:163-187.

Fuggles-Couchman, N. R. 1935. A parasitic weed of sorghums (*Rhamphicarpa veronicaefolia* Vatke). E. Afr. Agr. Jour. *1*:145-147.

Gadow, H. 1890-1899. Remarks on the structure of certain Hawaiian birds, with reference to their systematic position, *in:* S. B. Wilson and A. H. Evans, Aves Hawaiienses: The birds of the Sandwich Islands. London.

Gaertner, E. E. 1950. Studies of seed germination, seed identification and host relationships in dodders, *Cuscuta* sp. Cornell Univ. Agr. Expt. Sta. Mem. 294.

————. 1956. Dormancy in the seed of *Cuscuta europaea*. Ecology *37*:389.

Gäumann, E. 1950. Principles of plant infection. New York and London: Hafner.

Gagnepain, F., and E. Boureau. 1947. Une nouvelle famille de Gymnospermes: les Sarcopodacées. Bull. Soc. Bot. France *93*:313-320.

Garg, S. 1958. Embryology of *Atkinsonia ligustrina* (A. Cunn. ex F. Muell.) F. Muell. Nature (London) *182*:1615-1616.

Garman, H. 1903. The broom-rapes. Bull. Kentucky Agr. Expt. Sta. *105*:1-32.

Gavriliuk, V. A. 1965. A contribution to the biology of the parasitic plant *Boschniakia rossica* (Cham. et Schlecht.) B. Fedtsch. (in Russian) Bot. Zhurn. *50*: 523-528.

Gerarde, J. 1633. The Herball or Generall Historie of Plantes . . . London.

Gertz, O. 1915. Ueber die Schutzmittel einiger Pflanzen gegen schmarotzende *Cuscuta*. Jahrb. Wiss. Bot. *56*:123-154.

————. 1918. Ueber einige durch schmarotzende *Cuscuta* hervorgerufene Gewebeveränderungen bei Wirtspflanzen. Ber. Deuts. Bot. Ges. *36*:62-72.

Gibbs, R. D. 1954. Comparative chemistry and phylogeny of flowering plants. Trans. Roy. Soc. Canada, Sect. V, *48*:1-47.

Gill, L. S., and F. G. Hawksworth. 1961. The mistletoes: A literature review. U. S. Dept. Agr. Techn. Bull. *1242*:1-87.

Glück, H. 1911. Biologische und morphologische Untersuchungen über Wasser- und Sumpfgewächse. III. Die Uferflora. Jena.

Goebel, K. 1897. Morphologische und biologische Bemerkungen, 7. Ueber die biologische Bedeutung der Blatthöhlen bei *Tozzia* und *Lathraea*. Flora *83*:444-453.

————. 1908. Einleitung in die experimentelle Morphologie der Pflanzen. Leipzig and Berlin: Teubner.

————. 1932. Organographie der Pflanzen. 3d ed. *3*(1). Jena: Fischer.

Goeppert, H. R. 1847. Zur Kenntniss der Balanophoren insbesondere der Gattung *Rhopalocnemis* Jungh. Nov. Act. Acad. Caes.-Leopol.-Carol. Nat. Cur. *22*:117-158.

Govier, R. N., and J. L. Harper. 1965. Hemiparasitic weeds. 7th Brit. Weed Contr. Conf., pp. 577-582.

Govier, R. N., M. D. Nelson, and J. S. Pate. 1967. Hemiparasitic nutrition in angiosperms. I. The transfer of organic compounds from host to *Odontites verna* (Bell.) Dum. (Scrophulariaceae). New Phytol. *66*:285-297.

Graham, B. F., and F. H. Bormann. 1966. Natural root grafts. Bot. Rev. *32*:255-292.

Grant, V. 1950. The protection of the ovules in flowering plants. Evolution *4*:179-201.

Grant, V., and K. A. Grant. 1966. Records of hummingbird pollination in the western American flora. Aliso *6*: 51-66.

Gray, A. 1854. Plantae Novae Thurberianae. Mem. Amer. Acad. Arts and Sci., N.S. *5*:297-328.

Graziano, M. N., G. A. Widmer, J. D. Coussio, and R. Juliani. 1967. Isolation of tyramine from five Argentine species of Loranthaceae. Lloydia *30*:242-244.

Griffis, F. C. 1956. Mistletoe myth develops industry. Jour. Geogr. *55*:251-254.

Griffith, W. 1845. On the root-parasites referred by authors to Rhizantheae; and on various plants related to them. Trans. Linn. Soc. London *19*:303-347.

Guettard, —. 1744. Mémoire sur l'adhérence de la Cuscute aux autres plantes. Hist. de l'Acad. Roy. Sci. Paris, pp. 170-190.

Haberlandt, G. 1897. Zur Kenntniss der Hydathoden. Jahrb. Wiss. Bot. *30*:511-528.

————. 1918. Physiologische Pflanzenanatomie. 5th ed. Leipzig.

Haccius, B., and W. Troll. 1961. Ueber die sogenannten Wurzelhaare an den Keimpflanzen von *Drosera*- und

Cuscuta-Arten. Beitr. Biol. Pfl. *36*:139-157.

Hartel, O. 1956. Der Wasserhaushalt der Parasiten, *in:* W. Ruhland, Handb. Pfl. Physiol. *3*:951-960.

————. 1959. Der Erwerb von Wasser und Mineralstoffen bei Hemiparasiten, *in:* W. Ruhland, *idem,* *11*:33-45.

Hafsten, U. 1957. Om mistelteinens og bergflettens historie i Norge (On the history of mistletoe and ivy in Norway; Engl. summary). Blyttia *1957*:43-60.

Hallier, H. 1901. Ueber die Verwandtschaftsverhaltnisse der Tubifloren und Ebenalen, etc. Abhandl. Naturw. Ver. Hamburg *16*:1-112. (Not seen)

————. 1923. Ueber die Lennoeen, eine zu Linné's Bicornes verirrte Sippe der Borraginaceae. Beih. Bot. Centralbl. *40*:1-19.

Hambler, D. 1958. Some taxonomic investigations on the genus *Rhinanthus*. Watsonia *4*:101-116.

Hamilton, S. G., and B. A. Barlow. 1963. Studies in Australian Loranthaceae. II. Attachment structures and their interrelationships. Proc. Linn. Soc. N. S. W. *88*: 74-90.

Harms, H. 1935. Reihen Santalales, Aristolochiales, Balanophorales. Geschichtliche Entwicklung der Ansichten über die Umgrenzung der Reihen und ihre Zusammensetzung, *in:* Engler & Prantl, Die Nat. Pfl. Fam., 2d ed., *16*b: 1-4; Rafflesiaceae, *idem,* 243-281; Hydnoraceae, *idem,* 282-294; Balanophoraceae, *idem,* 296-339. Duncker & Humblot.

Harris, J. A. 1918. On the osmotic concentration of the tissue fluids of desert Loranthaceae. Mem. Torrey Bot. Club *17*:307-315.

————. 1924. The tissue fluids of *Cuscuta*. *Idem, 51*: 127-131.

Harris, J. A., and J. V. Lawrence. 1916. On the osmotic pressure of the tissue fluids of Jamaican Loranthaceae parasitic on various hosts. Amer. Jour. Bot. *3*:438-455.

Harris, J. A., T. A. Pascoe, and G. H. Harrison. 1930. Osmotic concentration and water relations in the mistletoes, with special reference to the occurrence of *Phoradendron californicum* on *Covillea tridentata*. Ecology *11*:687-702.

Harris, J. A., A. P. Truman, and D. J. Jones. 1930-1931. Note on the tissue fluids of *Phoradendron junipernum* parasitic on *Juniperus utahensis*. Bull. Torrey Bot. Club *58*:113-116.

Harris, J. A., and A. T. Valentine. 1921. The specific electrical conductivity of the tissue fluids of desert Loranthaceae. Proc. Soc. Exp. Biol. and Med. *18*:95-97.

Hart, T. S. 1925. The Victorian species of *Cassytha*. Vict. Natural. *42*:79-83.

Harvey-Gibson, R. J. 1913. Observations on the morphology and anatomy of the genus *Mystropetalon*. Trans. Linn. Soc. London *8*:143-154.

Hauman, L., and L. H. Irigoyen. 1923-1925. Catalogue des Phanérogames de l'Argentine. 2ieme Part, Dicotyledones, I. Anal. Mus. Nac. Buenos Aires *32*:1-314.

Hauri, H. 1912. *Anabasis aretioides* Moq. et Coss., eine Polsterpflanze der algerischen Sahara. Beih. Bot. Zentralbl. *28*:323-420.

Hawksworth, F. G. 1965. Life tables for two species of dwarf mistletoe. I. Seed dispersal, interception, and movement. Forest Sci. *11*:142-151.

Hawksworth, F. G., and D. Wiens. 1965. *Arceuthobium* in Mexico. Brittonia *17*:213-238.

————. 1966. Observations on witches'-broom formation, autoparasitism, and new hosts in *Phoradendron*. Madroño *18*:218-224.

Heckard, L. R. 1962. Root parasitism in *Castilleja*. Bot. Gaz. *124*:21-29.

Heckel, E. 1899. Sur le processus germinatif dans la graine de *Ximenia americana* L. et sur la nature des écailles radiciformes propre a cette espece. Rev. Gén. Bot. *11*:401-408.

————. 1901. Sur le processus germinatif dans les genres *Onguekoa* et *Strombosia* de la famille des Olacacées. Ann. Mus. Col. Marseilles *8*:17-27.

Hedayetullah, S., and J. C. Saha. 1942. A new phanerogamic parasite of sugarcane in Bengal. Curr. Sci. *11*: 109-110.

Hegi, G. 1907-1931. Illustrierte Flora von Mittel-Europa. Munich: Lehmann.

Heil, H. 1926. Haustorialstudien an *Struthanthus*arten. Flora *121*:40-76.

Heinricher, E. 1893. Biologische Studien an der Gattung *Lathraea*. Ber. Deuts. Bot. Ges. *11*:1-18.

————. 1894. Die Keimung von *Lathraea*. *Idem, 12*: 117-132.

————. 1895. Anatomischer Bau und Leistung der Saugorgane der Schuppenwurz-Arten (*Lathraea clandestina* Lam. und *L. squamaria* L.). Beitr. Biol. Pfl. *7*:315-406.

————.1898a. Die grünen Halbschmarotzer. I. *Odontites, Euphrasia,* und *Orthantha*. Jahrb. Wiss. Bot. *31*:77-124.

————. 1898b. *Ibid*. II. *Euphrasia, Alectorolophus,* und *Odontites*. *Idem, 32*:389-452.

————. 1901a. *Ibid*. III. *Bartschia* und *Tozzia,* nebst Bemerkungen zur Frage nach der assimilatorischen Leistungsfahigkeit der grünen Halbschmarotzer. *Idem, 36*:665-752.

————. 1901b. *Ibid*. IV. Nachtrage zu *Euphrasia, Odontites* und *Alectorolophus*. Kritische Bemerkungen zur Systematik letzterer Gattung. *Idem, 37*:264-337.

————. 1905. Beitrage zur Kenntnis der Rafflesiaceae. I. Denkschr. K. Akad. Wiss. Wien, Math.-Naturw. Kl. *78*:1-25.

————. 1907. Beitrage zur Kenntnis der Gattung *Balanophora*. Sitz-ber. K. Akad. Wiss. Wien, Math.-Naturw. Kl. *116*:439-465.

————. 1908a. Die grünen Halbschmarotzer. V. *Melampyrum*. Jahrb. Wiss. Bot. *46*:273-376.

————. 1908b. Van Tieghem's Anschauungen über den Bau der *Balanophora*-Knolle. Sitz-ber. K. Akad. Wiss. Wien, Math.-Naturw. Kl. *117*:337-346.

————. 1913. Einige Bemerkungen zur Rhinantheengattung *Striga*. Ber. Deuts. Bot. Ges. *31*:238-242.

————. 1917a. Die erste Aufzucht einer Rafflesiacee, *Cytinus hypocistis* L., aus Samen. *Idem, 35*:505-512.

————. 1917b. Zur Physiologie der schmarotzenden Rhinantheen, besonders der Halbparasiten. Naturwiss. *5*:113-119.

————. 1930. Wie steht es mit dem Beweis für die Behauptung . . . das eine chemische Reizung . . . für die Samen von *Lathraea clandestina* unnötig sei? Beitr. Biol. Pfl. *18*:1-16.

————. 1931. Monographie der Gattung *Lathraea*. Jena: Fischer.

Hendrych, R. 1963. *Austroamericium* genero nuevo (Santalaceae). Bol. Soc. Arg. Bot. *10*:120-128.

Henry, J. K. 1915. Flora of southern British Columbia. Toronto: Gage & Co., Ltd.

Herbert, D. A. 1918-1919. The West Australian Christmas tree. *Nuytsia floribunda* (The Christmas tree): its structure and parasitism. Jour. Roy. Soc. W. Austr. *5*:72-88.

————. 1925. The root parasitism of Western Australian Santalaceae. *Idem, 11*:127-149.

Heurn, W. C. van. 1922. Bladvulling. Club v. Nederl. Vogelk. Jaarber. *12*:84.

Heyne, K. 1950. De nuttige planten van Indonesië. The Hague: W. v. Hoeve.

Hieronymus, G. 1889. Santalaceae, *in:* Engler & Prantl, Die Nat. Pfl. Fam. 3(1):202-227.

Hill, H. 1926. *Dactylanthus taylori.* Order Balanophoreae; tribe Cynomorieae. Trans. N. Z. Inst. *56:* 87-90.

Hinds, T. E., and F. G. Hawksworth. 1965. Seed dispersal velocity in four dwarfmistletoes. Science *148:* 517-519.

Hinds, T. E., F. G. Hawksworth, and W. J. McGinnies. 1963. Seed discharge in *Arceuthobium:* a photographic study. Science *140:*1236-1238.

Hitchcock, C. L., A. Cronquist, M. Ownbey, and J. W. Thompson. 1959. Vascular plants of the Pacific Northwest. Part 4. Seattle: Univ. Washington Press.

Hodge, W. H., and D. Taylor. 1957. The ethnobotany of the Island Caribs of Dominica. Webbia *12:*513-644.

Hoehne, F. C. 1931. Algo sobre a ecologia do *Phrygilanthus eugenioides* (H.B.K.) Eichl. Secr. da Agric., Ind. e Commerc. Est. São Paulo, Brazil, Bol. Agric. *32:*258-290.

Hofmeister, W. 1858. Neuere Beobachtungen über Embryobildung der Phanerogamen. Jahrb. Wiss. Bot. *1:* 110.

Holdsworth, M., and P. S. Nutman. 1947. Flowering response in a strain of *Orobanche minor.* Nature (London) *160:*223-224.

Hooker, J. D. 1846. Flora Antarctica. Part II. Loranthaceae, pp. 289-302.

————. 1856. On the structure and affinities of Balanophoreae. Trans. Linn. Soc. London 22:1-68.

————. 1859. On a new genus of Balanophoreae from New Zealand, and two new species of *Balanophora. Idem,* 22:425-427.

————. 1886. The flora of British India. Vol. 5. London.

Hooker, W. J. 1840. *Langsdorffia indica.* Icon. Plant. 3:205-206.

————. 1844. *Pholisma. Idem,* 626.

————. 1848. *Aetanthus. Idem,* 734.

————. 1867. *Chaunochiton. Idem,* 1005.

————. 1871. *Buttonia. Idem,* 1080.

Hosford, R. M. 1967. Transmission of plant viruses by dodder. Bot. Rev. *33:*387-406.

Hosseus, C. C. 1907. Eine neue Rafflesiaceengattung aus Siam. Bot. Jahrb. *41:*55-61.

Howard, R. A., and C. E. Wood. 1955. Christmas plants in the Boston area. Arnoldia *15:*61-84.

Howell, J. T. 1966. *Viscum album* in California. Leafl. W. Bot. *10:*244.

Hürlimann, H., and H. U. Stauffer. 1957. Santalales-Studien II. *Daenikera,* eine neue Santalaceen-Gattung. Viertelj. Schr. Naturf. Ges. Zürich *102:*332-336.

Hull, R. J., and O. A. Leonard. 1964a. Physiological aspects of parasitism in mistletoes (*Arceuthobium* and *Phoradendron*). I. The carbohydrate nutrition of mistletoe. Plant Physiol. *39:*996-1007.

————. 1964b. *Ibid.* II. The photosynthetic capacity of mistletoe. *Idem, 39:*1008-1017.

Hultén, E. 1961. Two *Pedicularis* species from N. W. America, *P. albertae* n. sp. and *P. sudetica* sens. lat. Sv. Bot. Tidskr. 55:193-204.

Hunziker, A. T. 1949-1950. Las especies de *Cuscuta* (Convolvulaceae) de Argentina y Uruguay. Trab. Mus. Bot. Univ. Cordoba *1*(2):1-357.

Hunziker, J. 1926. Beiträge zur Anatomie von *Rafflesia patma* Bl. Thesis, Zürich.

Hutchinson, J. 1959. The families of flowering plants. Vol. 1, Dicotyledons. Oxford: Clarendon Press.

Irmisch, T. 1855. Bemerkungen über einige Pflanzen der deutschen Flora. Flora 38:625-638.

Jacob, F. 1966. Zur Auslösung des Blühvorganges bei der Kurztagpflanze *Cuscuta reflexa* Roxb. Flora, Abt. A. *156:*558-572.

Jaeger, P. 1961. The wonderful life of flowers. London: Harrap & Co.

Jensen, H. W. 1951. The normal and parthenogenetic forms of *Orobanche uniflora* in the eastern United States. Cellule *54:*135-142.

Jochems, S. C. J. 1928. Die Verbreitung der Rafflesiaceengattung *Mitrastemon.* Rec. Trav. Bot. Néerl. 25A: 203-207.

Johri, B. M., J. S. Agrawal, and S. Garg. 1957. Morphological and embryological studies in the family Loranthaceae. I. *Helicanthes elastica* (Desr.) Dans. Phytomorph. 7:336-354.

Johri, B. M., and Y. P. S. Bajaj. 1962. Behaviour of mature embryo of *Dendrophthoe falcata* (L.f.) Ettingsh. in vitro. Nature (London) *193:*194-195.

————. 1964. Growth of embryos of *Amyema, Amylotheca,* and *Scurrula* on synthetic media. *Idem, 204:* 1220-1221.

————. 1965. Growth responses of globular proembryos of *Dendrophthoe falcata* (L.f.) Ettingsh. in culture. Phytomorph. *15:*292-300.

Johri, B. M., and S. P. Bhatnagar. 1960. Embryology and taxonomy of the Santalales. I. Proc. Nat. Inst. Sci. India *26*B (Suppl.):199-220.

Johri, B. M., and S. Prakash. 1965. Morphological and embryological studies in the family Loranthaceae. XI. *Tapinostemma acaciae* (Zucc.) Van Tiegh. Phytomorph. *15:*150-158.

Johri, B. M., and B. Raj. 1965. Embryo sac development in *Moquiniella.* Nature (London) 205:415-416.

Johri, B. M., and B. Tiagi. 1952. Floral morphology and seed formation in *Cuscuta reflexa* Roxb. Phytomorph. 2:162-180.

Jones, B. L., and C. C. Gordon. 1965. Embryology and development of the endosperm haustorium of *Arceuthobium douglasii.* Amer. Jour. Bot. 52:127-132.

Jonker, F. P. 1938. A monograph of the Burmanniaceae. Meded. Bot. Mus. Rijksuniv. Utrecht *51:*1-279.

Juel, O. 1902. Zur Entwicklungsgeschichte des Samens von *Cynomorium.* Beih. Bot. Centralbl. *13:*194-202.

————. 1910. *Cynomorium* und *Hippuris.* Sv. Bot. Tidskr. 4:151-159.

Juliano, J. B. 1935. Anatomy and morphology of the Bunga, *Aeginetia indica* Linnaeus. Philip. Jour. Sci. 56:405-451.

Jumelle, H., and H. Perrier de la Bâthie. 1912. Quelques phanérogames parasites de Madagascar. Rev. Gén. Bot. 24:321-328.

Justesen, P. T. 1922. Morphological and biological notes on *Rafflesia* flowers observed in the Highlands of Mid-Sumatra (Padangsche Bovenlanden). Ann. Jard. Bot. Buitenz. 32:64-87.

Kadambi, K. 1954. Instances of fusion of the tissues of different trees. Indian For. *80:*726.

Kadry, A. 1952. The development of microsporangium and pollen grains in *Cistanche tinctoria* (Forssk.) G. Beck. Bot. Notiser *105:*46-57.

————. 1955. The development of endosperm and embryo in *Cistanche tinctoria* (Forssk.) G. Beck. *Idem, 108:*231-243.

Kadry, A., and H. Tewfic. 1956. A contribution to the morphology and anatomy of seed germination in *Orobanche crenata.* Bot. Notiser *109:*385-399.

Kamensky, K. W. 1928. Anatomische Struktur der Samen von einigen *Cuscuta*-Arten und deren systematischer

Wert. Angew. Bot. *10*:387-406.

Kamerling, Z. 1914. Ueber die Wachsthumsweise und über den Dimorphismus der Blätter von *Struthanthus flexicaulis* Mart. Rec. Trav. Bot. Néerl. *11*:342-352.

Kapil, R. N., and I. K. Vasil. 1963. Ovule, *in:* P. Maheshwari (ed.), Recent advances in the embryology of angiosperms. Intern. Soc. Plant Morphol., Delhi.

Keck, D. D. 1940. Notes on *Orthocarpus*. Madroño *5*: 164-165.

Kerner v. Marilaun, A. 1888. Ueber die Bestäubungseinrichtigungen der Euphrasieen. Verhandl. Zool.-Bot. Ges. Wien *38*:563-566.

————. 1896-1898. Pflanzenleben. 2d ed. Leipzig and Vienna: Bibl. Inst.

Kindermann, A. 1928. Haustorialstudien an *Cuscuta*-Arten. Planta *5*:769-783.

King, L. J. 1966. Weeds of the world. New York: Interscience Publ., Inc.

Kingsbury, J. M. 1965. Deadly harvest. New York: Holt, Rinehart and Winston.

Kirchmayr, H. 1908. Die extrafloralen Nektarien von *Melampyrum* vom physiologisch-anatomischen Standpunkt. Sitz-ber. K. Akad. Wiss. Wien, Math.-Naturw. Kl., Abt. 1, *117*:439-452.

Kirk, T. 1887. On the naturalized dodders and broomrapes of New Zealand. Trans. N. Z. Inst. *20*:182-186.

Klepac, D. 1956. (Effect of *Viscum album* on the increment of silver-fir stands; original in Croatian). Forest Abstr. *17*:2947.

Koch, L. 1874. Untersuchungen über die Entwicklung der Cuscuteen. Bot. Abhandl. 2(3):1-137.

————. 1878. Ueber die Entwicklung des Samens der Orobanchen. Jahrb. Wiss. Bot. *11*:218-261.

————. 1887a. Ueber die direkte Ausnutzung vegetabilischer Resten durch bestimmte chlorophyllhaltige Pflanzen. Ber. Deuts. Bot. Ges. *5*:352-364.

————. 1887b. Entwicklungsgeschichte der Orobanchen. Heidelberg.

————. 1889. Zur Entwicklungsgeschichte der Rhinanthaceen (*Rhinanthus minor* Ehrh.). Jahrb. Wiss. Bot. *20*:1-37.

————. 1891. Zur Entwicklungsgeschichte der Rhinanthaceen. II. *Euphrasia officinalis* L. *Idem*, 22:1-34.

Kostermans, A. J. G. H. 1957. Lauraceae. Communication Forest Res. Inst. Indonesia *57*:1-64.

Kostytschew, S. 1924. Untersuchungen über die Ernährung der grünen Halbschmarotzer. Beih. Bot. Centralbl. *40*:351-373.

Kozlowski, T. T. 1964. Water metabolism in plants. Harper & Row.

Kramer, P. J., and T. T. Kozlowski. 1960. Physiology of trees. New York: McGraw-Hill.

Krause, K. 1932. Loranthaceae, *in:* A. A. Pulle, Flora of Surinam, *1*:4-24.

Kribben, F. J. 1951. Die Blütenbildung von *Orobanche* in Abhängigkeit von der Entwicklungsphase des Wirtes. Ber. Deuts. Bot. Ges. *64*:353-355.

Krishna Iyengar, C. V. 1942. Development of seed and its nutritional mechanism in Scrophulariaceae. Part I. *Rhamphicarpa longifolia* Benth., *Centranthera hispida* Br., and *Pedicularis zeylanica* Benth. Proc. Nat. Inst. Sci. India *8*:249-261.

Kuijt, Job. 1955. Dwarf mistletoes. Bot. Rev. *21*:569-628.

————. 1959. A study of heterophylly and inflorescence structure in *Dendrophthora* and *Phoradendron* (Loranthaceae). Acta Bot. Neerl. *8*:506-546.

————. 1960. Morphological aspects of parasitism in the dwarf mistletoes (*Arceuthobium*). Univ. Calif. Publ. Bot. *30*:337-436.

————. 1960-1961. Distribution of dwarf mistletoes and their fungus hyperparasites in western Canada. Nat. Mus. Canada Bull. *186*, Contr. Bot. 1960-1961: 134-148 (1963).

————. 1961a. Observations on the life cycle in *Arceuthobium campylopodum*. Leafl. W. Bot. *9*:133-134.

————. 1961b. Notes on the anatomy of the genus *Oryctanthus* (Loranthaceae). Can. Jour. Bot. *39*:1809-1816.

————. 1961c. A revision of *Dendrophthora* (Loranthaceae). Wentia *6*:1-145.

————. 1963a. On the ecology and parasitism of the Costa Rican tree mistletoe, *Gaiadendron punctatum* (Ruiz and Pavón) G. Don. Can. Jour. Bot. *41*:927-938.

————. 1963b. *Dendrophthora*: additions and changes. Acta Bot. Neerl. *12*:521-524 (1964).

————. 1964a. Vegetative propagation of dwarf mistletoes. Forest Sci. *10*:78-79.

————. 1964b. A new *Antidaphne* from Ecuador. Brittonia *16*:331-333.

————. 1964c. Critical observations on the parasitism of New World mistletoes. Can. Jour. Bot. *42*:1243-1278.

————. 1964d. A revision of the Loranthaceae of Costa Rica. Bot. Jahrb. *83*:250-326.

————. 1964e. A peculiar case of hemlock mistletoe parasitic on larch. Madroño *17*:254-256.

————. 1965a. The anatomy of haustoria and related organs of *Gaiadendron* (Loranthaceae). Can. Jour. Bot. *43*:687-694.

————. 1965b. On the nature and action of the Santalalean haustorium, as exemplified by *Phthirusa* and *Antidaphne* (Loranthaceae). Acta Bot. Neerl. *14*: 278-307.

————. 1966a. Parasitism in *Pholisma* (Lennoaceae). I. External morphology of subterranean organs. Amer. Jour. Bot. *53*:82-86.

————. 1966b. Greenhouse culture of two neotropical mistletoes. Jour. Roy. Hort. Soc. *91*:400-402.

————. 1967a. The genus *Ixocactus* (Loranthaceae, s.s.): description of its first species. Brittonia *19*:61-66.

————. 1967b. Parasitism in *Pholisma* (Lennoaceae). II. Anatomical aspects. Can. Jour. Bot. *45*:1155-1162.

————. 1967c. On the structure and origin of the seedling of *Psittacanthus schiedeanus* (Loranthaceae). *Idem*, *45*:1497-1506.

————. 1968. Mutual affinities of Santalalean families. Brittonia *20*:136-147.

Kummerow, J. 1962. *Pilostyles berterii* Guill., eine wenig bekannte Rafflesiacee in Mittelchile. Zeits. Bot. *50*: 321-337.

Kummerow, J., and V. Matte. 1960. Anatomische Beobachtungen an *Phrygilanthus tetrandrus* (Ruiz et Pav.) Eichl., einer semiparasitischen Loranthacee. Phytopath. Zeits. *39*:321-326.

Kunkel, L. O. 1952. Transmission of alfalfa witch's broom to non-leguminous plants by dodder, and cure in periwinkle by heat. Phytopath. *42*:27-31.

Kunz, M. 1913. Die systematische Stellung der Gattung *Krameria* unter besonderer Berücksichtigung der Anatomie. Beih. Bot. Zentralbl. *30*:412-427.

Kusano, S. 1902. Studies on the parasitism of *Buckleya quadriala*, B. et H., a Santalaceous parasite, and on the structure of its haustorium. Jour. Coll. Sci., Imp. Univ. Tokyo *17*(10): 1-42.

————. 1908a. On the parasitism of *Siphonostegia* (Rhinantheae). Bull. Coll. Agr. Tokyo *8*:51-57.

————. 1908b. Further studies on *Aeginetia indica*.

Beih. Bot. Centralbl. *24*(1):286-299.

Kust, C. A. 1963. Dormancy and viability of witchweed seeds as affected by temperature and relative humidity during storage. Weeds *11*:247-250.

Kuwada, Y. 1928. An occurrence of restitution-nuclei in the formation of the embryosacs in *Balanophora japonica*, Mak. Bot. Mag. Tokyo *42*:117-129.

Lackey, C. F. 1953. Attraction of dodder and beet leafhopper to vascular bundles in the sugar beet as affected by curlytop. Amer. Jour. Bot. *40*:221-225.

Laing, R. M., and E. W. Blackwell. 1964. Plants of New Zealand. 7th ed. Christchurch: Whitcombe & Tombs, Ltd.

La Llave, P., and J. Lexarza. 1824. Novorum vegetabilum descriptum, fasc. *1*:7. Mexico. (Not seen)

Lam, H. J. 1945. Fragmenta Papuana. Sargentia *5*: 1-196.

————. 1948. Classification and the new morphology. Acta Biotheor. *8*:107-154.

Lane, H. C., and M. J. Kasperbauer. 1965. Photomorphogenic responses of dodder seedlings. Pl. Physiol. *40*:109-116.

La Rue, C. D. 1952. Root grafting in tropical trees. Science *115*:296.

Laudi, G. 1964. Ricerche infrastrutturali sui plastidi delle piante parassite. I. *Cuscuta*. Caryologica *17*: 139-152.

Leavitt, R. G. 1902. Subterranean plants of *Epiphegus*. Plant World *5*:114.

Lecomte, H. 1896. Sur une nouvelle Balanophorée du Congo français. Jour. de Bot. *10*:229-235.

Lee, A., and F. Goseco. 1932. Studies of the sugar-cane root parasite, *Aeginetia indica*. Intern. Soc. Sugar Cane Technol. Congr. (San Juan) Proc. *4*:1-12.

Lemon, P. C., and J. M. Voegeli. 1962. Anatomy and ecology of *Pieris phillyreifolia* (Hook.) DC. Bull. Torrey Bot. Club *89*:303-311.

Lewis, D. 1942. The evolution of sex in flowering plants. Biol. Rev. *17*:46-67.

Li, H.-L. 1948. A revision of the genus *Pedicularis* in China. Part I. Proc. Acad. Nat. Sci. Phila. *100*:205-378.

————. 1949. *Ibid.* Part II. *Idem, 101*:1-214.

————. 1951. Evolution in the flowers of *Pedicularis*. Evolution *5*:158-164.

Lindley, J. 1836. A natural system of botany. 2d ed. London.

————. 1853. The vegetable kingdom. London: Evans.

Ling, K. C. 1955. Bunga. Taiwan Sugar *2*:21.

Linnaeus, C. 1753. Species plantarum. 1st ed. Stockholm.

Littlefield, N. A., H. E. Pattee, and K. R. Allred. 1966. Movement of sugars in the alfalfa-dodder association. Weeds *14*:52-54.

Lloyd, F. E. 1902. The comparative morphology of the Rubiaceae. Mem. Torrey Bot. Club *8*:1-112.

Lo, T. T. 1955. N:Co 310 highly resistant to the root parasite bunga (*Aeginetia indica*). Taiwan Sugar *2*: 18-20.

Loo, S. W. 1946. Cultivation of excised stem tips of dodder *in vitro*. Amer. Jour. Bot. *33*:295-300.

Lotsy, J. P. 1899. *Balanophora globosa* Jungh., eine wenigstens örtlich verwitwete Pflanze. Ann. Jard. Bot. Buitenz., Sér. 2, *1*:174-184.

————. 1901. *Rhopalocnemis phalloides* Jungh., a morphological-systematical study. *Idem*, Sér. 2, *2*: 73-101.

Louis, J., and R. Boutique. 1947. Une espèce nouvelle

d'*Anacolosa* au Congo Belge. Bull. Jard. Bot. de l'Etat Bruxelles *18*:255-258.

Louis, J., and J. Léonard. 1948. Olacaceae, *in:* Flore du Congo Belge et du Ruanda-Urundi. Spermatophytes, *1*:249-278; Opiliaceae, *idem, 1*:279-287; Octoknemaceae, *idem, 1*:288-293.

Lundquist, G. 1915. Die Embryosackentwicklung von *Pedicularis sceptrum-carolinum* L. Zeits. Bot. *7*:545-559.

Lundströn, A. N. 1887. Pflanzenbiologische Studien. II. Die Anpassungen der Pflanzen an Thiere. Nova Acta Soc. Sci. Upsala, Ser. 3, *13*:1-88.

MacArthur, R. H. 1958. Population ecology of some warblers of northeastern coniferous forests. Ecology *39*: 599-619.

MacBride, J. F. 1937. Loranthaceae, *in:* Flora of Peru. Field Mus. Publ. Bot. *13*(II:2):375-416.

MacDougal, D. T. 1911. An attempted analysis of parasitism. Bot. Gaz. *52*:249-260.

MacKinney, G. 1935. On the plastid pigments of marsh dodder. Jour. Biol. Chem. *112*:421-424.

MacLeod, D. G. 1961. Some anatomical and physiological observations on two species of *Cuscuta*. Trans. Bot. Soc. Edinb. *39*:302-315.

————. 1963. The parasitism of *Cuscuta*. New Phytol. *62*:257-263.

McLuckie, J. 1924. Studies in parasitism. I. A contribution to the physiology of the genus *Cassytha*. Proc. Linn. Soc. N. S. W. *49*:55-79.

McWhorter, F. P. 1922. Concerning the sugar-cane parasite, *Aeginetia indica*. Philipp. Agr. *11*:89-90.

Maheshwari, P., and B. Baldev. 1961. Artificial production of buds from the embryos of *Cuscuta reflexa*. Nature (London) *191*:197-198.

Maheshwari, P., D. M. Johri, and S. N. Dixit. 1957. The floral morphology and embryology of the Loranthoideae (Loranthaceae). Jour. Madras Univ., B, *27*: 121-136.

Main, A. 1947. Artificial propagation of *Nuytsia floribunda*. W. Austr. Natural. *1*:25-31.

Makino, T. 1911. (*Mitrastemon*). Bot. Mag. Tokyo *25*: 251-257.

Malcolm, W. M. 1962. Culture of *Castilleja coccinea* (Indian paint-brush), a root-parasitic flowering plant. Mich. Bot. *1*:77-79.

————. 1966. Root parasitism of *Castilleja coccinea*. Ecology *47*:179-186.

Mangenot, G. 1947. Recherches sur l'organisation d'une Balanophoracée: *Thonningia coccinea*. Rev. Gén. Bot. *54*:201-244, 271-294.

Marloth, R. 1913. The flora of South Africa. Vol. 1.

————. 1932. *Idem*, Vol. 3.

Marudarjan, K. 1950. Note on *Orobanche cernua* Loefl. Curr. Sci. *19*:64-65.

Matuda, E. 1947. On the genus *Mitrastemon*. Bull. Torrey Bot. Club *74*:133-141.

May, V. 1941. A survey of the mistletoe of New South Wales. Proc. Linn. Soc. N. S. W. *66*:77-87.

Maybrook, A. C. 1917. On the haustoria of *Pedicularis vulgaris*. Ann. Bot. *31*:499-511.

Mayr, E., and D. Amadon. 1947. A review of the Dicaeidae. Amer. Mus. Nov. *1360*:1-32.

Mekel, J. C. 1935. Der Blütenstand und die Blüte von *Korthalsella dacrydii*. Blumea *1*:312-319.

Melchior, H. 1921. Ueber den anatomischen Bau der Saugorgane von *Viscum album* L. Beitr. Allgem. Bot. *2*:55-87.

Melvin, L. 1956. *Nestronia*, parasitic on roots of *Pinus*. Elisha Mitchell Sci. Soc. Jour. *72*:137-138.

Menke, W., and B. Wolfersdorf. 1968. Ueber die Plastiden von *Neottia nidus-avis*. Planta 78:134-143.

Menzies, B. P., and H. S. McKee. 1959. Root parasitism in *Atkinsonia ligustrina* (A. Cunn. ex F. Muell.) F. Muell. Proc. Linn. Soc. N. S. W. 84:118-127.

Metcalfe, C. R. 1935. The structure of some sandalwoods and their substitutes and of some other little known scented woods. Kew Bull. 1935:165-195.

Metcalfe, C. R., and Chalk, L. 1950. Anatomy of the dicotyledons. Oxford: Clarendon Press.

Meyen, J. 1829. Ueber das Herauswachsen parasitischer Gewächse aus den Wurzeln anderer Pflanzen. Flora 12:49-64.

Micheli, P. A. 1723. Relazione dell'erba detta da' Botanici Orobanche e volgarmente Succiamele, Fiamma, Emald'Occhio, etc. 2d ed. Florence: Tartini & Franchi.

————. 1729. Nova plantarum genera.

Michell, M. R. 1915. The embryo sac and embryo of *Striga lutea*. Bot. Gaz. 59:124-135.

Mildbread, J. 1935. Octoknemaceae, *in:* Engler & Prantl, Die Nat. Pfl. Fam., 2d ed., 16b:42-45.

Miller, L., and F. W. Woods. 1965. Root grafting in loblolly pine. Bot. Gaz. 126:252-255.

Millspaugh, C. F., and L. W. Nuttall. 1923. Flora of Santa Catalina Island (California). Field Mus. Nat. Hist. Publ. Bot. 5:1-413.

Miquel, F. 1856. Viscaceae. Fl. Ind. Bat. 1(1): 803-804.

Mirande, M. 1898. Sur les lacticifères et les tubes criblés des Cuscutes monogynées. Jour. de Bot. 12:70-90.

————. 1900. Recherches physiologiques et anatomiques sur les Cuscutacées. Thesis, Fac. Sci., Paris.

————. 1905a. Recherches sur le développement et l'anatomie des Cassythacées. Ann. Sci. Nat., Sér. 9, Bot., pp. 181-285.

————. 1905b. Sur l'origine pluricarpellaire du pistil des Lauracées. C. R. Acad. Sci. Paris 145:570-572.

Mitten, W. 1847. On the economy of the roots of *Thesium linophyllum* Hook. London Jour. Bot. 6:146-148.

Mohan Ram, H. Y. 1964. Famous plants: the sandalwood. The Botanica 14:164-167.

Monachino, J. 1961. A new *Liriosma* from Roraima. Brittonia 13:113-115.

Moore, L. B. 1940. The structure and life-history of the root parasite *Dactylanthus taylori* Hook. f. N. Z. Jour. Sci. and Technol. 21:206B-224B.

Moran, R. 1962. Muérdago en los Cactos. Cact. y Suc. Mex. 7:82-84.

————. 1966. ¿Cholla parásita? *Idem, 11*:81-84.

Moss, E. H. 1926. Parasitism in the genus *Comandra*. New Phytol. 25:264-276.

Munz, P. A., and D. D. Keck. 1959. A California flora. Berkeley: Univ. Calif. Press.

Narasimha, V. L., and V. Rabindranath. 1964. A further contribution to the host range of *Dendrophthoe falcata* (L.f.) Ettingsh. Bull. Bot. Survey India 6:103.

Narayana, R. 1958. Morphological and embryological studies in the family Loranthaceae. III. *Nuytsia floribunda* (Labill.) R. Br. Phytomorph. 8:306-323.

Nash, S. M., and S. Wilhelm. 1960. Stimulation of broomrape seed germination. Phytopath. 50:772-774.

Nayar, B. K., and P. N. Nayar. 1952. On the range of the hosts of *Cassytha filiformis* Linn. Sci. and Cult. 17:383-384.

Nevlin, L. I. 1960. Opiliaceae and Olacaceae, *in:* Flora of Panama, Ann. Miss. Bot. Garden 47:291-302.

Oehler, E. 1927. Entwicklungsgeschichtlich-zytologische Untersuchungen an einigen saprophytischen Gentianaceen. Planta 3:641-733. Springer-Verlag.

Offord, H. R. 1964. A new record of dwarfmistletoe on planted Monterey pine in California. Plant Dis. Rept. 48:912.

Okonkwo, S. N. C. 1967. Studies on *Striga senegalensis* Benth. I. Mode of host-parasite union and haustorial structure. Phytomorph. 16:453-463.

Ortiz, G. J. 1936. Es obligatoria la destruccion del "Quintral" *Phrygilanthus (Loranthus) tetrandrus*. Bol. Soc. Agric. Norte 24:87.

Paine, L. A. 1950. The susceptibility of pear trees to penetration and toxic damage by mistletoe. Phytopath. Zeits. 17:305-327.

Paliwal, R. L. 1956. Morphological and embryological studies in some Santalaceae. Agra Univ. Jour. Res. (Sci.) 5:193-284.

Pant, D. D., and R. Banerji. 1965. Epidermal structure and development of stomata in some Convolvulaceae. Senck. Biol. 46:155-173.

Pattee, H. E., K. R. Allred, and H. H. Wiebe. 1965. Photosynthesis in dodder. Weeds 13:193-195.

Patvardhan, G. B. 1924. Note on *Loranthus* on *Eucalyptus* in Poona. Jour. Indian Bot. Soc. 4:71-72.

Pearson, H. H. W. 1912. On the Rooibloem (Isona or Witchweed). Agr. Jour. Union S. Afr. 2(3):1-7.

Peé-Laby, M. E. 1904. La passiflore parasite sur les racines du fusain. Rev. Gén. Bot. 16:453-457.

Peirce, G. J. 1893. On the structure of haustoria of some phanerogamic parasites. Ann. Bot. 7:291-327.

Pennell, F. W. 1925. The genus *Afzelia*: a taxonomic study in evolution. Proc. Acad. Nat. Sci. Phila. 77: 335-373.

————. 1928. *Agalinis* and allies in North America. I. *Idem, 80*:339-449.

————. 1929. *Agalinis* and allies in North America. II. *Idem, 81*:111-249.

————. 1935. The Scrophulariaceae of eastern temperate North America. Acad. Nat. Sci. Phila. Monogr. 1:1-650.

————. 1948. The taxonomic significance of an understanding of floral evolution. Brittonia 6:301-308.

Pennell, F. W., and G. N. Jones. 1937. A new Indian paintbrush from Mount Rainier. Proc. Biol. Soc. Wash. 50:207-210.

Pennell, F. W., and M. Ownbey. 1950. *Castilleja nivea*, snowy Indian paint-brush (Scrophulariaceae), a new species from the northern Rocky Mountains. Notulae Naturae 227.

Penzig, O. 1901. Beiträge zur Kenntniss der Gattung *Epirrhizanthes* Bl. Ann. Jard. Bot. Buitenz. 17:142-170.

Percival, W. C. 1931. The parasitism of *Conopholis americana* on *Quercus borealis*. Amer. Jour. Bot. 18: 817-837.

Philipson, W. R. 1959. Some observations on root-parasitism in New Zealand. Trans. Roy. Soc. N. Z. 87:1-3.

Piehl, M. A. 1962a. The parasitic behavior of *Melampyrum lineare* and a note on its seed color. Rhodora 64: 15-23.

————. 1962b. The parasitic behavior of *Dasistoma macrophylla*. *Idem, 64*:331-336.

————. 1963. Mode of attachment, haustorium structure, and hosts of *Pedicularis canadensis*. Amer. Jour. Bot. 50:978-985.

————. 1965a. Studies of root parasitism in *Pedicularis lanceolata*. Mich. Bot. 4:75-81.

————. 1965b. Observations on the parasitic behavior of *Buckleya distichophylla* (Santalaceae) (Abstr.) Amer. Jour. Bot. 52:626.

————. 1965c. The natural history and taxonomy of *Comandra* (Santalaceae). Mem. Torrey Bot. Club 22: 1-97.

————. 1966. The root parasitism of *Cordylanthus* and some of its ecological implications (Abstr.). Amer. Jour. Bot. *53:*622.

Pilger, R. 1935. Santalaceae, *in:* Engler & Prantl, Die Nat. Pfl. Fam., 2d ed., *16b:*52-91.

Pitra, A. 1861. Ueber die Anheftungsweise einiger phanerogamen Parasiten an ihre Nahrpflanzen. Bot. Zeit. *19:*53-58, 61-67, 69-74.

Pora, E., E. Pop, D. Rosca, and A. Radu. 1957. Der Einfluss der Wirtspflanze auf den Gehalt an hypotensiven und herzwirksamen Prinzipen der Mistel (*Viscum album*). Die Pharmacie *12:*528-538.

Porsch, O. 1929. Vogelblumenstudien. II. Jahrb. Wiss. Bot. *70:*181-277.

Prain, D. 1891. The species of *Pedicularis* of the Indian Empire and its frontiers. Ann. Roy. Bot. Garden Calcutta *3:*1-196.

Prakash, S. 1963. Morphological and embryological studies in the family Loranthaceae. X. *Barathranthus axanthus* (Korth.) Miq. Phytomorph. *13:*97-103.

Quisumbing, E. 1940. On *Christisonia wrightii* Elmer, a parasite of sugar cane. Philipp. Jour. Agr. *11:*397-401.

Raj, B. 1963. Female gametophyte of *Buckleya lanceolata* Sieb. et Zucc. (Abstr.). Proc. 50th Ind. Sci. Congr. (Delhi) *3:*389-390.

Ram, M. 1957. Morphological and embryological studies in the family Santalaceae. I. *Comandra umbellata* (L.) Nutt. Phytomorph. *7:*24-35.

————. 1959. *Ibid.* II. *Exocarpus,* with a discussion on its systematic position. *Idem, 9:*4-19.

Rangan, T. S. 1965. Morphogenesis of the embryo of *Cistanche tubulosa* Wight in vitro. Phytomorph. *15:*180-182.

Rangaswamy, N. S. 1963. Studies on culturing seeds of *Orobanche aegyptiaca* Pers., *in:* P. Maheshwari and N. S. Rangaswamy, Plant tissue and organ culture, pp. 345-354. Intern. Soc. Plant Morphol., Delhi.

Rangaswamy, N. S., and T. S. Rangan. 1966. Effects of seed germination-stimulants on the witchweed *Striga euphrasioides* (Vahl) Benth. Nature (London) *210:*440-441.

Rao, P. S. 1965. In vitro induction of embryonal proliferation in *Santalum album* L. Phytomorph. *15:*175-179.

Rao, R. S. 1957. A revision of the Indo-Malayan species of *Viscum* Linn. Jour. Indian Bot. Soc. *36:*113-168.

Rattray, J. M. 1932. A parasite on cowpeas (*Alectra vogelii* Benth.). Rhod. Agr. Jour. *29:*791-794.

Rau, M. A. 1961. Occurrence of *Alectra parasitica* A. Rich. in India. A new variety from Banda District, U. P. Bull. Bot. Survey India *3:*25-27.

Record, S. J., and R. W. Hess. 1943. Timbers of the New World. New Haven: Yale Univ. Press.

Rediske, J. H., and K. R. Shea. 1961. The production and translocation of photosynthate in dwarfmistletoe and lodgepole pine. Amer. Jour. Bot. *48:*447-452.

Reed, C. F. 1955. The comparative morphology of the Olacaceae, Opiliaceae and Octoknemaceae. Mem. Soc. Brot. *10:*29-79.

Reiche, C. 1904. Bau und Leben der chilenischen Loranthacee *Phrygilanthus aphyllus.* Flora *93:*271-297.

Rendle, A. B. 1925. The classification of flowering plants. Vol. 2. Dicotyledons. Cambridge.

Rendle, A. B., F. G. Baker, and S. L. M. Moore. 1921. A systematic account of the plants collected in New Caledonia and the Isle of Pines by Prof. R. H. Compton, M. A., in 1914. Jour. Linn. Soc. London, Bot., *45:*245-417.

Renner, O. 1934. Guttation bei *Cuscuta* und *Casuarina.* Ann. Jard. Bot. Buitenz. *44:*90-93.

Ridley, H. N. 1930. The dispersal of plants throughout the world. Kent: Ashford.

Rigby, J. F. 1959. Light as a control in the germination and development of several mistletoe species. Proc. Linn. Soc. N. S. W. *84:*335-337.

Rizzini, C. T. 1960. Loranthaceae, *in:* Flora of Panama, Ann. Miss. Bot. Garden *47:*263-290.

Robinson, E. L. 1960. Growth of witchweed (*Striga asiatica*) as affected by soil types and soil and air temperatures. Weeds *8:*576-581.

————. 1961. Soil-incorporated pre-planting herbicides for witchweed control. Weeds *9:*411-428.

Robinson, E. L., and C. A. Kust. 1962. Distribution of witchweed seeds in the soil. Weeds *10:*335.

Rock, J. F. 1916. The sandalwoods of Hawaii: A revision of the Hawaiian species of the genus *Santalum.* Terr. Hawaii, Board Agr. and For., Bot. Bull. 3.

Rogers, W. E., and R. R. Nelson. 1962. Penetration and nutrition of *Striga asiatica.* Phytopath. *52:*1064-1070.

Rose, J. N. 1906. Studies of Mexican and Central American plants. No. 5. Contr. U. S. Nat. Herb. *10:*79-132.

————. 1909. *Ibid.* No. 6. *Idem, 12:*259-302.

Roth, L. F. 1959. Natural emplacement of dwarfmistletoe seed on Ponderosa pine. For. Sci. *5:*365-369.

Ruiz, A., and F. A. Roig. 1958. Los hospedantes de hemiparásitos y parásitos fanerogámicos en el valle de Atuél (Mendoza). Bol. Soc. Arg. Bot. *7:*116-119.

Rutishauser, P. 1935. Entwicklungsgeschichtliche und zytologische Untersuchungen an *Korthalsella dacrydii* (Ridl.) Dans. Ber. Schweiz. Bot. Ges. *44:*389-436.

————. 1937. Blütenmorphologische und embryologische Untersuchungen an den Viscoïdeen *Korthalsella opuntia* Merr. und *Ginalloa linearis* Dans. *Idem, 47:*5-28.

Sablon, L. du. 1887. Organes d'absorption des plantes parasites (Rhinanthées et Santalacées). Ann. Sci. Nat., Sér. 7, *6:*90-117.

Sahni, B. 1933. Explosive fruits in *Viscum japonicum,* Thunb. Jour. Indian Bot. Soc. *12:*96-101.

St. John, H. 1947. The history, present distribution, and abundance of sandalwood on Oahu, Hawaiian Islands: Hawaiian plant studies 14. Pac. Sc. *1:*5-20.

Sakimura, K. 1947. Virus transmission by *Cuscuta sandwichiana.* Phytopath. *37:*66.

Samuelsson, G. 1913. Studien über die Entwicklungsgeschichte der Blüte einiger Bicornes-Typen. Sv. Bot. Tidskr. *7:*69-188.

————. 1958. Phytochemical and pharmacological studies on *Viscum album* L. Sv. Farmaceut. Tidskr. *62:*169-189.

————. 1959. *Ibid. Idem, 63:*412-425; 545-553.

Sands, W. N. 1924. Mistletoes attacking cultivated trees in Malaya. Malay. Agr. Jour. *12:*64-76.

Sastri, R. L. N. 1956. Embryo sac haustoria in *Cassytha filiformis* Linn. Curr. Sci. *25:*401-402.

————. 1962. Studies in Lauraceae. III. Embryology of *Cassytha.* Bot. Gaz. *123:*197-206.

Saunders, A. R. 1933. Studies in phanerogamic parasitism with particular reference to *Striga lutea* Lour. Dept. Agr. S. Afr. Sci. Bull. 128.

————. 1934. *Melasma*—a dangerous parasite on legumes. Farming in S. Afr. *9:*342.

Schaar, F. 1898. Ueber den Bau des Thallus von *Rafflesia rochussenii* Teijsm. et Binn. Sitz-ber. K. Akad. Wiss. Wien, Math.-Naturw. Kl. *107:*1039-1056.

Schaeppi, H., and F. Steindl. 1945. Blütenmorphologische

und embryologische Untersuchungen an einigen Viscoideen. Beih. Viertelj. Schr. Naturf. Ges. Zürich *90:* 1-46.

Scharpf, R. F. 1963. Epidemiology and parasitism of the dwarfmistletoe *Arceuthobium campylopodum* Engelm., in California. Ph. D. Thesis, Univ. Calif., Berkeley.

Schellenberg, G. 1932. Ueber Systembildung und über die Reihe der Santalales. Festschr. Deuts. Bot. Ges. *50a:*136-145.

Scherffel, A. 1888. Die Drüsen in den Höhlen der Rhizomschuppen von *Lathraea squamaria* L. Mitth. Bot. Inst. Graz *1888:*185-212.

Schimper, A. F. W. 1880. Die Vegetationsorgane von *Prosopanche burmeisteri.* Abhandl. Naturf. Ges. Halle *15:*21-47.

Schmidt, A. T. 1902. Zur Anatomie von *Cassytha filiformis* L. Oesterr. Bot. Zeitschr. *52:*173-177.

Schmidt, O. C. 1935. Aristolochiaceae, *in:* Engler & Prantl, Die Nat. Pfl. Fam., 2d ed., *16b:*204-242.

Schmucker, T. 1959. Höhere Parasiten, *in:* W. Ruhland, Handb. Pfl. Physiol. *11:*480-529.

Schnarf, K. 1931. Vergleichende Embryologie der Angiospermen. Berlin: Borntraeger.

Scholl, R. 1957. Weitere Untersuchungen über Veränderungen der Reaktionslage des Birnbaumes (*Pirus communis* L.) gegenüber der Mistel (*Viscum album* L.). Phytopath. Zeitschr. *28:*237-258.

Schrenk, H. 1894. Parasitism of *Epiphegus virginiana.* Proc. Amer. Micr. Soc. *15:*91-128.

Schumacher, W. 1934. Die Absorptionsorgane von *Cuscuta odorata* und der Stoffübertritt aus den Siebröhren der Wirtspflanze. Jahrb. Wiss. Bot. *80:*74-91.

Schumacher, W., and W. Halbsguth. 1938. Ueber den Anschluss einiger höheren Parasiten an die Siebröhren der Wirtspflanzen. Jahrb. Wiss. Bot. *87:*324-355.

Seeger, R. 1910. Versuche über die Assimilation von *Euphrasia* (sens. lat.) und über die Transpiration der Rhinantheen. Sitz-ber. K. Akad. Wiss. Wien, Math.-Naturw. Kl. *119*(1):987-1004.

Selawry, O. S., F. Vester, W. Mai, and M. R. Schwartz. 1961. Zur Kenntnis der Inhaltsstoffe von *Viscum album.* II. Tumorhemmende Inhaltsstoffe. Hoppe-Seyler's Zeitschr. Physiol. Chem. *324:*262-281.

Senn, G. 1913. Der osmotische Druck einiger Epiphyten und Parasiten. Verhandl. Naturf. Ges. Basel *42:*179-183.

Setchell, W. A. 1935. *Acroblastum* vs. *Polyplethia:* a complex of the Balanophoraceae. Univ. Calif. Publ. Bot. *19:*141-158.

Sharma, S. L., D. Rao, and K. N. Trivedi. 1956. Taxonomy of three species of *Striga* parasitic on sugarcane. Proc. Indian Acad. Sci. Sect. B *43:*67-71.

Shaw, W. C., D. R. Shepherd, E. L. Robinson, and P. F. Sand. 1962. Advances in witchweed control. Weeds *10:*182-192.

Singh, T. C. N. 1933. *Cuscuta* as a parasite on a fern. Ann. Bot. *47:*423-425.

Skottsberg, C. 1910. Juan Fernandez—örnas sandelträd. Sv. Bot. Tidskr. *4:*167-173.

————. 1913a. A botanical survey of the Falkland Islands. K. Sv. Vetensk. Akad. Handl. *50*(3):1-129.

————. 1913b. Morphologische und embryologische Studien über die Myzodendraceae. *Idem, 51*(4):1-34.

————. 1915. Ett par fall af heterostyli i Patagoniens flora. Bot. Notiser *1915:*195-204.

————. 1916. Zur Morphologie und Systematik der Gattung *Arjona* Cav. Sv. Bot. Tidskr. *10:*520-528.

————. 1930. The geographical distribution of the sandal-woods and its significance. Proc. 4th Pac. Sci. Congr. (Java) *3:*435-442.

Sleumer, H. 1935. Olacaceae and Opiliaceae, *in:* Engler & Prantl, Die Nat. Pfl. Fam., 2d ed., *16b:*5-41.

————. 1942. Erythropalaceae, *in:* Engler & Prantl, *idem, 20b:*401-403.

Smart, C. 1952. The life-history of *Tupeia* Cham. et Schl. Trans. Roy. Soc. N. Z. *79:*459-466.

Smith, A. C. 1901. The structure and parasitism of *Aphyllon uniflorum* Gray. Trans. and Proc. Bot. Soc. Penn. *1:*111-121.

Smith, F. H., and E. C. Smith. 1943. Floral anatomy of the Santalaceae and related forms. Ore. State Monogr. Stud. Bot. *5:*1-93.

Smith, G. G. 1951. New records of distribution of *Pilostyles hamiltonii.* W. Austr. Natural. *3:*21-24.

Smith, H. 1933. Plantae Sinenses: XXVI. Orobanchaceae. Acta Hort. Gothoburg. *8:*127-146.

Smith, J. 1878. Bible plants, their history, with a review of the opinions of various writers regarding their identification. 265 pp.

Solms-Laubach, H. 1867-1868. Ueber den Bau und Entwicklung der Ernährungsorgane parasitischer Phanerogamen. Jahrb. Wiss. Bot. *6:*509-638.

————. 1870. Die Familie der Lennoaceen. Abhandl. Naturf. Ges. Halle *11:*121-178.

————. 1874a. Ueber den Thallus von *Pilostyles haussknechtii.* Bot. Zeit. *32:*49-59, 65-74.

————. 1874b. Ueber den Bau des Samens in den Familien der Rafflesiaceae und Hydnoraceae. *Idem, 32:*337-342, 353-358, 369-374, 385-389.

————. 1875. Das Haustorium der Loranthaceen und der Thallus der Rafflesiaceen und Balanophoreen. Abhandl. Naturf. Ges. Halle *13:*238-276.

————. 1876. Die Entwicklung der Blüten bei *Brugmansia zippelii* und *Aristolochia clematitis.* Bot. Zeit. *34:*449-461, 465-470, 481-489, 497-503.

————. 1898. Die Entwicklung des Ovulum und des Samens bei *Rafflesia* und *Brugmansia.* Ann. Jard. Bot. Buitenz., Suppl. *2:*11-22.

Solomon, S. 1952. Studies in the physiology of phanerogamic parasites with special reference to *Striga lutea* Lour. and *S. densiflora* Benth. on *Andropogon sorghum.* 1. The osmotic pressure of the host and parasite in relation to the nutrition of the host. Proc. Indian Acad. Sci. 35B:122-131.

Sota, E. R. de la. 1960. *Mamorea singeri,* un nuevo género y especie de Burmanniaceae. Darwiniana *12:* 43-47.

Soyer-Poskin, D., and A. Schmitz. 1962. Phanérogames parasites et hémiparasites des arbres des environs d'Elisabethville (Katanga). Lejeunia, N.S. 7:3-52.

Sperlich, A. 1908. Ist bei grünen Rhinanthaceen ein von einem pflanzlichen Organismus ausgehender äusserer Keimungsreiz nachweisbar? Ber. Deuts. Bot. Ges. 26A: 574-587.

————. 1925. Die Absorptionsorgane der parasitischen Samenpflanzen, *in:* K. Linsbauer, Handbuch der Pflanzenanatomie 9(2A):1-53.

Spisar, K. 1910. Beiträge zur Physiologie der *Cuscuta gronovii.* Ber. Deuts. Bot. Ges. 28:329-334.

Sprague, E. F. 1962a. Parasitism in *Pedicularis.* Madroño *16:*192-200.

————. 1962b. Pollination and evolution in *Pedicularis* (Scrophulariaceae). Aliso 5:181-209.

Sprague, T. A., and V. Summerhayes. 1927. *Santalum, Eucarya* and *Mida.* Kew Bull. *1927:*193-202.

Srivastava, L. M., and K. Esau. 1961a. Relation of

dwarfmistletoe (*Arceuthobium*) to the xylem tissue of conifers. I. Anatomy of parasite sinkers and their connection with host xylem. Amer. Jour. Bot. *48*:159-167.

—————. 1961*b*. *Ibid*. II. Effect of the parasite on the xylem anatomy of the host. *Idem, 48*:209-215.

Standley, P. C. 1937. Flora of Costa Rica. Field Mus. Publ. Bot. *18*(1-2).

Standley, P. C., and J. A. Steyermark. 1946. Flora of Guatemala. Fieldiana Bot. *24*(4).

Stauffer, H. U. 1956. *Petrusia madagascariensis* Baillon, eine Zygophyllacee. Mitt. Bot. Mus. Univ. Zürich *66*:433-446.

—————. 1957. Santalales-Studien I. Zur Stellung der Gattung *Okoubaka* Pellegrin et Normand. Ber. Schweiz. Bot. Ges. *67*:422-427.

—————. 1959. *Ibid*. IV. Revisio Anthobolearum. Mitt. Bot. Mus. Univ. Zürich *213*:1-260.

—————. 1961*a*. *Ibid*. V. Afrikanische Santalaceae. I. *Osyris, Colpoon* und *Rhoiacarpos*. Viertelj. Schr. Naturf. Ges. Zürich *106*:387-400.

—————. 1961*b*. *Ibid*. VI. Afrikanische Santalaceae. II. *Osyridicarpos*. *Idem, 106*:400-406.

—————. 1961*c*. *Ibid*. VII. Südamerikanische Santalaceae. I. *Acanthosyris, Cervantesia* und *Jodina*. *Idem, 106*:406-412.

—————. 1961*d*. *Ibid*. VIII. Zur Morphologie und Taxonomie der Olacaceae-Tribus Couleae. *Idem, 106*:412-418.

Stauffer, H. U., and H. Hürlimann. 1957. *Ibid*. III. *Amphorogyne*, eine weitere Santalaceen-Gattung aus Neukaledonien. *Idem, 102*:337-349.

Stebbins, G. L. 1950. Variation and evolution in plants. New York: Columbia Univ. Press.

Steenis, C. G. G. J. van. 1931. Some remarks on the genus *Rhopalocnemis* Junghuhn. Handel. 6. Nederl.-Ind. Natuurw. Congr. (Bandoeng), pp. 464-475 (1932).

—————. 1934. Determinatie-tabel voor de Nederlandsch Indische bladgroenlooze Phanerogamen. Trop. Natuur *23*:45-57.

Steffen, K. 1956. Endomitosen im Endosperm von *Pedicularis palustris* L. Planta *47*:625-652.

Stephens, E. L. 1912. The structure and development of the haustorium of *Striga lutea*. Ann. Bot. *26*:1067-1076.

Steude, H. 1935. Beiträge zur Morphologie und Anatomie von *Mourera aspera*. Beih. Bot. Zentralbl. *53*:627-650.

Stevenson, G. B. 1934. The life history of the New Zealand species of the parasitic genus *Korthalsella*. Trans. Proc. Roy. Soc. N. Z. *64*:175-190.

Stocken, C. 1964. *Viscum cruciatum*—an unusual mistletoe. Jour. Roy. Hort. Soc. *89*:303.

Stojanovich, D. 1959. (A contribution to the study of the biology of *Cuscuta epithymum* Murray — vegetative hibernation). Plant Protection (Zashtita Bilja) *51*:47-49.

Story, R. 1958. Some plants used by the Bushmen in obtaining food and water. Mem. Bot. Survey S. Afr. *30*:1-115.

Stout, G. L. 1938. A recurrence of broomrape, *Orobanche ramosa* L., on tomato plants in California. Bull. Dept. Agr. Calif. *27*:166-170.

Straka, H. 1966. Palynologia madagassica et mascarenica. Fam. 5—59bis. Pollen et Spores *8*:241-264.

Strigl, M. 1907. Der anatomische Bau der Knollenrinde von *Balanophora* und seine mutmassliche funktionelle Bedeutung. Sitz-ber. K. Akad. Wiss. Wien, Math.-Naturw. Kl. *116*(1):1041-1060.

—————. 1908. Der Thallus von *Balanophora*, anatomisch-physiologisch geschildert. *Idem, 117*:1127-1175.

Subramanian, C. L., and A. R. Srinivasan. 1960. A review of the literature on the phanerogamous parasites. Indian Counc. Agr. Res., New Delhi.

Suessenguth, K. 1927. Ueber die Gattung *Lennoa*. Ein Beitrag zur Kenntnis exotischer Parasiten. Flora *122*:264-305.

Sunderland, N. 1960*a*. The production of the *Striga* and *Orobanche* germination stimulants by maize roots. I. The number and variety of stimulants. Jour. Expt. Bot. *11*:236-245.

—————. 1960*b*. Germination of the seeds of angiospermous root parasites. Brit. Ecol. Soc. Symp. (1959) *1*:83-98.

—————. 1960*c*. The production of the *Striga* and *Orobanche* germination stimulants by maize roots. II. Conditions of synthesis in the root. Jour. Expt. Bot. *11*:356-366.

Sutton, G. M. 1951. Mistletoe dispersal by birds. Wilson Bull. *63*:235-237.

Swamy, B. G. L. 1949. The comparative morphology of the Santalaceae: node, secondary xylem, and pollen. Amer. Jour. Bot. *36*:661-673.

—————. 1960. Contributions to the embryology of *Cansjera rheedii*. Phytomorph. *10*:397-409.

Swamy, B. G. L., and J. Dayanand Rao. 1963. The endosperm in *Opilia amentacea* Roxb. Phytomorph. *13*:423-428.

Szczawinski, A. F., and G. A. Hardy. 1962. Guide to common edible plants of British Columbia. B. C. Prov. Mus. Handb. No. 20. Victoria.

Taubert, P. 1894. Leguminosae, in: Engler & Prantl, Die Nat. Pfl. Fam. *3*(3):70-396.

Taylor, W. R. 1950. Plants of Bikini and other northern Marshall Islands. Univ. Michigan Press.

Teijsmann, J. E. 1856. Nadere bijdrage tot de kennis van de voortteling van *Rafflesia arnoldii* R. Br. in 's Lands Plantentuin te Buitenzorg. Natuurk. Tijdschr. Nederl. Indië *12*:277-281.

Temme, F. 1883. Ueber den Chlorophyllgehalt und die Assimilation von *Cuscuta europaea*. Ber. Deuts. Bot. Ges. *1*:485.

Templeton, B. C. 1962. A morphological comparison of *Pholisma arenarium* Nuttall and *Pholisma paniculatum* Templeton (Lennoaceae). L. A. Co. Mus. Contr. Sci. *57*:3-29.

Thackeray, F. A. 1953. Sand food of the Papagos. Desert Mag. *16*:22-24.

Thieret, J. W. 1967. Supraspecific classification in the Scrophulariaceae: a review. Sida *3*:87-106.

Thoday, D. 1951. The haustorial system of *Viscum album*. Jour. Expt. Bot. *2*:1-19.

—————. 1956. Modes of union and interaction between parasite and host in the Loranthaceae. I. Viscoideae, not including Phoradendreae. Proc. Roy. Soc. B. *145*:531-548.

—————. 1957. *Ibid*. II. Phoradendreae. *Idem, 146*:320-338.

—————. 1958*a*. *Ibid*. III. Further observations on *Viscum* and *Korthalsella*. *Idem, 148*:188-206.

—————. 1958*b*. *Ibid*. IV. *Viscum obscurum* on *Euphorbia polygona*. *Idem, 149*:42-57.

—————. 1960. *Ibid*. V. Some South African Loranthoideae. *Idem, 152*:143-162.

—————. 1961. *Ibid*. VI. A general survey of the Loranthoideae. *Idem, 155*:1-25.

—————. 1963. *Ibid*. VII. Some Australian Loranthoideae with exceptional features. *Idem, 157*:507-516.

Thoday, D., and E. F. Johnson. 1930. On *Arceuthobium pusillum*, Peck. I. The endophytic system. Ann. Bot. 44:393-413. II. Flowers and fruit. *Idem, 44*:813-824.

Thomasson, K. 1959. Nahuel Huapi. Plankton of some lakes in an Argentine National Park with notes on terrestrial vegetation. Acta Phytogeogr. Suec. *42*:1-83.

Thomson, J. 1925. Studies in irregular nutrition. No. 1. The parasitism of *Cuscuta reflexa* (Roxb.). Trans. Roy. Soc. Edinb. *54*:343-356.

Thun, H. 1943. Pharmakologische Untersuchungen über die Eichenmistel und die Amerikanische Mistel. Arch. Expt. Pathol. Pharmakol. *202*:642-645.

Thunberg, C. P. 1775. Beskrifning paa en ganska besynnerlig och obekant svamp, *Hydnora africana* . . . Handl. K. Vetensk. Acad. *36*:69-75.

Thurman, L. D. 1965. Genecological studies in *Orthocarpus*, subgenus *Triphysaria* (Scrophulariaceae). Ph. D. Thesis, Univ. Calif. Berkeley.

Tiagi, B. 1951*a*. A contribution to the morphology and embryology of *Cuscuta hyalina* Roth. and *C. planiflora* Tenore. Phytomorph. *1*:9-21.

————. 1951*b*. Studies in the family Orobanchaceae. III. A contribution to the embryology of *Orobanche cernua* Loeffl. and *O. aegyptiaca* Pers. *Idem, 1*:158-169.

————. 1952*a*. *Ibid*. I. A contribution to the embryology of *Cistanche tubulosa* Wight. Lloydia *15*:129-148.

————. 1952*b*. *Ibid*. A contribution to the embryology of *Aeginetia indica* Linn. Bull. Torrey Bot. Club *79*:63-78.

————. 1956. A contribution to the embryology of *Striga orobanchioides* Benth. and *Striga euphrasioides* Benth. *Idem, 83*:154-170.

————. 1963. Studies in the family Orobanchaceae. IV. Embryology of *Boschniakia himalaica* Hook., and *B. tuberosa* (Hook.) Jepson, with remarks on the evolution of the family. Bot. Notiser *116*:81-93.

————. 1965. *Ibid*. VI. Development of the seed in *Conopholis americana* (L. fil.) Wallr. Acta Bot. Hung. *11*:253-261.

————. 1966*a*. Floral morphology of *Cuscuta reflexa* Roxb. and *C. lupuliformis* Krocker with a brief review of the literature on the genus *Cuscuta*. Bot. Mag. Tokyo *79*:89-97.

————. 1966*b*. Development of the seed and fruit in *Rhinanthus major* and *R. serotinus*. Amer. Jour. Bot. *53*:645-651.

Tiagi, B., and N. S. Sankhla. 1963. Studies in the family Orobanchaceae. V. A contribution to the embryology of *Orobanche lucorum*. Bot. Mag. Tokyo *76*:81-88.

Tingey, D. C., and K. R. Allred. 1961. Breaking dormancy in seeds of *Cuscuta approximata*. Weeds *9*:429-436.

Toledo, J. F. 1952. Species Brasilienses *Agonandrae* Miers. Arq. Bot. Est. São Paulo, N.S. *3*:11-18.

Torrey, J. 1867. On *Ammobroma*, a new genus of plants, etc. Ann. Lyc. Nat. Hist. N. Y. *8*:51-56.

Trattinick, *in*: Schlechtendal, D. F. L. 1828. Nachtrag zu der *Ichthyosma wehdemanni*. Linnaea *3*:194-198.

Trelease, W. 1916. The genus *Phoradendron*. A monographic revision. Urbana: Univ. Illinois.

Treub, M. 1883. Sur une nouvelle catégorie de plantes grimpantes. Ann. Jard. Bot. Buitenz. *3*:44-75.

————. 1898. L'organe femelle et l'apogamie du *Balanophora elongata* Bl. *Idem, 15*:1-25.

Tronchet, J. 1958. Mise en évidence de l'hydrotropisme des plantules de *Cuscuta gronovii* Willd. Ann. Sci. Univ. Besançon, Sér. 2, Bot. *10*:11-13.

Truscott, F. H. 1958. On the regeneration of new shoots from isolated dodder haustoria. Amer. Jour. Bot. *45*: 169-177.

————. 1966. Some aspects of morphogenesis in *Cuscuta gronovii*. *Idem, 53*:739-750.

Tsoong, P. C., and K. T. Chang. 1965. Palynological study of *Pedicularis* and its relation with the taxonomic systems of the genus (in Chinese). Acta Phytotax. Sinica *10*:257-282.

Tubeuf, C. von. 1923. Monographie der Mistel. Berlin: Oldenbourg.

Turner, B. L. 1958. Chromosome numbers in the genus *Krameria*: evidence for familial status. Rhodora *60*: 101-106.

Twisselmann, E. C. 1956. A flora of the Temblor range and the neighboring part of the San Joaquin Valley. Wasmann Jour. Biol. *14*:161-320.

Ulbrich, E. 1907. Ueber europäische Myrmekochoren. Verhandl. Bot. Ver. Prov. Brandenburg *48*:214-241.

Ule, E. 1907. Beiträge zur Flora der Hylaea nach den Sammlungen von Ule's Amazonas-Expedition. Loranthaceae. Verhandl. Bot. Ver. Prov. Brandenburg *48*: 117-208.

————. 1915. Ueber brasilianische Rafflesiaceen. Ber. Deuts. Bot. Ges. *33*:468-478.

Ultée, A. J. 1926. Ueber das sogenannte Balanophorin. Bull. Jard. Bot. Buitenz., Sér. 3, *8*:32-34.

Umiker, O. 1920. Entwicklungsgeschichtlich-cytologische Untersuchungen an *Helosis guyanensis* Rich. Arb. Inst. Allgem. Bot. Pfl. Physiol. Univ. Zürich, No. 23.

Unger, F. 1833. Ueber das Einwurzeln parasitischer Pflanzen auf der Mutterpflanze. Isis von Oken *4*:373-374.

Uttaman, P. 1950. A study of the germination of *Striga* seed and on the mechanism and nature of parasitism of *Striga lutea* on rice. Proc. Indian Acad. Sci. 32B:133-142.

Vaccaneo, R. 1934. Hydnoracea Africae. Mem. R. Accad. dei Lincei, Roma, Ser. 6, *5*:413-458.

Vallance, K. B. 1951. Studies on the germination of the seeds of *Striga hermontica*. III. On the nature of pretreatment and after-ripening. Ann. Bot., N.S. *15*:109-128.

————. 1952. The germination of the seeds of *Rhinanthus crista-galli*. *Idem, 16*:409-420.

Vallentin, E. F. 1921. Illustrations of the flowering plants and ferns of the Falkland Islands. London: Reeve & Co. Ltd.

Van Tieghem, P. 1896. Sur l'organisation florale des Balanophoracées et sur la place de cette famille. Bull. Soc. Bot. France *43*:295-310.

————. 1897. Sur la structure de l'ovule et de la graine chez les Hydnoracées. Jour. de Bot. *11*:233-238.

Vasey, G., and J. N. Rose. 1890. List of plants collected by Dr. Edward Palmer in Lower California and western Mexico in 1890. Contr. U. S. Nat. Herb. *1*:63-90.

Vaucher, J. P. 1823. Mémoire sur la germination des orobanches. Mém. Mus. Hist. Nat. Paris *10*:261-273.

Venkata Rao, C. 1964. On the morphology of the calyculus. Jour. Indian Bot. Soc. *42*:618-628.

Walldén, B. 1961. Misteln vid dess nordgräns. Sv. Bot. Tidskr. *55*:428-549.

Walls, F. 1962. Dodder—serious parasitic weed. Agr. Gaz. N. S. Wales *73*:133-135.

Walzel, G. 1952*a*. *Cuscuta* auf *Nicotiana* Nikotin-frei. Phyton (Austria) *4*:121-123.

————. 1952*b*. Colchicinierte *Cuscuta*. *Idem, 4*:137-143.

Warburg, O. 1913. Pflanzenwelt *1*:505.

Watanabe, K. 1936-1937. Morphologisch - biologische Studien über die Gattung *Mitrastemon*. Jour. Jap. Bot. *12*:603-618, 698-711, 759-773, 847-858; *13*:14-24, 75-86, 154-162.

Watt, G. 1889-1896. A dictionary of the economic products of India. 7 vols. London and Calcutta: W. H. Allen & Co.

Watt, J. M., and M. G. Breyer-Brandwijk. 1962. The medical and poisonous plants of southern and eastern Africa. Edinburgh: Livingstone.

Weberbauer, A. 1901. Ueber die Frucht-Anatomie der Scrophulariaceae. Beih. Bot. Zentralbl. *10*:393-457.

Weddell, H. A. 1850. Considérations sur l'organe reproducteur femelle des Balanophorées et des Rafflesiacées. Ann. Sci. Nat., Sér. 3, *14*:166-187.

————. 1858-1861. Mémoire sur le *Cynomorium coccineum*, parasite de l'ordre des Balanophorées. Arch. Mus. Hist. Nat. *10*:269-308.

Weeraratna, W. G. 1960. The ecology and biology of parasitism of the Loranthaceae of Ceylon, *in:* J. L. Harper, The biology of weeds, Brit. Ecol. Soc. Symp. *1*:189-202.

Wetmore, A. 1914. The development of the stomach in the Euphonias. Auk *31*:458-461.

Wettstein, R. 1895. Scrophulariaceae, *in:* Engler & Prantl, Die Nat. Pfl. Fam. *4*(3b):39-107.

————. 1896. Monographie der Gattung *Euphrasia*. Leipzig.

Wiens, D. 1964a. Chromosome numbers in North American Loranthaceae (*Arceuthobium*, *Phoradendron*, *Psittacanthus*, *Struthanthus*). Amer. Jour. Bot. *51*:1-6.

————. 1964b. Revision of the acataphyllous species of *Phoradendron*. Brittonia *16*:11-54.

————. 1967. Chromosomal relationships of the New World mistletoes. Amer. Jour. Bot. *54*:654.

Wiens, H. E. 1962. Atoll environment and ecology. New Haven: Yale Univ. Press.

Wiesner, J. 1872. Untersuchungen über die Farbstoffe einiger für chlorophyllfrei gehaltenen Phanerogamen. Jahrb. Wiss. Bot. *8*:575-594.

————. 1894. Pflanzenphysiologische Mittheilungen aus Buitenzorg. IV. Vergleichende physiologische Studien über die Keimung europaischer und tropischer Arten von *Viscum* und *Loranthus*. Sitz-ber. K. Akad. Wiss. Wien, Math.-Naturw. Kl. *103*(1):1-37.

————. 1897. Ueber die Ruheperiode und über einige Keimungsbedingungen der Samen von *Viscum album*. Ber. Deuts. Bot. Ges. *15*:503-516.

Wild, H. 1948. A suggestion for the control of tobacco witchweed (*Striga gesnerioides* (Willd.) Vatke) by leguminous crops. Rhod. Agr. Jour. *45*:208-215.

Wilhelm, S., L. C. Benson, and J. E. Sagen. 1958. Studies on the control of broomrape on tomatoes. Soil fumigation by methyl bromide is a promising control. Plant Dis. Rept. *42*:645-651.

Wilhelm, S., R. C. Storkan, J. E. Sagen, and T. Carpenter. 1959. Large-scale soil fumigation against broomrape. Phytopath. *49*:530-531.

Williams, C. N. 1958. The parasitism of witchweed: a review. W. Austr. Jour. Biol. Chem. *2*:57-73.

————. 1959. Resistance of *Sorghum* to witchweed. Nature (London) *184*:1511-1512.

————. 1960a. *Sopubia ramosa*, a perennating parasite on the roots of *Imperata cylindrica*. Jour. W. Afr. Sci. Assoc. *6*:137-141.

————. 1960b. Growth movements of the radicle of *Striga*. Nature (London) *188*:1043-1044.

————. 1961a. Growth and morphogenesis of *Striga* seedlings. *Idem, 189*:378-381.

————. 1961b. Tropism and morphogenesis of *Striga* seedlings in the host rhizosphere. Ann. Bot., N.S., *25*:407-415.

————. 1961c. Effect of inoculum size and nutrition on the host-parasite relations of *Striga senegalensis* on *Sorghum*. Plant and Soil. *15*:1-12.

————. 1962. Non-geotropic growth curvature in the *Striga* radicle. Ann. Bot., N.S. *26*:647-655.

Williams, C. N., and G. H. Caswell. 1959. An insect attacking *Striga*. Nature (London) *184*:1668.

Wilson, L. L. W. 1898. Observations on *Conopholis americana*. Contr. Bot. Lab. Univ. Penn. *2*:3-19.

Winkler, H. 1927. Ueber eine *Rafflesia* aus Zentralborneo. Planta *4*:1-79.

Winter, A. G. 1961. New physiological and biological aspects in the interrelationships between higher plants. Soc. Exp. Biol. Symp. *15*:229-244.

Witkus, E. R. 1945. Endomitotic tapetal cell divisions in *Spinacia*. Amer. Jour. Bot. *32*:326-330.

Woods, F. W., and K. Brock. 1964. Interspecific transfer of Ca-45 and P-32 by root systems. Ecology *45*:886-889.

Worsham, A. D., D. E. Moreland, and G. C. Klingman. 1964. Characterization of the *Striga asiatica* (Witchweed) germination stimulant from *Zea mays* L. Jour. Exp. Bot. *15*:556-567.

Yarwood, C. E. 1956. Obligate parasitism. Ann. Rev. Plant Physiol. *7*:115-142.

Yeo, P. F. 1961. Germination, seedlings, and the formation of haustoria in *Euphrasia*. Watsonia *5*:11-22.

York, H. H. 1913. The origin and development of the embryo sac and embryo of *Dendrophthora opuntioides* and *D. gracilis*. I and II. Bot. Gaz. *56*:89-111, 200-216.

Youngken, H. W. 1921. Observations on Muira-puama. Amer. Jour. Pharm. *93*:625-627.

Yuncker, T. G. 1920. Revision of the North American and West Indian species of *Cuscuta*. Ill. Biol. Monogr. *6*(2/3):1-141.

————. 1932. The genus *Cuscuta*. Mem. Torrey Bot. Club *18*:113-331.

————. 1934. *Cuscuta japonica* Choisy, an Asiatic species new to America. Torreya *44*:34-35.

————. 1942. A note on the precocious germination of dodder seed. Proc. Indiana Acad. Sci. *51*:114-115.

————. 1961. *Cuscuta cassythoides*, a South African species, in North Carolina. Brittonia *13*:144.

Zakharov, B. S. 1940. Der Einfluss der Tageslange auf den Befall der Sonnenblume durch *Orobanche cumana*. C. R. Acad. Sci. URSS, N.S. *27*:267-270.

Zender, J. 1924. Les haustoriums de la Cuscute et les réactions de l'hote. Bull. Soc. Bot. Geneve, Sér. 2, *16*:189-264.

Ziegler, H. 1955. *Lathraea*, ein Blutungssaftschmarotzer. Ber. Deuts. Bot. Ges. *68*:311-318.

Zietz, H. 1954. Neue Beitrage zur Entwicklungsphysiologie von *Cuscuta gronovii* Willd. Biol. Zentralbl. *73*:129-155.

Zimmerman, C. E. 1962. Autotropic development of dodder (*Cuscuta pentagona* Englm.) in vitro. Crop Sci. *2*:449-450.

Index